Vol. 29. **The Analytical Chemistry of Sulfur and Its Compc** Karchmer

Vol. 30. **Ultramicro Elemental Analysis.** By Günther Tölg

Vol. 31. **Photometric Organic Analysis** *(in two parts).* By Eugene Sawicki

Vol. 32. **Determination of Organic Compounds: Methods and Procedures.** By Frederick T. Weiss

Vol. 33. **Masking and Demasking of Chemical Reactions.** By D. D. Perrin

Vol. 34. **Neutron Activation Analysis.** By D. De Soete, R. Gijbels, and J. Hoste

Vol. 35. **Laser Raman Spectroscopy.** By Marvin C. Tobin

Vol. 36. **Emission Spectrochemical Analysis.** By Morris Slavin

Vol. 37. **Analytical Chemistry of Phosphorus Compounds.** Edited by M. Halmann

Vol. 38. **Luminescence Spectrometry in Analytical Chemistry.** By J. D. Winefordner, S. G. Schulman, and T. C. O'Haver

Vol. 39. **Activation Analysis with Neutron Generators.** By Sam S. Nargolwalla and Edwin P. Przybylowicz

Vol. 40. **Determination of Gaseous Elements in Metals.** Edited by Lynn L. Lewis, Laben M. Melnick, and Ben D. Holt

Vol. 41. **Analysis of Silicones.** Edited by A. Lee Smith

Vol. 42. **Foundations of Ultracentrifugal Analysis.** By H. Fujita

Vol. 43. **Chemical Infrared Fourier Transform Spectroscopy.** By Peter R. Griffiths

Vol. 44. **Microscale Manipulations in Chemistry.** By T. S. Ma and V. Horak

Vol. 45. **Thermometric Titrations.** By J. Barthel

Vol. 46. **Trace Analysis: Spectroscopic Methods for Elements.** Edited by J. D. Winefordner

Vol. 47. **Contamination Control in Trace Element Analysis.** By Morris Zief and James W. Mitchell

Vol. 48. **Analytical Applications of NMR.** By D. E. Leyden and R. H. Cox

Vol. 49. **Measurement of Dissolved Oxygen.** By Michael L. Hitchman

Vol. 50. **Analytical Laser Spectroscopy.** Edited by Nicolo Omenetto

Vol. 51. **Trace Element Analysis of Geological Materials.** By Roger D. Reeves and Robert R. Brooks

Vol. 52. **Chemical Analysis by Microwave Rotational Spectroscopy.** By Ravi Varma and Lawrence W. Hrubesh

Vol. 53. **Information Theory As Applied to Chemical Analysis.** By Karel Eckschlager and Vladimir Stepánek

Vol. 54. **Applied Infrared Spectroscopy: Fundamentals, Techniques, and Analytical Problem-solving.** By A. Lee Smith

Vol. 55. **Archaeological Chemistry.** By Zvi Goffer

Vol. 56. **Immobilized Enzymes in Analytical and Clinical Chemistry.** By P. W. Carr and L. D. Bowers

Vol. 57. **Photoacoustics and Photoacoustic Spectroscopy.** By Allan Rosencwaig

Vol. 58. **Analysis of Pesticide Residues.** Edited by H. Anson Moye

Vol. 59. **Affinity Chromatography.** By William H. Scouten

Vol. 60. **Quality Control in Analytical Chemistry.** By G. Kateman and F. W. Pijpers

Vol. 61. **Direct Characterization of Fineparticles.** By Brian H. Kaye

Vol. 62. **Flow Injection Analysis.** By J. Ruzicka and E. H. Hansen

Receptor Modeling in Environmental Chemistry

Receptor Modeling in Environmental Chemistry

Philip K. Hopke
Professor of Environmental Chemistry
Institute for Environmental Studies
University of Illinois at Urbana-Champaign

A WILEY-INTERSCIENCE PUBLICATION

JOHN WILEY & SONS

New York / Chichester / Brisbane / Toronto / Singapore

Library of Congress Cataloging in Publication Data:

Hopke, Philip K., 1944–
Receptor modeling in environmental chemistry.

(Chemical analysis, ISSN 0069-2883; v. 76)
"A Wiley-Interscience publication."
Bibliography: p.
Includes index.
1. Environmental chemistry–Methodology. I. Title. II. Series.

TD193.H67 1985 628.5′3 84-19568
ISBN 0-471-89106-1

Printed in the United States of America

10 9 8 7 6 5 4 3 2 1

PREFACE

During the past few years, there has been a rapid increase in the sophistication of environmental analytical techniques. With the more readily available automation that laboratory computers afford, large volumes of data can be obtained to characterize a system. Improvements in computers have also made statistical methods accessible and more reliable. In this period, a number of people have been exploring ways to put these techniques to use to solve a variety of problems and have contributed to a rapidly expanding literature in what has now come to be called receptor modeling within the air pollution research community. Although the biochemists may not take kindly to our pirating of the term, it has now become the way of referring to the class of methods where the properties of the sample are used to infer the origins of its components.

The principal use of these models has been in trying to quantitatively apportion aerosol mass to particle sources but the methods are really generally applicable to any problem where the measured property can be considered to be a linear sum of independently contributing components. Thus the idea behind this book is that the time has come to try to have a coordinated, coherent presentation of these methods in a way that can introduce newcomers to these potent methods as well as warn them to some extent about the currently understood pitfalls.

In writing this book I drew upon the help of a large number of people who have contributed to the field, and to begin to list everyone from whom I have extracted information and obtained good ideas only invites omissions. They are all included in what I hope is a comprehensive literature review through the end of 1983.

Because this sort of volume attempts to present the current state of a rapidly growing and changing field, it may become obsolete in a few years. However, I have tried to incorporate some basic concepts that are not expected to change. It is hoped that this volume will prove useful in the attempt to set in perspective the current state of the art in receptor models.

It is necessary that I thank a number of individuals who have contributed significantly to the preparation of this work. Because of my limited understanding of microscopic methods, I have relied heavily on Russ Crutcher of Microlab Northwest, who has allowed me to use much of his written material in producing Chapter 3, and has provided some of the micrographs used to illustrate it. Ron

Draftz of IITRI has also helped greatly with discussions of microscopy and with other micrographs. Gary Casuccio of Energy Technology Consultants has kindly provided many of the electron micrographs as well as a very useful introduction to automated scanning electron microscopy. Rich Linton of the University of North Carolina has also kindly provided some of the micrographs from his work. Without the assistance of these individuals, it would not have been possible to include microscopic techniques, and their loss would have been a serious one to the utility of this volume. Finally, I want to thank Victoria Corkery for the difficult task of typing all of my cutting, pasting, and chicken scratches into a comprehensible volume.

PHILIP K. HOPKE

Urbana, Illinois
January 1985

CONTENTS

Receptor Modeling in Environmental Chemistry

CHAPTER

1

INTRODUCTION TO RECEPTOR MODELING

In order to develop strategies to manage the quality of environmental systems, it is necessary to understand the flow of materials through the system, the sources and sinks, transformations, and transport pathways of the species of interest. An approach that has been used extensively in air pollution control models the emission and dispersion of materials from point, line, or area sources after making a series of assumptions regarding the nature of the atmospheric mixing processes. These dispersion models are then used to predict the results of possible control scenarios or to predict the effects of siting new sources in the airshed.

The development of any model requires that a number of assumptions be made as to how to approximate reality. There is always a question about how accurate the models are in their predictions and their ability to develop the most effective management strategy. The basic concept of air quality models has been reviewed by Budiansky (1980). It is clear that these models have been very useful, particularly in areas with substantial air quality problems.

However, the implementation of strategies based on these models has not always brought about the degree of improvement expected. For example, in 1970, efforts began in Portland, Oregon to solve an ambient air quality problem. The airshed had too high a level of airborne particulate matter. Examination of emission factors and testing of industrial sources led to an emissions inventory that was then used in a proportional rollback model to predict air quality improvement. Additional control equipment was installed, leading to a 60,000 tons per year reduction in industrial emissions in the region. However, the progress toward cleaner air was substantially less than anticipated, since air quality violations continued to occur, caused by unknown sources that had to account for over half of the measured particulate concentration levels. The inability to accurately predict the impact of existing sources threatened to force curtailment of economic growth in the region, because permitting new sources was predicted to exacerbate the existing air quality problem.

There are substantial problems in predicting short-term source impacts as well as in identifying all of the sources, particularly nonducted sources. The problems associated with complex, multisource urban areas are substantial and it would, therefore, be extremely useful to have alternative approaches to identifying sources of materials and quantitatively assessing their impacts on environmental samples. Most of the discussion will focus on suspended particulate matter, since this is the area that has had the greatest attention over the last 15 years. The basic principles and the techniques presented in this volume will be applicable to a wide variety of environmental research problems.

Receptor models are based on the concepts of conservation of mass and a mass balance analysis. For example, suppose that in an airborne particle monitoring program, samples of particulate matter are collected and analyzed for several elements including lead. We can assume that the total volumetric lead concentration (ng/m^3) measured at the site is a linear sum of contributions from independent source types such as motor vehicles, incinerators, smelters, and so on,

$$\mathrm{Pb}_T = \mathrm{Pb}_{\mathrm{auto}} + \mathrm{Pb}_{\mathrm{incin.}} + \mathrm{Pb}_{\mathrm{smelter}} + \cdots \tag{1.1}$$

where the sum would include all of the sources of airborne particles that contain lead. We can extend the analysis by considering further the concentration of airborne lead contributed by a specific source. For example, a motor vehicle burning leaded gasoline emits particles containing materials other than lead. Therefore, the atmospheric concentration of lead from automobiles in ng/m^3, $\mathrm{Pb}_{\mathrm{auto}}$, can be considered to be the product of two cofactors: the gravimetric concentration (ng/mg) of lead in automotive particulate emissions, $a_{\mathrm{Pb,\,auto}}$, and the mass concentration (mg/m^3) of automotive particles in the atmosphere, f_{auto}:

$$\mathrm{Pb}_{\mathrm{auto}} = a_{\mathrm{Pb,\,auto}} f_{\mathrm{auto}} \tag{1.2}$$

A similar partitioning can be made for each source type. The normal approach to obtaining a data set for receptor modeling is to determine a number of elements in a number of samples. The mass balance equation can thus be extended to account for all m elements in the n samples as contributions from p independent sources,

$$x_{ij} = \sum_{k=1}^{p} a_{ik} f_{kj}, \qquad \begin{array}{l} i = 1, m \\ j = 1, n \end{array} \tag{1.3}$$

where x_{ij} is the ith elemental concentration measured in the jth sample, a_{ik} is the gravimetric concentration of the ith element in material from the kth source, and f_{kj} is the airborne mass concentration of particles. This kind of reasoning can be extended to a variety of properties regarding the sample and to different kinds of samples. The basic premise of receptor models is that properties of material collected in the environment can be used to infer their origins. These

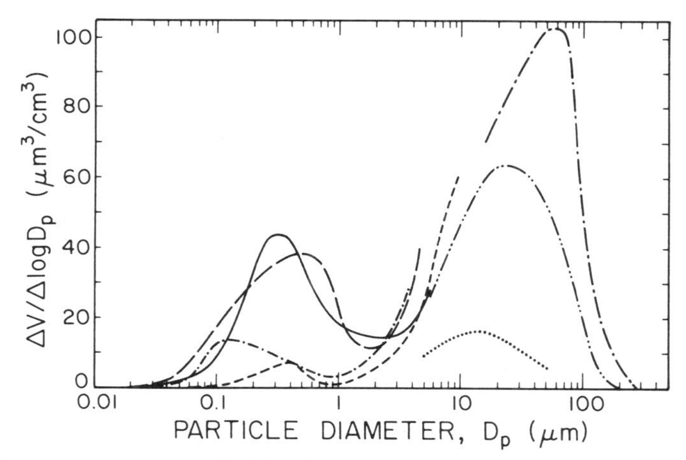

Figure 1.1. A comparison by Whitby, Husar, and Liu (1972) of several volume distribution functions. The volume increment per log size interval, $dV/d(\log D_p)$, is plotted on a linear axis versus size on a log axis. The area under the curve represents the total volume. The locations are: (———) Pasadena, California; (— —) and (— · —) Minneapolis, Minnesota; (- - -) Fort Collins, Colorado; (— - —) Seattle, Washington; (— ·· —) Germany (various locations); (· · ·) Japan.

properties include characteristics of individual particles, such as the physical parameters of particle size, shape, color, density, or optical properties, and the chemical composition parameters for individual particles or collections of particles, including elemental distribution, identification of specific organic or inorganic compounds, or the determination of isotopic or radionuclide abundances for specific elements.

The interest in characterization of airborne particles has led to the development of a variety of improved sampling methods that help to separate samples into less complex mixtures. For example, urban airborne particles typically demonstrate a multimodel size distribution as shown in Figure 1.1. Different types of sources primarily produce particles with one of the particular modes. Different particle formation processes lead to different size airborne particles. Thus, separation of the ambient particles into size classes before further examination for specific properties will tend to group particles from a limited number of sources, whereas examining the whole size range distribution forces the simultaneous study of a much larger number of source types. The development of instruments like the dichotomous sampler (Dzubay and Stevens, 1975) that collects samples of fine (<2.5 μm) and coarse ($2.5 \leq d \leq 15$ μm) particles makes the use of receptor models for source apportionment of aerosols more practical.

One of the other reasons for the rapid development of receptor models has been the rapid, concurrent development of a variety of analytical methods that permit greatly improved analyses of samples to provide the needed input data. These methods include multielemental techniques such as energy-dispersive X-ray fluorescence with excitation by photons (XRF) or charged particles (PIXE), instrumental activation analysis with neutron- or photon-induced radioactivity, and inductively coupled plasma emission spectroscopy. The coupling of improved scanning electron microscopes with X-ray fluorescence analyzers and computer control and image processing has led to the development of computer-controlled scanning electron microscopy. Improvements have taken place in X-ray diffraction methods that have also increased its utility for receptor modeling. In addition to these methods, optical and electron microscopy has been used as extensively in particle identification studies.

These new analytical techniques have caused modifications to the sampling methods and materials in order to take advantage of the full analytical capabilities of a method. For example, an ideal sample for energy-dispersive X-ray fluorescence analysis is a thin, uniform deposit of material on a nonexcited backing material. The dichotomous sampler was designed not only to separate samples into two size ranges, but also to collect the samples on filter media such that they could readily be analyzed (Dzubay and Stevens, 1975). In addition, new mass measurement methods were also developed so it was possible to automatically collect particle samples, and determine the mass and up to 27 elements in large numbers of samples. For example, over 34,000 samples of airborne particles were collected and analyzed in the Regional Air Pollution Study of St. Louis, Missouri (Nelson, 1979; Goulding, Jaklevic, and Loo, 1981). Other large-scale programs have been conducted as a result of the advances in sampling and analysis techniques.

The field of receptor modeling has been in existence informally for as long as efforts have been made to identify the origins of materials of interest. Its use in environmental studies has recently been developing rapidly into a formalized area of study with particular emphasis on airborne particulate matter problems. The U.S. Environmental Protection Agency has now begun to formally recognize the utility of these methods by the development of a series of guidelines, the first of which gives an overview of these models (O.A.Q.P.S., 1981a). An extensive literature has grown from a few key initial papers of 10-20 years ago. Three major symposia and two EPA-sponsored workshops have been held in the last 4 years to bring together the people developing various aspects of this field and those researchers interested in initiating such studies. The proceedings of the symposia (Kneip and Lioy, 1980; Macias and Hopke, 1981; Dattner and Hopke, 1983) have provided useful compilations of papers to both summarize the current state of developments in the field and show potential new directions for research. The 1980 EPA workshop was summarized by Watson (1981) and many of the results of the 1983 workshop are in the Dattner and Hopke volume, with more detailed reports to be published in the journal literature. There is thus a large, diverse literature in the field. There have been several summary papers (Gordon, 1980; Cooper and Watson, 1980), but they have rapidly gone out of date with the new developments and there is still no unified presentation of the body of material relating to the analytical and mathematical procedures that have been developed and applied to various environmental problems. It is the purpose of this volume to outline the analytical methods so that their capabilities and limitations for receptor modeling purposes are presented and to then describe how the data that result can be employed to identify and quantitatively apportion environmental materials to their sources.

CHAPTER

2

SAMPLING AND ANALYTICAL METHODOLOGIES

In the discussions of the various mathematical models to follow, the input data to those calculations are the results of the characterization of ambient and/or source material samples. These data are obtained by collecting environmental samples and determining their characteristics through the application of some set of analytical protocols. In some cases, the analytical procedure consists of detailed microscopic examination of the sample by a light or an electron microscope. These topics are dealt with directly in subsequent chapters. As with most of the discussions in this volume, the primary focus is on the collection and characterization of airborne particulate matter samples. However, many of the analytical methods can be applied to determination of the composition of other kinds of samples and would yield similar data for the receptor models. We first discuss airborne particle sampling and then analytical methodologies.

SAMPLING

The ability to obtain a sample of material that is, in fact, representative of the ambient environment is an extremely difficult task. In general, there will always be some aspects of the sample that prevent it from being ideal. However, if those defects are understood, the results obtained from its analysis can be placed in the proper perspective and possible systematic biases can be estimated.

The general method of sampling airborne particles is to use a pump to pull a known volume of air through a filter so as to collect the particles. The airstream can be processed on its way to the filter so that only particles fulfilling some set of predetermined aerodynamic characteristics are allowed to reach the filter. The characteristics of the filter govern the efficiency of particle collection and the presentation of a sample suitable for a particular analytical technique. The accurate quantitation of the volume of air sampled is also an important consideration if the ultimate data set is to be reliable and fully useful to the purpose of quantitative source apportionment. All of these aspects of sampling must be considered in the development of the experimental design of the sampling and analytical program.

SAMPLERS

The most widely used airborne particle sampler is the high-volume sampler that is the designated standard method for sampling total suspended particulates (TSP) and determining compliance or noncompliance with the Ambient Air Quality Standard for Particulate Matter (CFR, 1975). This very simple device was first developed and put into wide use in early 1960s (Robson and Foster, 1962). There is a housing that shelters the filter holder from direct precipitation and does to some extent act to keep very large particles from reaching the filter. The exact maximum size particle collected by the sampler strongly depends on the ambient wind speed and direction (Wedding et al., 1977). Thus, although this is the "standard method" for collecting airborne particulate samples, it is not a very desirable sampler in that it does not provide samples with a well-defined range of particle sizes. Another problem with the sampler is that it permits wind-blown dust to accumulate on the filter even when the pump is not running. This "passive loading" can lead to an overestimate of total airborne particulate levels. Modifications to the samplers are now available which eliminate this concern, but it does represent a problem for most of the historical samples collected with this device.

In addition, the standard medium for particle collection is glass fiber filters. Although these filters do have a high retention capacity for particles greater than 0.3 μm, they have a substantial ability to produce artifact sulfate on the filter (Pierson et al., 1980) that may distort the TSP measurement. These filters are also not suitable for analysis by instrumental neutron activation or X-ray fluorescence analysis. The particulate matter can be leached and the dissolved material analyzed by atomic absorption or inductively coupled plasma emission spectroscopy. However, the quantitative recovery of the material from the filter is difficult and time consuming. Thus, this combination of sampler and filter material is not well suited to receptor modeling studies.

An initial approach to improved sampling is to utilize an inlet to provide an upper size cutoff. Current designs are for 50% efficiency collection at 10 μm (Wedding and Weigand, 1982; McFarland and Ortiz, 1982). These samplers eliminate most of the wind speed and direction problems as well as being able to prevent most of the passive loading difficulties. These inlets represent a significant improvement in the collection of a well-defined sample.

To examine the sources of particles, however, it is useful to recognize that different types of source processes give rise to different size particles. Figure 2.1 illustrates that fine particles (<2 μm) arise primarily from condensation of vapor or chemical conversion or gaseous compounds to low volatility vapors that can nucleate. These initially very small particles coagulate into the relatively stable accumulation range particles. Coarse particles (>2 μm) are generated principally by mechanical action. Thus, separation of particles into different size categories

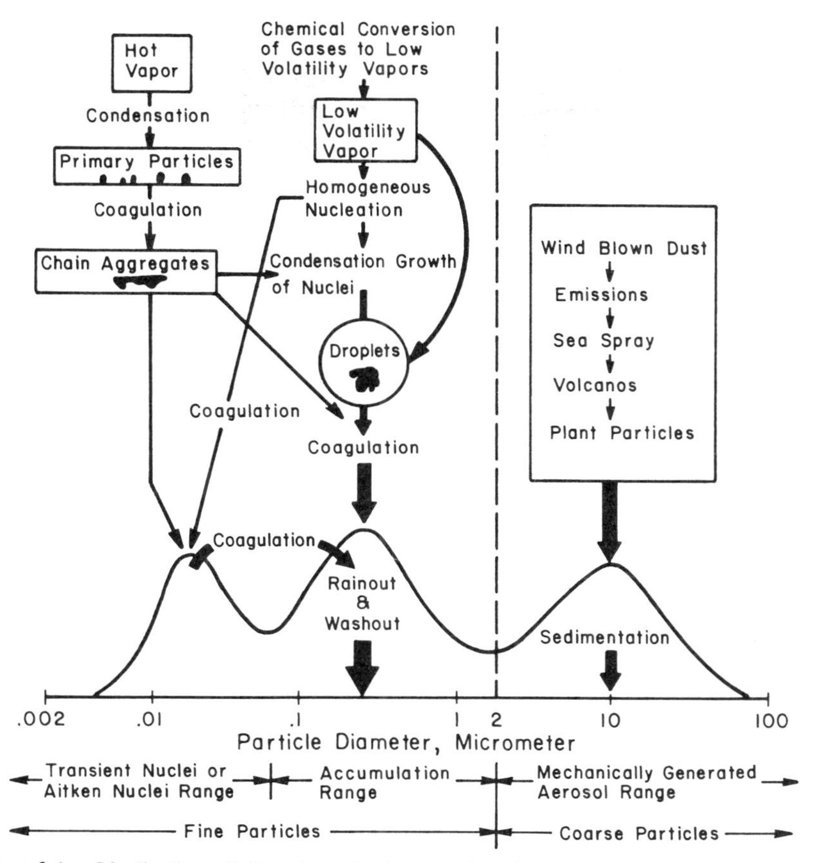

Figure 2.1. Idealization of the atmospheric aerosol surface area distribution showing the principal modes, main sources of mass for each mode, and the principal processes involved in injecting mass in each mode as well as primary removal mechanisms. Taken from Whitby and Cantrell (1976) and used with permission of IEEE. Copyright 1976 IEEE.

may provide samples that are affected by fewer sources and that are easier to analyze and interpret. A device developed to provide such a separation is the dichotomous sampler alluded to in Chapter 1.

The originally designed dichotomous sampler is shown schematically in Figure 2.2. It was designed to separate samples at a 50% efficiency for a 2.5-μm aerodynamic diameter particles and to exclude large particles by having a 50% efficient inlet for 15-μm-diameter particles. The values of 2.5 and 15 μm were chosen primarily to match the important separation sizes that occur in the human respiratory tract, but they are generally useful size ranges for source separation as well.

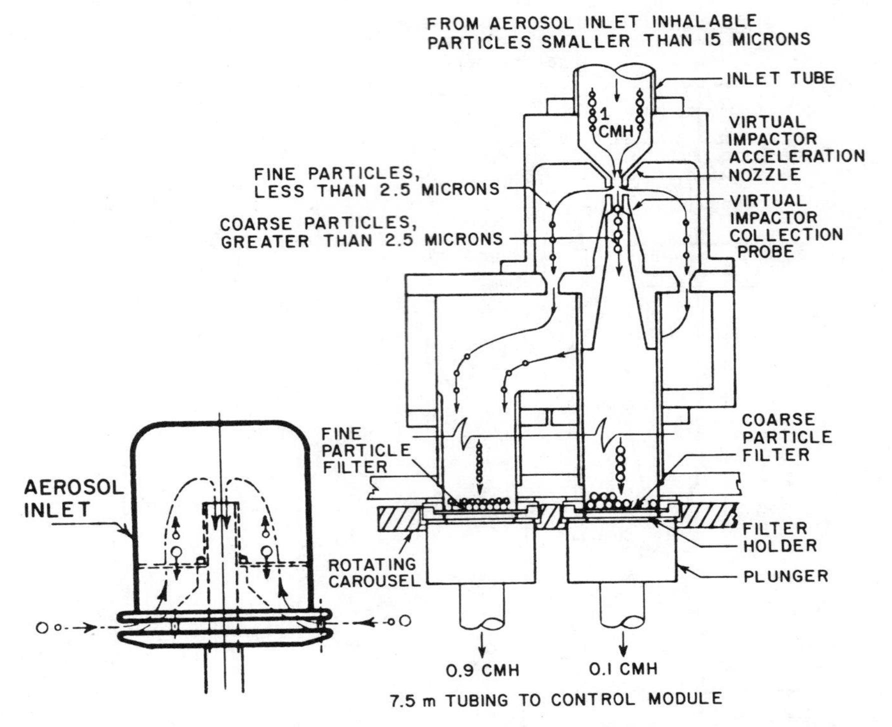

Figure 2.2. Design of a dichotomous particle sampler for separating airborne particulate matter into two size fractions. Drawing courtesy of Sierra Instruments, Inc., Carmel Valley, CA.

There have been several intensive evaluations of these samplers to investigate the accuracy of their aerodynamic sizing and their sensitivity to wind speed and direction effects. Wedding et al. (1980) showed that the size cutoff was strongly dependent on wind speed and that there could be considerable variation in the size particle that could penetrate the sampler to the course size sample. John et al. (1983) have also shown that nozzle concentricity was a difficult parameter to measure but when determined, samplers could be found to be significantly out of specified tolerance ranges. This lack of tolerance affects the rate of wall loss for particles 3 μm and larger.

Recently there has been a shift in interest to a sampler with a 50% efficient size cutoff to 10 μm and new inlets have been designed to provide the separation and also address the wind speed dependency of the previous designs (Liu and Pui, 1981). The dichotomous sampler thus is a very useful device to provide well-characterized size range samples.

The particles are collected in the sampler on membrane filters. Currently, the most commonly used are Teflon filters. In the past, Nucleopore polycarbonate and cellulose acetate filters have been employed. These filters provide samples that are well suited for automated mass measurement using beta gauge techniques (Jaklevic et al., 1981a) and for elemental analysis using X-ray fluorescence and/or instrumental neutron activation analysis.

Other sampling devices can be used for particle collection. There has been considerable use of cascade impactors that provide multiple samples with various size range particles collected on each stage. These samplers can be single- or multiple-jet devices that sample air over a wide range of flow rates. The impactor is designed so that the air stream carrying the entrained particle is forced to change direction rapidly. Particles with sufficient inertia detach from the streamline and impact on the collection surface. A major difficulty with impactors is the bounce of particles after collision with the collection surface. Problems of this nature have been reviewed by Esmen and Lee (1980). It is necessary to exert some care and effort to minimizing this problem in order to yield properly size-segregated samples. One approach has been the use of a virtual impaction surface as in the dichotomous samples. However, the virtual impactor concept has not been extended beyond the one-stage separation in the dichtomous sampler. Coatings have often been applied to lower bounce in conventional multistage impactors (Lawson, 1980). In some cases with low-volume flow systems, the impactor has been used upside down so that particles that bounce are forced back onto the same impactor stage. The individual particle size-range samples can then be analyzed. They are generally quite suitable for instrumental analysis although there is usually only a small amount of mass collected. Typical urban size distributions have been given by Gladney et al. (1974), Paciga and Jervis (1976), and Kowalczyk, Gordon, and Rheingrover (1982). The additional information available in the elemental size distributions has not yet been employed effectively in receptor models that rely on collected particle composition data.

In addition to ambient air sampling, there is the problem of taking source samples that are representative of the emissions from that plant. For ducted emissions, it is possible to obtain a gas stream containing the materials about to be released. However, even after passing through the pollution control devices, the material in the stack may be physically and chemically different than that plant's emissions will be when they get to the sampling site.

Pollution control devices such as electrostatic precipitators and venturi scrubbers are quite efficient for large-particle collection. However, each large particle uncollected carries very much more mass than a small, uncollected particle. Yet the large particles, when they reach the ambient atmosphere, may settle out before before reaching the ambient sampling site while the small particles do not. Thus, the size distribution must be taken into account when performing the stack sampling. Often particle samples are taken using an in-stack cascade impactor in the place of the filter on an EPA Method 5 system

(Rom, 1972). A review of the properties of the most commonly used source-test impactors is given by Cushing, McCain, and Smith (1979). Although the particle sizing techniques based on aerosol dynamics can be used to estimate the physical properties of the stack emissions at the collection site, chemical considerations should be included in developing an effective stack sampling strategy.

The gases and associated particles in a stack are typically still at an elevated temperature after the control devices and before emission. Thus, volatile species will tend to be in the gas phase and not associated with the particulate phase. Yet when the emissions in the plume begin to cool, those species will then adsorb onto the particle surface and change the composition of the particles from what they were in the stack. Pillay and Thomas (1971) have found that a substantial fraction of selenium in coal-fired power plant stacks is in the gas phase. Kowalcyzk, Choquette, and Gordon (1978) suggest similar problems may exist for arsenic and antimony and that the stack samples may substantially underestimate these concentrations. Similarly, organic compounds such as polycyclic aromatic hydrocarbons may also be underestimated using hot flue gas samples.

Another important aspect of the volatilization-condensation phenomenon is that the concentration of the volatile elements that then adsorb onto the particle surface will show a particle size dependence to their concentration. Although the details of the enrichment mechanism are the subject of debate in the literature, it is clear the smaller particles tend to be more highly enriched in volatile elements (Fisher and Natusch, 1979; Ondov, 1981; Hansen and Fisher, 1980; Ondov, Ragaini, and Biermann, 1979). Ondov, Ragaini, and Biermann (1978) have used a reduced pressure cascade impactor to subdivide in-stack coal-fired power plant particles below 1 μm and have found very substantial arsenic and selenium enrichments on these particles (Biermann and Ondov, 1980). Thus, not only may a cascade impactor be needed, but one that will allow fractionation to below 0.1-μm diameters may provide extremely useful data on source characteristics. Unfortunately very few such data are currently available.

In response to these considerations, there has been some substantial improvements in source sampling methods. Houck, Core, and Cooper (1982) have developed a system that extracts a stream of flue gas from the stack, dilutes and cools it, and then samples it with a dichotomous sampler. Thus, the stack samples obtained are similar in nature to those ambient samples obtained with the same device. Alternatively, plume samples can be obtained. Previously, aircraft have been used to sample plumes (e.g., Small et al., 1981a). However, it is difficult to spend long periods of time in the plume and the plane must be carefully modified to permit representative sampling without contamination from propellar wash or engine exhaust. It is expensive to modify aircraft and difficult to get the modifications certified by the FAA. Thus, only limited aircraft sampling has been done.

Another approach is to use tethered balloons to hold a sampler in the plume (Armstrong et al., 1981). Current systems do not permit sufficiently large payloads to allow for large enough samples or the use of size segregating samplers. This approach has been useful in collecting samples for microscopic examination and it may be possible to develop it further for source sampling.

It is much more difficult to obtain representative materials from fugative and area sources. There has been much discussion over the validity of collection of soils and their suspension by ambient sampling devices. This approach is certainly better than bulk soil analysis since there will be size dependence to the composition of the material. Similar problems exist in obtaining source emission samples from a variety of materials handling operations such as ore crushing and moving. Heavy truck traffic over unpaved lots yields a large airborne particle mass loading with nontypical source composition when nonsoil materials such as ore is being transported in the area. Size segregation helps since these mechanical processes primarily produce large-particle aerosols. However, the acquisition of truly representative source samples poses some of the greatest problems associated with a number of the receptor models requiring such data.

In other situations it may be of interest to determine the sources of a settled

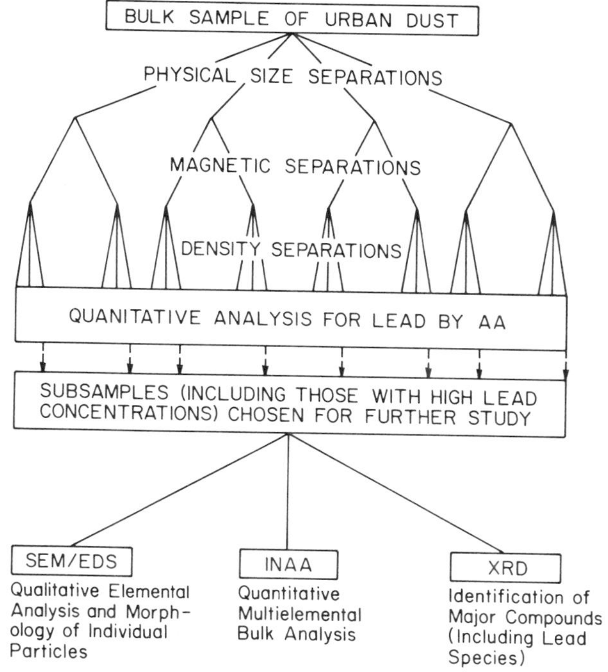

Figure 2.3. Flowchart for the physical fractionation of settled dust samples. Reprinted with permission from Linton et al. (1980b). Copyright 1980 American Chemical Society.

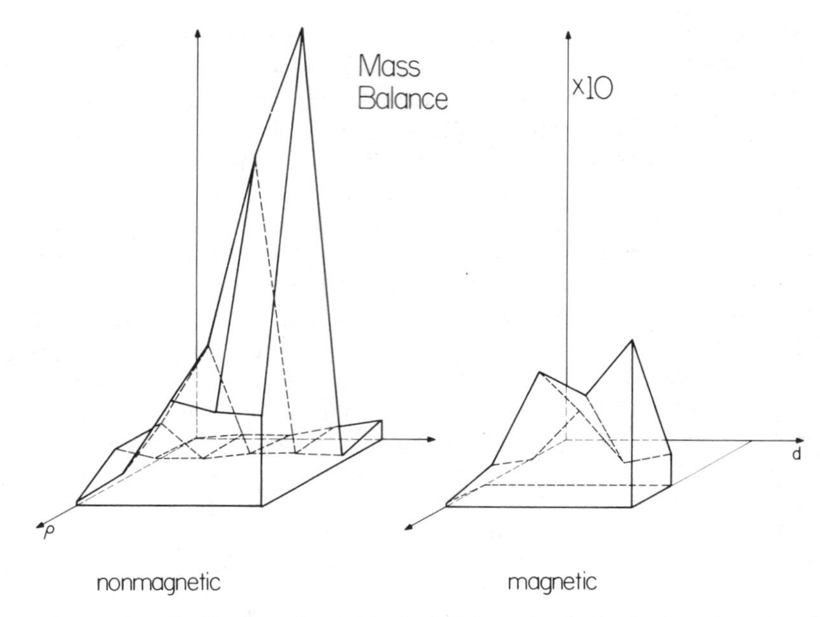

Figure 2.4. Mass distribution of a physically fractionated sample of urban street dust. The mass of each sample is displayed as a function of particle diameter and density for the nonmagnetic and magnetic fractions. Reprinted with permission from Hopke, Lamb, and Natusch (1980). Copyright 1980 American Chemical Society.

dust. For example, if the sources of lead and other toxic trace metals in an urban street dust are to be determined, it is helpful to collect and fractionate the sample in such a way as to provide more information on the individual particles. A system of particle separations is outlined in Figure 2.3 and has been applied to several urban street dusts (Hopke, Lamb, and Natusch, 1980; Linton et al., 1980a,b) and indoor dusts (Hopke, 1978). In this approach the particles are first dry sieved for particle sizing and passed by a magnet to collect magnetic materials. The sized, magnetic/nonmagnetic particles are then density separated by a float/sink technique. Mixtures of heavy liquids such as chloroform and methylene iodide can be prepared to give a known specific gravity. Particles that float in the mixture have a density less than that specific gravity and those that sink have greater. Adequate time must be given to allow sedimentation to take place. There is the strong possibility that some of the material in the sample will dissolve. However, it does provide an approach to separate particles into a number of different categories that can be individually examined; particularly characteristic particles can there be selectively collected in a given fraction by proper choice of conditions. The mass distribution of an urban street dust sample is given in Figure 2.4 (Hopke, Lamb, and Natusch, 1980).

FILTERS

Another important consideration in obtaining a useful sample is the material on which it is collected. The filter should be retentive of particles but permit ready air flow. It should provide the sample in a manner that makes it easy to quantify both the total mass of collected material and the chemical composition of the sample. Although these conditions sound straightforward, they present several inherent and unresolvable conflicts.

There are two general types of filters, fibrous and pore-type. The fibrous filters include the glass fiber filters already mentioned, quartz fiber filters, and paper. In effect, they capture particles as the air moves through and around the randomly positioned fibers that have been pressed together into a mat often with a binding material to help keep the filter together. These filters have relatively low pressure drops across them and are, therefore, well suited to high-volume sampling. The glass and quartz fiber filters have a high retention of particles with sizes above 0.3 μm. Paper filters have somewhat lower retention when sampling begins and improve in their collection capability as the filter begins to develop a loading of particles. Neustadter et al. (1975) found that Whatman 41 paper could be employed in ambient monitoring with insignificantly lower collected mass if proper care were taken in the humidity equilibration that must be conducted before taking exposed or unexposed filter weights. The glass fiber filter has a now well-known positive artifact mass owing to the in-situ conversion of sulfur dioxide to sulfate. The quartz fiber and paper filters do not show appreciable sulfate artifacts (Pierson et al., 1980). The use of quartz fiber filters can provide accurate mass values that neither paper nor glass fiber filters can provide, although extra care must be taken in handling them because they tend to be friable.

An important use of glass and quartz fiber filters is in the collection of particles for organic compound analysis. It is essential that the filter have low blank values of the compounds to be analyzed. For example, in the ATEOS Program in New Jersey (Lioy and Daisey, 1983), glass fiber filters are being used for collecting particulate samples for extractable organic material, polycyclic aromatic hydrocarbons (PAH), alkylating agents, and mutagenic activity. The ability to bake these filters at high temperatures before sampling also helps in lowering organic blank values. The samples collected on quartz or high-purity glass fibers can aso be analyzed for trace metals by acid leaching and spectrophotometric methods such as atomic absorption (Kneip et al., 1971) or inductively coupled plasma optical emission (Broekaert, Wopenka, and Puxbaum, 1982). These methods entail the preparation of the liquid extracts and, therefore, are destructive of the sample.

For receptor modeling purposes, it is more common to collect the sample on pore-type filters that provide samples better suited for neutron activation, X-ray fluorescence, or microscopic analysis. The filters commonly used for particle sampling include Nucleopore polycarbonate and polytetrafluoroethylene filters produced by several manufacturers. Spurny et al. (1969) provided the first review of these relatively uniform size pore filters. There have been a number of subsequent tests of the filtration efficiency of these membrane type filters (Liu and Lee, 1976; John and Reischl, 1978). The collection efficiency depends on the pore size, particularly for the Nucleopore filters. For example, John and Reischl found 0.8-μm Nucleopore filters had only a 72% efficiency for submicron particles observable with a condensation nuclei counter. In contrast, 1–3 μm pore Ghia fluorocarbon filters and 1-μm Fluoropore filters are >99.9% efficient under the same experimental conditions.

Although there can be some difficulties in handling the fluorocarbon filters when preparing them for neutron activation, they seem to be the filter of choice for inorganic analysis. They also have proved to be a useful substrate for X-ray diffraction analysis of particle samples (Davis and Johnson, 1982). One problem that has arisen with the fluorocarbon filters is the loss of coarse particles from the filters in handling and transport from the sampling site to the laboratory. Recently, Dzubay and Barbour (1983) have suggested the use of an oil coating on the filters to prevent such shakeoff effects.

The Teflon filters, since they are chemically inert, are particularly good for sampling with minimal artifact production of sulfates on the filter. However, there may be negative artifacts for ammonium nitrate and particulate nitric acid. These species have sufficiently high-vapor pressure, particularly at high ambient temperatures, that substantial nitrogenous material can violatilize from the filter after collection. Recently, new methods for nitrate collection on nylon filters and analysis have been developed and tested (Grosjean, 1982; Spicer et al., 1982). It seems necessary to utilize alternative sampling methods to properly determine the concentrations of airborne particulate nitrate species.

ANALYTICAL METHODS

Mass

One of the primary measurements that can be made for a sample of airborne particles is the total mass loading. The traditional approach is to equilibrate the filter at <50% relative humidity and weigh it, then collect the sample, reequilibrate it, and weigh it. This procedure provides good results for larger samples where sufficient materials can be collected. It can be done using microbalances for smaller filters and air volumes. However, it is not a process that is easily automated. Therefore, alternative methods have been developed.

The most widely used alternative is β-gauging. A radioactive source of an isotope like ^{14}C is used to provide a constant flux of energetic electrons to the filter. The presence of the filter between the radioactive source and the detector lowers the total detected radiation. The reduction is proportional to the mass loading on the filter. Although the exact relationship includes average atomic number and particle size effects, the procedure yields a good estimate of the gravimetric mass (Husar, 1974). The technique has been automated and refined by Jaklevic et al. (1981a). The precision and accuracy have been assessed by Courtney, Shaw, and Dzubay (1982) and found to be of the order of 20–25 μg with a bias between gravimetric and β-gauge values of less than 5%. Thus, the β-gauge is capable of providing sufficiently accurate mass measurements and is readily incorporated in an automated sampling and analysis protocol.

Energy Dispersive X-Ray Fluorescence

One of the most commonly used methods of elemental concentration analysis is energy dispersive X-ray fluorescence (XRF) analysis using either photons or charged particles to induce the X-ray emission. The basic concept of X-ray fluorescence is that an inner electron orbital electron is knocked out of the atom by the action of an incoming photon or charged particle. The excited atom then fills the inner electron vacancy by the transition of an outer shell electron to the inner shell vacancy. The release of the excess binding energy may occur as a characteristic X-ray. The X-ray has an energy dependent only on the energies of the two bound electron orbitals that are involved in the transition and it, therefore, has an energy that is characteristic of the element. For thin samples, the intensity of the emitted X-rays depend on the number of excited atoms in the viewing range of the detector. The number of excited atoms depends on the mode and strength of excitation radiation and the number of atoms of the particular element which are available for excitation. The theoretical development and methodology of X-ray fluorescence analysis is described in great detail by Bertin (1975). The applications of X-ray fluorescence methods to environmental samples are described in the collection of reports edited by Dzubay (1977).

Samples of atmospheric particles collected on membrane-type filters are very well analyzed by XRF methods. These samples are generally thin compared to the range of the exciting radiation (photons or charged particles) in the sample. It is relatively easy to automate the analysis procedure such that fairly large numbers of samples can be analyzed in a reasonable time. For example, in the Regional Air Pollution Study (RAPS) of the St. Louis, Missouri airshed, over 34,000 samples were obtained with automated dichtomous samples, analyzed for mass using a β-gauge, and for 27 elements by X-ray fluorescence analysis.

The production of the X-rays using photons as the primary excitation radiation requires a source of X-rays. To maximize the signal to background with energy dispersive detection, monoenergic photons are employed. However, the sensitivity of detecting an element is highest when the energy of the excitation radiation is just above the binding energy of the electron and drops off rapidly as the difference between exciting and binding energies increases. Thus, to cover a range of elements, it is necessary to obtain fluoresced spectra with several different excitation energies. For example, in the RAPS study, excitation radiations were obtained from the secondary fluorescence of Ti (4.5 keV), Mo (17.4 keV), and Sm (40 keV) (Jaklevic et al., 1981b). Typical spectra for fine ($<2\ \mu m$) and coarse ($>2\ \mu m$) particles with molybdenum X-ray excitation radiation are shown in Figure 2.5.

An alternative method of X-ray excitation is particle-induced X-ray fluorescence. This approach has a more uniform efficiency of X-ray production as a function of atomic number. Thus, a broad range of elements can be measured with a single bombardment count. Particles used to produce X-rays include protons (Nelson, 1977) and alpha particles (Cahill, 1975). A typical spectrum with proton excitation is shown in Figure 2.6. This sample was deposited on a Nucleopore filter and was shown to yield a relatively high background in the low-energy portion of the spectrum caused by bremsstrahlung from the decelerating protons. The magnitude of the background is dependent on the thickness and composition of the material the sample is deposited on. Thus, the proper removal of the background is an important part of the analysis of the X-ray spectral data.

Particle-induced X-ray fluorescence has the advantage that the particle beam can be sharply focused to give good spatial resolution. If samplers can be used that spread the collected particles over a collection surface as a function of time, then this spatial resolution of analysis can yield a time resolution of the aerosol composition. The group at Florida State University developed a moving orifice device (Nelson, 1977) to provide such samples while the group at the University of California at Davis used a Lundgren impactor to provide similar samples (Cahill, 1975). About 2-hr time resolution can be obtained with these combinations of samplers and analytical methods yielding useful time sequences of elemental concentrations.

Another feature of particle-induced X-ray excitation is that it is possible to use the spectrum of scattered particles to measure the amounts of light elements present. Figure 2.7 shows a scattered proton spectrum taken for an air particulate filter. However, for best scattering results, a somewat higher energy particle beam is needed than would be optimum for X-ray production. More details regarding particle scattering are given by Cahill (1975) and Nelson (1977).

X-ray fluorescence is thus a very useful method for the analysis of airborne particle samples. It has the capability of detecting with sufficient sensitivity for most atmospheric samples, as shown in Figure 2.8 for photon-excited XRF and Figure 2.9 for particle-induced fluorescence. It can detect a variety of elements

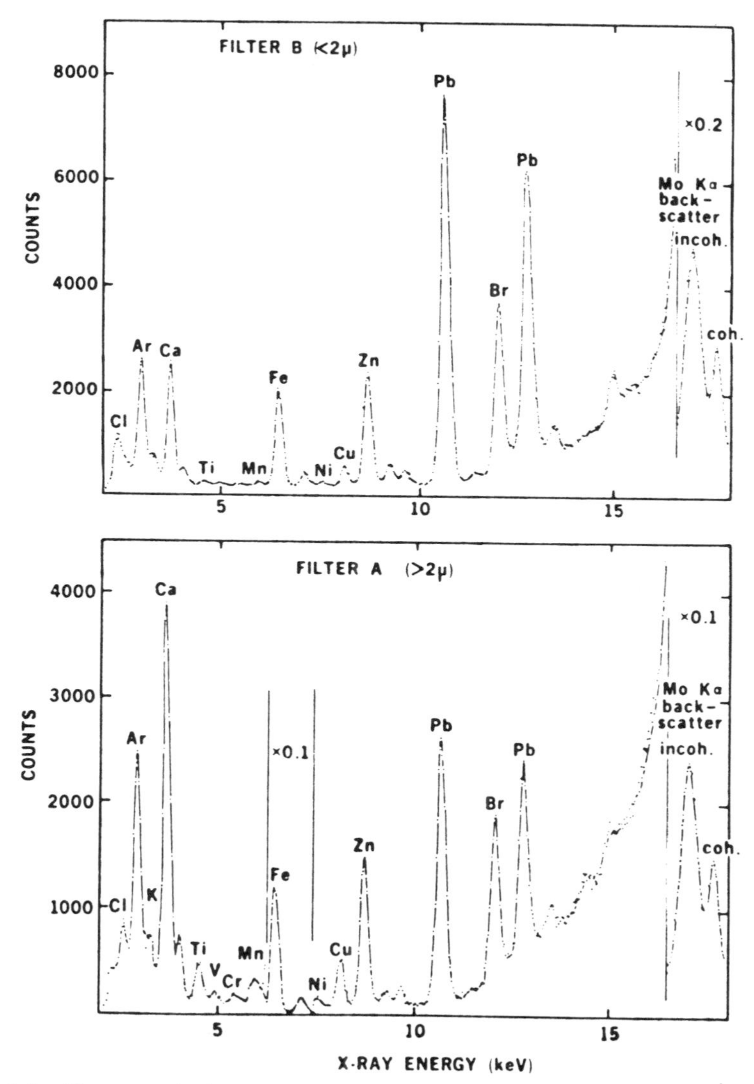

Figure 2.5. Photon-excited, energy-dispersive X-ray fluorescence spectra of a membrane filter containing the fine particle fraction (upper) and the coarse particle fraction (lower) from a dichotomous sampler. The figure is taken from Jaklevic, Loo, and Goulding (1977) and used with permission. Copyright 1977 Ann Arbor Science Publishers, Inc.

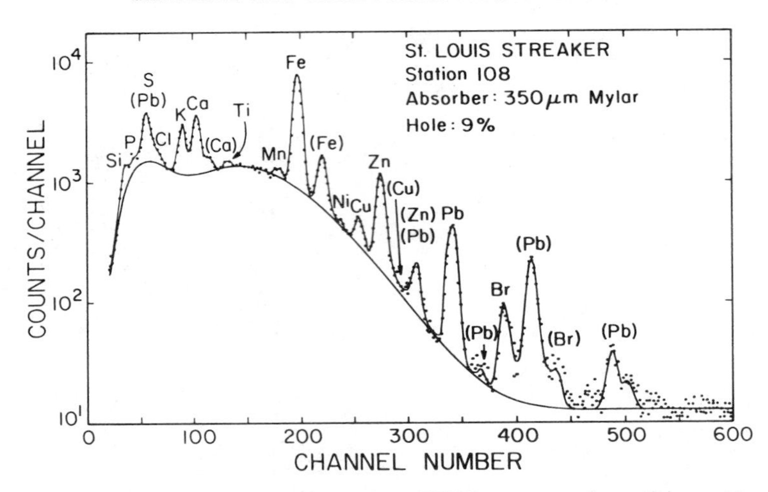

Figure 2.6. Particle-induced X-ray excitation (PIXE) spectrum for a 2-mm wide region (2-hr time interval) of a streaker sampler. The figure is taken from Nelson (1977) and used with permission. Copyright 1977 Ann Arbor Science Publishers, Inc.

that are useful in receptor modeling including silicon, sulfur, and lead. It does not typically permit the quantitative determination of elements lighter than aluminum and there are often problems with aluminum analyses, particularly in large-particle size samples. For a more complete elemental analysis, it is useful to have complementary methods, such as instrumental neutron activation.

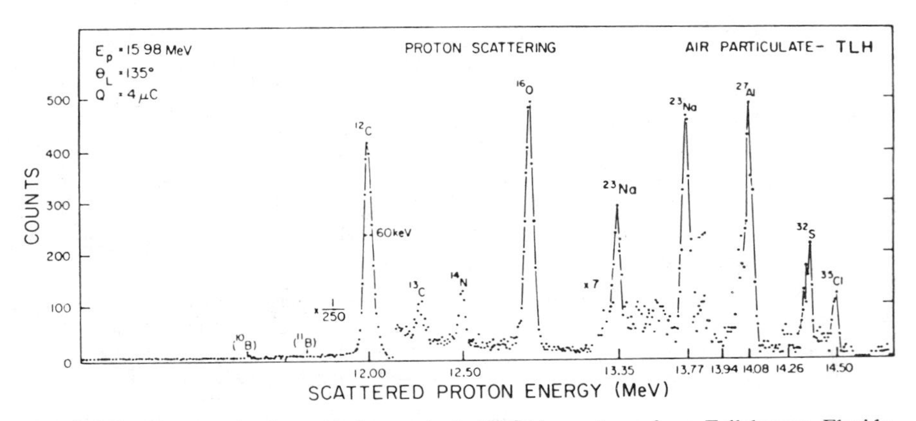

Figure 2.7. Proton elastic scattering analysis (PESA) spectrum for a Tallahassee, Florida impactor sample. Taken from Nelson (1977) and used with permission. Copyright 1977 Ann Arbor Science Publishers, Inc.

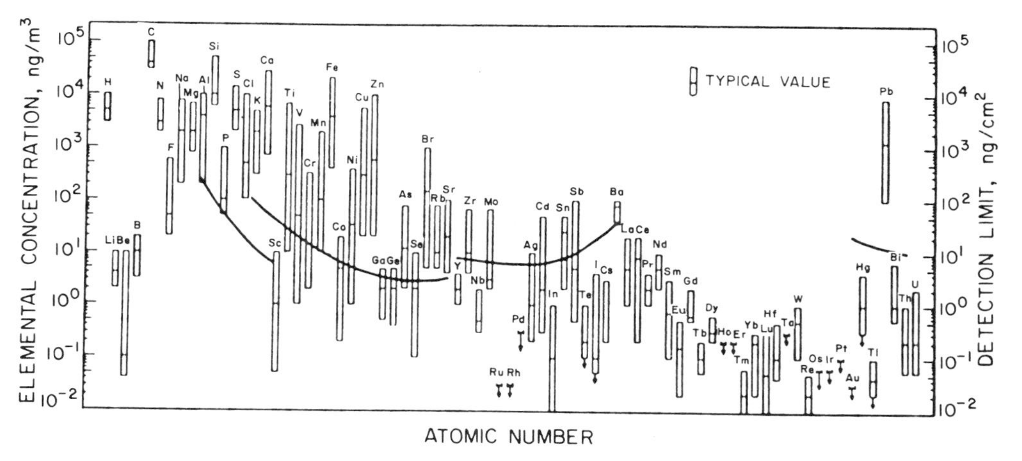

Figure 2.8. Minimum detectable limits in ng/cm² for an energy-dispersive X-ray fluorescence system. The curves are for the secondary fluorescers titanium, molybdenum, and samarium (left to right) and for molybdenum (far right). The rectangles represent typical upper and lower ranges of concentrations for urban aerosols. Common values are represented by the horizontal bars inside the rectangles. Figure taken from Jaklevic and Walter (1977) and used with permission. Copyright 1977 Ann Arbor Science Publishers, Inc.

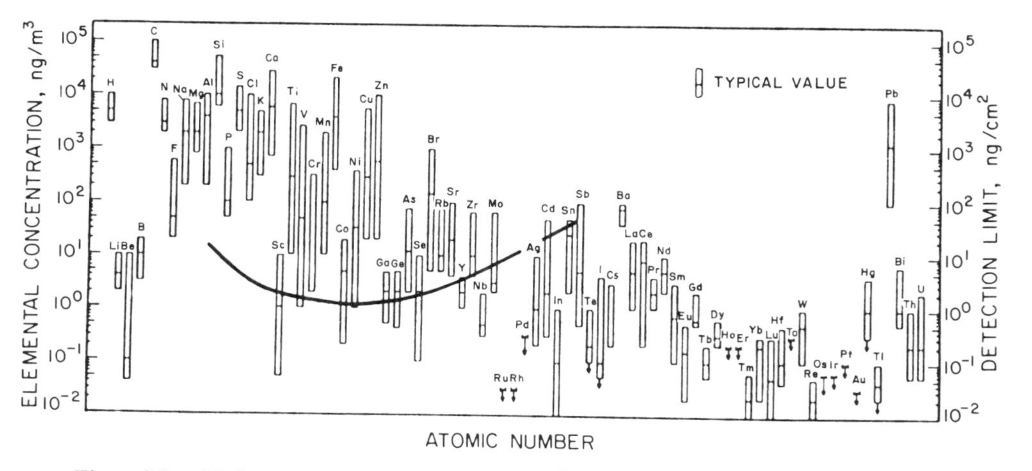

Figure 2.9. Minimum detectable limits in ng/cm² for the Florida State PIXE system described by Nelson (1977) using Nucleopore filters. See caption of Figure 2.8 for additional explanation. Figure taken from Jaklevic and Walter (1977) and used with permission. Copyright 1977 Ann Arbor Science Publishers, Inc.

Instrumental Activation Analysis

Activation analysis is the technique of converting a small fraction of the stable nuclei present in a sample to being radioactive and then relating the measured level of radioactivity to the initial concentration of an element in the sample. The nuclear transformation can be induced by irradiating the sample with thermal neutrons, high-energy neutrons, high-energy photons, or charged particles. For most nuclei, reactions can occur for each type of bombarding particle leading to a radioactive nucleus. The measurement and quantification of the resulting radioactivity leads to the elemental analysis.

The introduction of high-resolution, gamma-ray spectroscopy permitted the development of nondestructive instrumental activation analysis. Instrumental activation analysis is capable of trace multielement determinations with a minimum of sample preparation and postirradiation operations. There are four types of activation methods for converting stable nuclei to radioactive nuclei. These reactions are with thermal neutrons, fast neutrons, high-energy photons, and positively charged ions. Each type of projectile gives a different set of product nuclei. Since not all product nuclei will be radioactive and not all of the radioactive nuclei are analytically useful, the availability of more than one activation technique increases the number of elements in a sample that can be determined by activation analysis. The most widely used irradiation technique is with reactor or thermal neutrons. These neutrons have a low kinetic energy (0.025 eV). Because of the absence of an electrostatic barrier for the electrically neutral neutron, they can move to a distance within the range of the nuclear force (10^{-13} cm). The neutron can then be added to the stable nucleus to form a nucleus of the same element but heavier by one atomic mass unit. This new isotope may be radioactive. The flux of neutrons, that is, the number of neutrons crossing a unit area from any direction in unit time, range from 10^{11} to 10^{15} neutrons/cm^2-sec with typical values being 10^{12} to 10^{13} neutrons/cm^2-sec for most reactors. The probability of a nucleus capturing a thermal neutron is expressed as a cross-sectional area and is generally on the order of 10^{-24} cm^2.

The major disadvantage of thermal neutrons is that several elements of environmental interest such as silicon, sulfur, and lead do not activate to nuclei that decay by gamma emission, or have sufficiently low gamma emission rates to make the technique impractical. Thus, the combination of XRF and INAA provides complementary elemental analyses that can be very useful in obtaining a complete source resolution. These elements must be activated by other means, since gamma-ray spectroscopy is the only detection mode for instrumental analysis.

It is possible to produce neutrons with about 14 MeV of kinetic energy by using a charged particle accelerator (Nargolwalla and Przybylowicz, 1973). These energetic neutrons can then be used as projectiles to initiate reactions in the sample to be analyzed. The types of reactions involve the inelastic scattering of the neutron, leaving the residual nucleus in an excited state, the ejection of two neutrons from the nucleus, the ejection of a proton, or the ejection of more complex particles such as deuterons or alpha particles. Thus one of the advantages as well as the disadvantages of fast neutron reactions is that they lead to a number of possible products from a single initial nucleus. This is advantageous because it is likely that at least one product will emit gamma rays. It is a disadvantage in that if there is more than one way to form a given radioactive nucleus, there is greater difficulty in relating the number of radionuclides to the initial weight of various elements in the sample. For example, $^{27}Al(n, \alpha)^{24}Na$ and $^{24}Mg(n, p)^{24}Na$ yield the same product nucleus, ^{24}Na. However, by observing the decay of other reaction products, an analysis can be made. Another disadvantage of fast neutron activation is the relatively low flux of neutrons that can be obtained. Typical fluxes are on the order of 10^9-10^{10} neutrons/cm^2-sec. Finally, the cross sections for fast neutron reactions are generally 1-3 orders of magnitude smaller than thermal neutron cross sections. Thus, the sensitivities are generally 3-4 orders of magnitude poorer than with thermal neutrons.

It is also possible to activate samples with high-energy photons. If a high-energy electron is stopped in a target containing a high atomic number element such as tantalum, a continuous distribution of photons is generated with energy up to that of the electron. This bremsstrahlung radiation can then be used to irradiate samples. The photons can knock out a neutron, a proton, or complex particles. As in the fast neutron case, it is likely that there is at least one radioactive product that decays via gamma emission. The sensitivities for photon activation are generally much lower than those thermal neutron activation sensitivities obtainable for the same element. However, since there is zero thermal neutron sensitivity for some elements, photon activation does permit the analysis of lead and several other elements. The number of electron accelerator facilities where such activation experiments can be conducted is unfortunately limited and the expense of construction and operation of such a device is large. It is, therefore, unlikely to become a routine activation analysis method.

In general, charged particle activation is not very satisfactory. Multiple products can result from a bombardment and the sensitivities obtained as much poorer than those obtained with thermal neutrons. Thus, charged particle activation has only a very limited utility in analytical applications. Most of the activation analyses of ambient airborne particles that have been made have used thermal neutrons. Some photon activation has been employed.

The initial work on applying neutron activation analysis to airborne particles was by Zoller and Gordon (1970) and Dams et al. (1970). A typical set of gamma spectra for ambient air samples is shown in Figure 2.10. Subsequently, a number of groups have employed neutron activation analysis (Gladney et al., 1974; Moyers et al., 1977), although there have not been any INAA studies mounted

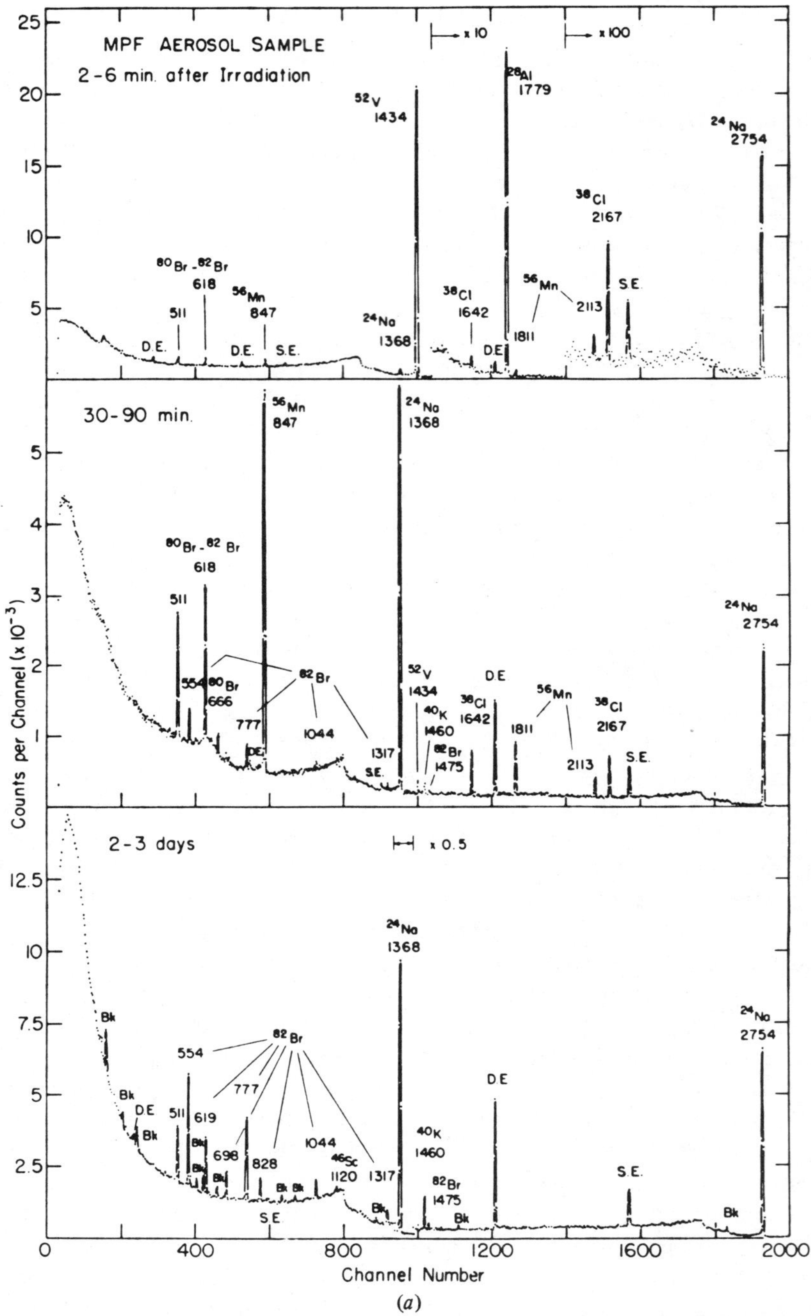

(a)

Figure 2.10. *(a)* Gamma-ray spectra at three different times following a 5-min irradiation of a Millipore filter sample. S.E. and D.E. designate single and double annihilation-photon escape peaks. Bk indicates peaks occurring in the background. Figure taken from Zoller and Gordon (1970) and used with permission. Copyright 1970 American Chemical Society.

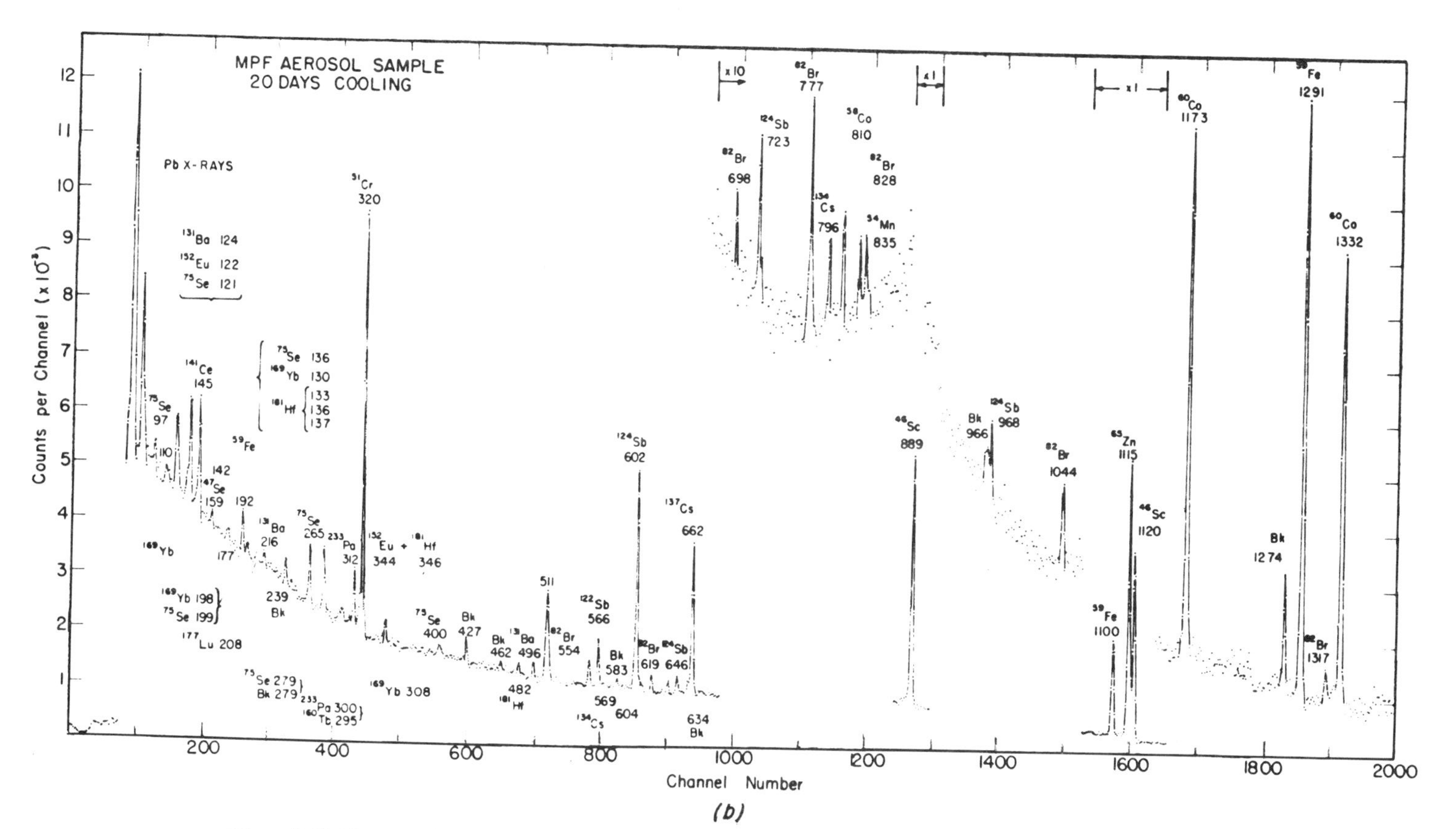

Figure 2.10 (*Continued*). (b) Gamma-ray spectrum of an irradiated Millipore filter sample 20 days after a 9-hr irradiation. The ^{137}Cs was present as an energy calibration standard. Figure taken from Zoller and Gordon (1970) and used with permission. Copyright 1970 Copyright American Chemical Society.

on airborne particles on the scale of the RAPS program. Photon activation has been used for the analysis of samples taken from particle sources (for example, see Gladney et al., 1976), and has been used in the analysis of a few ambient samples (Aras et al., 1973; Olmez et al., 1974).

ATOMIC SPECTROSCOPY

A large number of studies have employed atomic spectroscopy as an elemental analytical method. The most widely used methods have been atomic absorption spectroscopy and arc emission spectroscopy, that was used for many years to analyze filters from the National Air Sampling Network (NASN). These methods have been primarily used to determine trace metals in samples collected on glass fiber filters. Because of the previously discussed problems with these filters, it is necessary to quantitatively leach the airborne particulate matter from the filters and to determine the amount of the elements of interest in the leachate. A state-of-the art review of atomic spectroscopy methods for trace element analysis has recently been produced by Parsons, Major, and Forster (1983). Atomic absorption provides good quantitative analysis, but because only one element at a time is measured, the number of elements determined is usually low (a maximum of six to ten in most studies). Arc emission can provide semiquantitative results on a wider suite of elements, but obtaining accurate quantitative analyses is difficult.

An alternative emission method that has become more widely available recently is the inductively coupled plasma (ICP) emission spectrometer. Such a system does permit the simultaneous analysis of a large number of elements. The elements observable by ICP are shown in Figure 2.11 and cover the species of interest to receptor modeling except for the halogens. Although many elements can nominally be determined, typical instruments are limited to 30–40 elements. Many of the elements will be present in airborne samples in amounts below detection limits. In the study of cascade impactor samples by Broekaert,

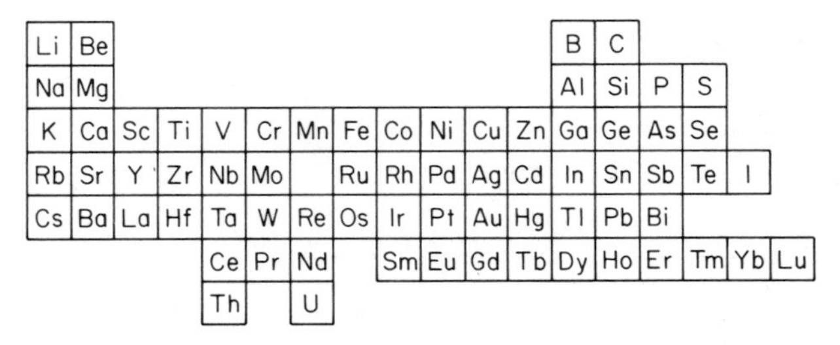

Figure 2.11. Elements in the periodic table that can be detected using inductively coupled plasma emission spectroscopy. The specific instrumental configuration may limit the number of elements that can be simultaneously analyzed.

Wopenka, and Puxbaum (1982), only 11 elements (Al, Ca, Cr, Cu, Fe, Mg, Mn, Pb, Sr, V, and Zn) were determined. Analyses for a number of elements are theoretically possible but this method has not been extensively used since it does require sample dissolution before an analysis can be made. For other systems where receptor models might be applied such as tracing water pollution sources, ICP could serve as an extremely important tool. However, such studies have not yet been reported.

X-RAY DIFFRACTION

In addition to the use of X-rays for elemental analysis, there have been recent applications of X-ray diffraction (XRD) analysis to the determination of compounds in airborne particle samples. XRD is a well-known phenomenon that has been widely employed in examining the structure of solid, crystalline materials (Klug and Alexander, 1974). In general, airborne particle samples pose severe difficulties for XRD because of the thinness of the deposit and the scattering due to the filter itself.

When X-rays of a fixed wavelength enter a crystalline material, the lattice can act as a three-dimensional diffraction grating. The constructive and destructive interferences in the propagation of the electromagnetic radiation are such that the angle θ, at which the waves show reinforcement, occurs when the wavelength λ and the crystal spacings are related by

$$n\lambda = 2d \sin \theta \tag{2.1}$$

commonly called Bragg's law. Since different chemical compounds have different lattice spacings d, the measurement of the angles at which a given wavelength X-ray is scattered yields qualitative information on the nature of the compounds present. X-ray diffraction has been known for a long time and a considerable library of diffraction patterns are available for many species of potential interest in airborne particulate identification.

The intensity of the XRD will depend on the distribution of electrons in a given atom as well as the configuration of atoms in the lattice. The presence of impurity atoms in the crystal can substantially affect the intensity of scattering although generally not the angle. Thus different samples of the same mineral or compound can have significant differences in the number of scattering X-rays per unit time in the same experimental system. Thus, quantitative analysis can often be difficult. Qualitative and relative quantitative analysis for a given set of samples can usually be accomplished straightforwardly. A detailed discussion of XRD theory and practice is given by Klug and Alexander (1974).

The first notable use of XRD for identification of crystalline phases in airborne particulate samples was by Brosset, Andreasson, and Ferm (1975) in which they identified a number of sulfate compounds in Swedish particle samples. However, in order to make the analysis, the particles were scraped from the filter on which they were collected and remounted in the manner necessary to perform the analysis. Several other groups have also removed the particles from the filters in order to make the analysis. Biggins and Harrison (1979) have examined particles size-fractionated using an Andersen impactor. They have analyzed each stage by ultrasonically stripping the particles from the filter using hexane and then redepositing the material on a cellulose ester membrane filter for the XRD analysis. Foster and Lott (1980) removed the particles from glass fiber high-volume samples using the jet from a plastic wash bottle filled with 5% Duco cement in acetone to concentrate the sample onto a paper filter. Fukasawa et al. (1980) used an ultrasonic bath with mixtures of carbon tetrachloride or hexane to remove the particles from the filters. They subsequently separated various density fractions using mixtures of carbon tetrachloride and methylene iodide. They did analyze the filtrates from these separations by XRF and found that some material had been dissolved by the procedure. However, by separating into density fractions, they were able to identify more compounds than could be seen in the unseparated sample as shown in Figure 2.12. Fukasawa, Iwatsuki, and Tillekeratne (1983) have also looked at size-fractionated samples using an Anderson impactor. Here they removed the particles from the collection substrates with CCl_4 and redeposited them, but they did not perform a density separation on these samples. The separation of the particles from the filter allows the sample to be concentrated and to be separated further. Better identification of minor species can be obtained in this manner. However, the possible solubilization and other losses of particles raises serious questions regarding quantitative analysis.

To avoid the problems of removing the particles from the filter, there has been an effort to develop methods to examine the particles in situ on the filters. Davis and Cho (1977), Davis (1978, 1980, 1981), and Davis and Johnson (1982) have examined the standard high-volume sampler glass fiber filter. O'Connor and Jaklevic (1980, 1981) and Davis and Johnson (1982) have examined the use of membrane-type filters for XRD analysis.

In their works, Davis and collaborators focused on the development of quantitative methods for using XRD data to determine the amount of each of the identifiable mineral phases. Their approach is based on the development of multicomponent analysis by Chung (1974a, 1974b, 1975) using a single internal standard or "reference" material. In their initial studies, Davis and Cho (1977) and Davis (1978) suggested the need for adding this standard to the sample, a difficult procedure that changes the sample. Davis (1980) then showed that it is possible to make the analysis without adding any material to the sample as long as the relative intensities from a series of binary mixtures have been measured separately. Davis (1981) has examined the errors involved in such analyses.

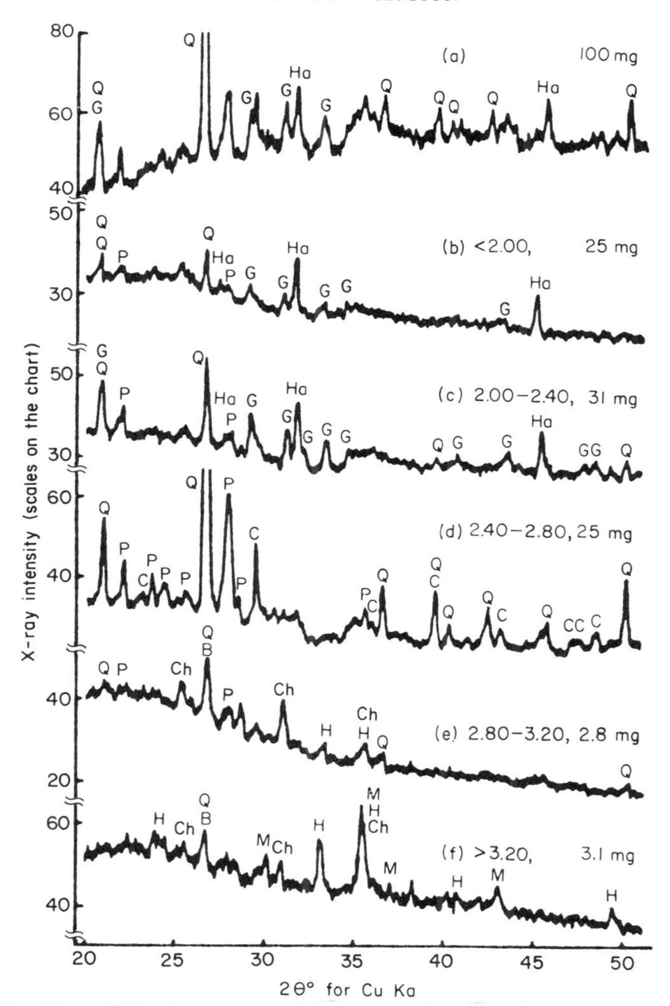

Figure 2.12. X-ray diffraction pattern of a sample of airborne particles and its fractions separated with heavy liquids. (*a*) Starting sample. (*b*)–(*f*) Fraction separated with heavy liquids. Numerical values without dimensions show densities (g/cm^3) of the liquids used. Weight in mg is the sample weight used for X-ray diffraction analysis. Q: α-quartz, G: gypsum, Ha: halite, C: calcite, P: plagioclase, H: hematite, M: magnetite, Ch: chlorite, B: biotite. Full scale: 8×10^2 cps. Figure taken from Fukasawa et al. (1980) and used with permission. Copyright 1980 American Chemical Society.

O'Connor and Jaklevic (1980, 1981) showed the utility of a qualitative analysis. The problem with typical membrane filter samples is that there is very little collected material and it is difficult to obtain good diffraction patterns. Twenty-three-hour scans were necessary to obtain the qualitative data with the cellulose acetate filter samples taken from the Regional Air Pollution Study program.

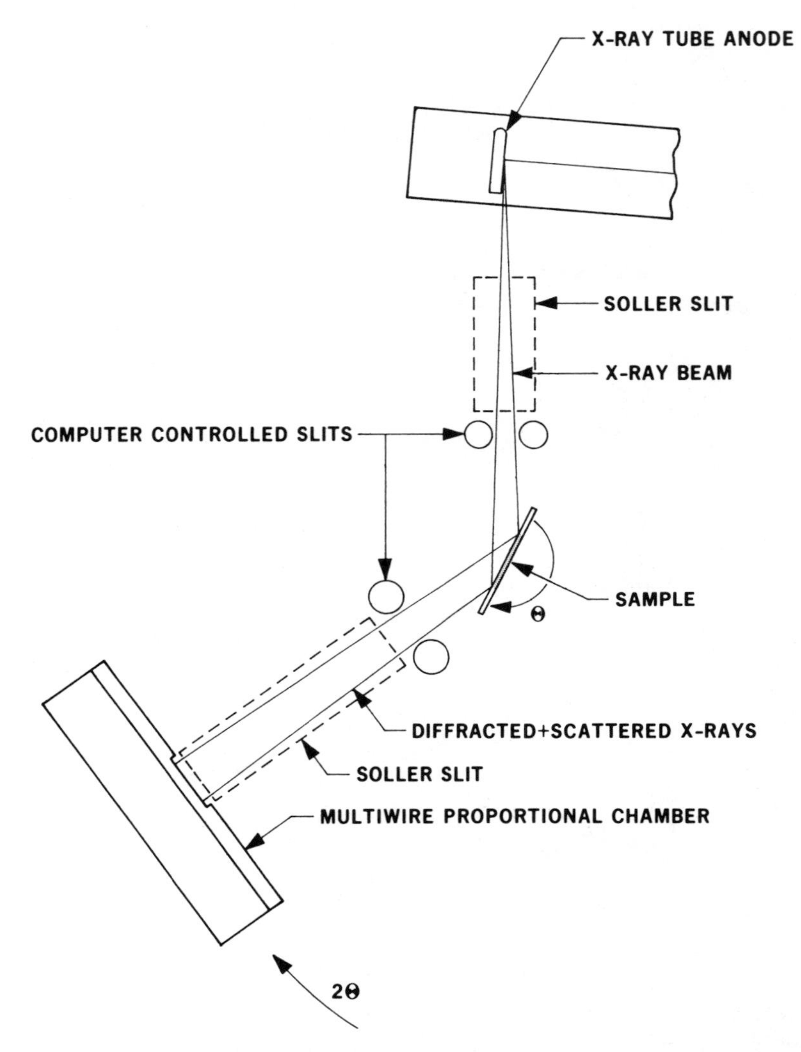

Figure 2.13. General layout of the LBL powder diffractometer system using a position sensitive detector. Figure taken from Thompson, Johnson, and Jaklevic (1983) and used with the permission of Lawrence Berkeley National Laboratory.

To permit quantitative analysis with shorter counting times, an improved diffractometer was developed using a position sensitive proportional wire chamber instead of the conventional slit/detector system (Thompson et al., 1982). This system is shown schematically in Figure 2.13. A 50-fold increase in counting rates is obtained for this system and it has been used to obtain quantitative results for Teflon filter samples from a dichotomous sampler that was used to study particulate visibility effects in Houston, Texas (Thompson, Johnson, and Jaklevic, 1983). Although the low level of collected material still limits the analysis to major species, it is now possible to obtain excellent qualitative and improving quantitative data on the presence of specific crystalline materials in airborne particle samples. The use of such data in receptor models is discussed in subsequent chapters.

INORGANIC SPECIATION METHODS

It is possible to examine samples for the presence of specific inorganic species. Braman (1983) has just provided an excellent review of the methodologies that are available for antimony, arsenic, copper, germanium, lead, mercury, sulfur, selenium, and tin. Only arsenic, mercury, lead, and sulfur compounds have been examined in the atmosphere. In the cases of arsenic and mercury, it is possible to selectively trap these compounds and then selectively analyze for them. Volatile arsenic species can be trapped on silvered glass beads, removed by a mild NaOH wash, and analyzed by the hydride method. Although the NaOH oxidizes the adsorbed arsines, it apparently leaves them such that they can be readily reduced by the $NaBH_4$ to their original form. These compounds can be cold trapped on a short column and separated by slow warming so that individual species can be identified.

Mercury can be trapped on a series of selective adsorbing media. It is possible to separate and analyze for $HgCl_2$-type compounds, CH_3HgCl-type compounds, elemental mercury, and dimethylmercury. Several studies have been made on the amounts of these various compounds, but no efforts have been made to relate their presence to specific mercury sources.

Organolead compounds have been collected by cold trapping as a preconcentration method. The various alkyl lead compounds can then be desorbed and analyzed by an atomic spectroscopy system. However, there is some disagreement in the literature as to what is separated by a filter and what is measured. Robinson, Rhodes, and Wolcott (1975) suggest that tetraalkyl leads are very unstable and decompose in a very short time period to inorganic salts. They suggest the presence of inorganic vapor-phase lead. Harrison and Perry (1977) argue that the presence of low-vapor-pressure inorganic lead compounds in the gaseous phase is very unlikely and that alkyl-lead has an atmospheric lifetime of several hours. Subsequently, Rohbock, Georgh, and Muller (1980) used a gas chromatographic step to show that the gaseous lead was principally in the form of tetramethyl lead with some tetramethyl lead and mixed ethyl–methyl lead compounds. In most cases, the gas to particle lead concentrations were less than 0.1. With the increasing effort to phase lead out of motor fuels, the utility of tracing automobiles with alkyl lead compounds will diminish. In areas where there are lead sources other than motor vehicles, the alkyl lead compounds provide a useful automotive tracer.

Sulfur speciation in particles has primarily focused on the differences in the amounts of sulfuric acid, ammonium bisulfate, and ammonium sulfate. Complex systems have been developed for such purposes (Weiss, Larson, and Waggoner, 1982; Huntzicker, Hoffman, and Ling, 1978). However, since these are principally secondary aerosols, this information is not very relevant to source identification in most circumstances. A thermometric method has been developed to examine the relative amounts of sulfur (IV) and sulfate in aerosols (Hansen et al., 1976). In some cases, this information might prove useful.

Finally there are gas-phase sulfur species that can be collected and analyzed. Natural sources and fluxes of H_2S, COS, dimethylsulfide, CS_2, dimethyldisulfide, and methyl mercaptan have been examined by Adams et al. (1981a,b) using cryogenic enrichment sampling and wall-coated, open tubular, capillary column, cryogenic gas chromatography with a flame photometric detector. Other recent studies have been made on preconcentration methods for volatile sulfur compounds (Torres et al., 1983).

The development of additional methods for the collection, separation, and identification of specific inorganic compounds could provide valuable additional receptor modeling tools. As will be discussed later, it is often found that the total elemental concentration values are not sufficient to provide all of the source resolution desired. Further work to develop practical methods that can be employed in large field studies would assist in obtaining that additional measure of resolution.

CARBON ANALYSIS

A major problem with both neutron activation and X-ray fluorescence analysis is that they cannot determine carbon. There are some very specialized nuclear methods to determine carbon. Deuteron activation has been used by Clemenson, Novakov, and Markowitz (1980) to convert ^{12}C to ^{13}N and measured the amount of 10.0-min ^{13}N. This method has a sensitivity of 0.5 $\mu g/cm^2$ on the filter which corresponds to 0.2% in a sample of total thickness 250 $\mu g/cm^2$. Another approach is the measurement of gamma rays induced in the sample by inelastic proton scattering (Macias et al., 1978). These methods can provide good estimates of total carbon in the sample. However, they require the availability of facilities that are not widely accessible. In atmospheric aerosol problems, carbon may represent a significant fraction of the particulate mass and the inability to measure its concentration presents a serious problem. Carbon can be present in a wide variety of forms and so there are a number of different ways that various investigators have undertaken the study of carbon in atmospheric aerosols.

ORGANIC AND ELEMENTAL CARBON

Several methods have been developed to determine the amounts of two gross carbon fractions, "organic" and "elemental" carbon (Cadle, Groblicki, and Stroup, 1980; Johnson et al., 1981). Organic carbon is that present primarily in discrete compounds adsorbed on particles or on the filter. Elemental carbon is the polymeric particulate matter. There is considerable debate as to whether the fine, carbonaceous particles are primary or secondary in origin (Chu and Macias, 1981). The actual materials being analyzed are not well characterized and these fractions are operationally defined. Typically, the organic carbon is evolved from the filter by heating to the order of 350°C in an inert atmosphere. The evolved gas is catalytically oxidized to CO_2 and the CO_2 determined by an infrared monitor. In other systems, the CO_2 is reduced to CH_4 and the carbon determined with a flame ionization detector. The elemental carbon is then oxidized in air at ca. 600°C and the evolved CO_2 is measured. During the evolution of the organic fraction, pryolysis can occur, converting part of the organic to elemental carbon. The ability to determine the amount of elemental carbon (soot) by examining the reflectance of the filter was suggested by Rosen et al. (1980). This optical absorption technique has been incorporate into the analytical process by Johnson et al. (1981) so that a correction can be made for the amount of pyrolyzed organic carbon in the elemental carbon value. This system has been field tested at a number of rural and urban sites around the United States (Shah et al., 1982). Some other modifications have been made by Tanner, Gaffney, and Phillips (1982) to produce another similar system.

Problems have been reported by several authors with the adsorption of gas-phase organic compounds on the filters (Cadle, Groblicki, and Mulawa, 1983; Appel, Tokiwa, and Kothny, 1983). The measurement of the carbonaceous particulate organic carbon is thus overestimated, depending on the amount of filter adsorption. Appel, Tokiwa, and Kothny (1983) suggest the use of a denuder tube to remove the organic vapors by diffusion to the walls of the tube. It appears that there are still problems to be resolved in the sampling of particles for proper carbon analysis.

EXTRACTABLE ORGANICS

An alternative to these thermal evolution methods is the extraction of organics from the filter. These methods will remove most of the organic compounds but will not remove much of the graphitic material. By extracting the filter with a series of solvents of increasing or decreasing polarity, a measure of the organic carbon fraction can be obtained as well as relative amounts of various polarity compounds. The results clearly depend on the solvents used and the order in which they are used and the quantities measured again represent operationally defined variables. Appel, Colodny, and Wesolowski (1976) suggested that cyclohexane was a selective solvent for the extraction of "primary" particle phase organics—those injected into the air in the particle phase. "Total" organics were those solubilized by successive extraction with benzene and 1:2 v/v methanol/chloroform. The residual insoluble carbon is then considered as elemental carbon. "Secondary" organics were then the "total" minus "primary" values. The insoluble carbon also included carbonate carbon. This procedure was then used to examine samples collected in Pasadena, Pomona, and Riverside (Appel et al., 1979).

Daisey, Hershman, and Kneip (1982) have used a sequence of cyclohexane, dichloromethene, and acetone to separate nonpolar, moderately polar, and polar organic compounds. This approach has also been applied to particles in New York City to examine samples for mutagenicity (Daisey et al., 1978; Daisey, Leyko, and Kneip, 1979). A similar approach is being applied to samples in New Jersey (Lioy and Daisey, 1983). As in the thermal methods, it is clear there are filter effects that are not well characterized and make the interpretation of the measured values difficult (Schwartz, Daisey, and Lioy, 1981). These operationally defined carbon values help to determine a larger fraction of the total sample mass. There are ways in which information that may be better used to relate the carbon to sources can be obtained.

CONTEMPORARY CARBON ANALYSIS

Another way to separate particulate carbon into useful receptor modeling categories is to measure the amount of contemporary and fossil fuel carbon. The combustion of wood, paper, or agricultural fields produces particles of carbon that have recently been in the atmosphere as CO_2. Fossil fuel combustion particles have carbon that has been in the ground for a long time. In the atmosphere ^{14}C is constantly being produced by the interaction of cosmic rays. Therefore, the measurement of ^{14}C gives a measure of current carbon content in the particles. The development of the counting methods to measure the ^{14}C content of small samples as found in airborne particulate samples is described by Currie et al. (1978). These methods have been applied to particles in Portland, Oregon (Cooper, Currie, and Klouda, 1981), Sydney, Australia (Court et al., 1981), and Bonanza, Utah (Voorhees et al., 1981). The ability to use contemporary carbon as an important variable in receptor modeling was tested in a simulated data set in an EPA-sponsored intercomparison study (Gerlach, Currie, and Lewis, 1983) and found to improve the resolution of sources that can be obtained by having this additional parameter. It is a difficult measurement and can be used in limited situations. It is currently unlikely to become a routinely measured parameter.

ORGANIC COMPOUND ANALYSIS

In the previous sections, the discussion has focused on the carbon analysis of airborne particulates. However, it is also possible to examine the gaseous phase organic compounds, and to utilize these compounds as well. Clearly, there is an extensive literature on organic compound separation and analysis utilizing gas chromatography, gas chromatography/mass spectrometry, liquid chromatography, and so on, that will not be reviewed here. Once the sample is taken and reduced to a mixture of compounds in an appropriate solvent, these separation and quantification methods can be employed. The key aspects with regard to atmospheric receptor modeling applications are the sample collection and extraction steps.

The first step is to separate the particulate phase organics from the gaseous phase compounds. As mentioned, this process is complicated by filter adsorption of gas-phase organics. It is a temperature-dependent process so that both phases should really be measured if a full balance of carbon in the air is to be determined. Several recently published volumes provide a number of papers on the collection and analysis of atmospheric organic species (Verner, 1980; Keith, 1984). Keith (1984) gives a very comprehensive review of the literature and the availability and applicability of current organic analytical methods and this volume represents an important resource for those interested in this area.

Typically, a glass fiber filter is used to remove particles and the gaseous material can be collected in an appropriate bed of adsorbent. In other cases, an air sample is collected in an airtight bag or in a glass gas pipette (Nelson and Quigley, 1982). The material can be thermally desorbed from the adsorbent or extracted. The air sample can by cryogenically preconcentrated and then flash evaporated into a gas chromatograph. Thus, depending on the nature of the compounds sought and the analytical methods employed, there are a number of alternatives to organic compound sampling.

The most commonly analyzed gaseous components have been aliphatic hydrocarbons; the most common sought particulate compounds have been polycyclic aromatic hydrocarbons. There are a number of such reports in the literature with the majority of them merely reporting the observation of the presence of various compounds. There have been only a few attempts to utilize organic compounds in receptor models (Daisey, Leyko, and Kneip, 1979; Daisey and Kneip, 1981; Daisey, 1983).

One major problem is that very little source data are available for organic compounds. Lee et al. (1977) examined some polycyclic aromatic hydrocarbons from the combustion of a variety of common fuels and found notable differences in the alkyl branching distributions from the different fuels. Hites and Howard (1978) reported on a more extensive study of flame production of particulate organic matter. Serth and Hughes (1980) looked at polycyclics in carbon black vent gas. Lee and Schuetzle (1983) reviewed the methods and results of examining internal combustion engine exhausts for polycyclic compounds. Eiceman and Vandiver (1983) examined the emissions of polycyclics on incinerator and coal-fired power plant particles. A broader range of compounds was examined by Eiceman, Clement, and Karasek (1979) in refuse incinerator emissions. Thus, there is far less information on the emission of organics from sources and far less knowledge about their behavior in transit from source to receptor site. For example, there is considerable uncertainty as to the reactivity of polycyclics adsorbed onto fly ash particles. Model studies have been made (Korfmacher et al., 1980), but these studies focused on particles that would not reside long in the atmosphere because of their size. Since there is a size dependence to the composition of emitted particles and considerable uncertainty regarding the composition of the surface, the exact nature of the reactions that might occur to adsorbed organics is very uncertain. A recent paper by Khalil, Edgerton, and Rasmussen (1983) has suggested CH_3Cl_3 in urban air as a tracer for wood smoke particles. The idea of using a gaseous tracer for particulate pollution is a novel idea, but further verification of the coupling of the particle and gas dispersion processes is needed.

The sampling, separation, and analysis of specific organic compounds as well as studies to elucidate the transformations between source and receptor will be receiving ever increasing attention in the future. The ability to examine a large number of specific species offers great promise for source resolution studies, although substantial problems must be overcome before they come into as widespread use as elemental analyses.

ISOTOPE RATIO METHODS

Most of the elements have more than one stable isotope and in some cases there are physical or chemical factors that can affect the relative isotope abundances. The measurement of these abundances can potentially be used to infer information on the processes that gave rise to the material analyzed. These measurements are made using mass spectrometers that are typically dedicated to isotope ratio measurements and great care must be taken to eliminate contamination. However, careful analysis can be done with quite high precision. Only very limited work has been done in this field.

For example, biological processes that involve elements like oxygen and sulfur tend to enrich the lighter isotopes in preference to the heavier ones. Thus biologically enriched sulfur will tend to have more ^{32}S per unit ^{34}S than will material of geological origin. Hitchchock and Black (1978) suggest the examination of sulfur isotope ratio defined by

$$\delta^{34}S\% = \frac{{}^{34}S/{}^{32}S_{sample} - {}^{34}S/{}^{32}S_{std}}{{}^{32}S/{}^{32}S_{std}} \times 1000 \qquad (2.2)$$

Fractionation occurring in the formation and decomposition of organic sulfides yields volatile sulfides with $\delta^{34}S$ values in the range +14 to +17% while that which occurs during bacterial sulfate reduction gives H_2S with $\delta^{34}S$ values of +5 to -10%. Seawater sulfate has a $\delta^{34}S$ value of +20% and fossil fuel sulfur values are typically in the range of -8 to +16%. Although there is overlap in these groups, some useful information can be obtained.

Oxygen isotope ratios can also be obtained examining the relative amounts of ^{18}O and ^{16}O. Holt, Kumar, and Cunningham (1982) have used this isotope ratio to examine the relative amounts of primary to secondary sulfate aerosol in the air at Argonne, Illinois. They suggest that ^{18}O is enriched in high-temperature-produced SO_4^{2+} compared to secondary sulfate by using laboratory studies and some stack samples (Holt, Kumar, and Cunningham, 1982). They report that 20-30% of the sulfate in collected precipitation was primary.

Although lead isotope ratios are not changed appreciably by the kinetic isotope effects as are lighter elements, lead isotopic ratios can be affected by the amount of radiogenic lead produced by the decay of naturally radioactive uranium and thorium. These radionuclides produce ^{206}Pb, ^{207}Pb, and ^{208}Pb but not ^{204}Pb. Thus, the examination of the ratio of ^{204}Pb to ^{206}Pb, ^{207}Pb, or ^{208}Pb will yield information on the amounts of U and Th in the lead deposit. It is found that in many cases specific ores will have unique lead isotopic ratio. Rabinowitz and Wetherill (1972) have shown that it is possible to clearly distinguish between lead from gasoline-fueled motor vehicles and smelter emissions in both southeastern Missouri and in an area near Benicia, California. In the latter example, they are able to make a strong case for the emissions from the smelter causing lead poisoning found in horses pastured near the smelter. However, there have not been many subsequent studies utilizing the lead isotope ratios. Shirahata et al. (1980) have used the changes in lead isotope ratio in sediments of a remote subalpine lake to date the sediments but little work on urban lead source problems using lead isotope ratios has been reported.

CHAPTER

3

OPTICAL MICROSCOPY

One of the earliest and most used methods to identify the source of airborne particles is optical microscopy. The optical microscope with a properly trained observer can utilize a variety of particle properties to identify the nature of particles and infer their probable source. The specificity of the light microscope may make it the single most powerful tool available for the qualitative characterization and identification of particles. The microscopist can not only identify the chemical composition but he or she can also distinguish between different types of particles having the same chemical compositions. For example, SiO_2 in a particle can be found as quartz, vitreous silica, opal, and tridymite, all of which have different optical properties and can be distinguished with a polarizing light microscope. Although there are physical limits to the size of the smallest particle that can be resolved and identified, virtually all particles with diameters greater than 1 μm can be characterized using light microscopy. In this chapter, the nature of these properties will be outlined and some typical applications of the light microscopic method presented.

A detailed discussion of physical optics is clearly out of place in this volume. A very useful introduction to optics and the light microscope is presented by McCrone and Delly (1973a) and the use of light microscopy in particle source identification studies has been described by Crutcher (1983). In light microscopy, two kinds of information can be obtained from the observed image. These properties are the shape and optical characteristics.

SHAPE

The initial type of data that can be obtained with the light microscope includes the physical size and shape of each particle. The form or morphology of the particle can often provide conclusive identification of its origin. Biological materials such as pollen are generally easy to identify by their size and shape alone. For example, pollen counts based on the identification of ragweed pollen are probably the most commonly reported specific airborne particle identifications in the United States.

The first step in any optical microscopic analysis of a particulate sample is the identification of the particle types in the sample. This phase of the study is qualitative. Individual particles are examined in great detail to establish their identity. There are good collections of photomicrographs describing a wide variety of individual particles (McCrone and Delly, 1973b; McCrone, Delly, and Palenik, 1979). However, it takes considerable experience to be able to immediately recognize specific particle types. The shape of any object is the result of the internal and external forces that have acted upon the object. The structural unit of a material is modified in a bulk sample by distortions such as voids or misalignments whose frequency is a function of the number of unit volumes. As linear dimensions increase, the number of defects increases as the cube of the form that indicates the structural units of their compositional material. These structural units may be a biological cell or may possess a characteristic crystal symmetry.

For individual particles, there is an increased importance of surface active forces over those dependent on volume or mass. Surface tension and surface chemistry become dominant over gravity and bulk chemistry. These forces have a profound effect on morphology as well as some analytical approaches.

Different types of objects often exhibit morphologies that have significance only for that type of object. Crystal cleavage planes have no place in the discussion of biological particles just as there are no cell wall structures for a crystal. The terms used to describe different type materials must be different. For this reason, the discussion is divided into three broad categories: crystals, biologicals, and morphological artifacts.

Crystals

Each crystal class has a set of symmetries associated with it. These symmetries are divided into axial, planar, and point type. Many of these symmetries are immediately apparent in a crystal and are useful in helping to confirm an identity as well as a source type. A crystal habit is an outward form that expresses the internal order of the compound. It exhibits many of the symmetry elements of the specific crystal class to which the compound belongs and occasionally possesses unique interfacial angles that are sufficient to establish the compound's identity. The shape of a crystalline solid may result from its molecular crystalline structure in two general ways. The first, and most characteristic, is the mode of its crystal growth. This is a reflection of the molecular arrangement in the compound and can be sufficient in and of itself to establish the identity of the compound. For example, Figure 3.1 shows a characteristic needle of NH_4NO_3 that has formed on a glass fiber filter after being immersed in mounting fluid used to clear the filter. These crystals are not initially present on the filter, but form about 5 to 10 min after the filter is immersed. The second way

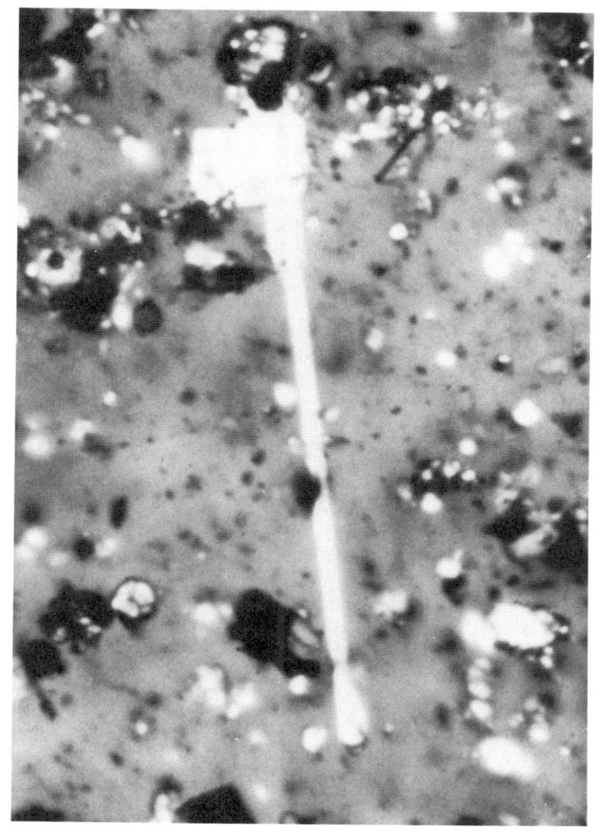

Figure 3.1. NH_4NO_3 needle observed with slightly uncrossed polarizers (20°) at 51× magnification. Micrograph courtesy of R. Draftz of IITRI.

is through the arrangement of the cleavage planes in the material. Cleavage planes reflect the relative bonding energies through the lattice and are generally less characteristic of specific compounds.

The observed crystalline morphology may help to identify the crystal's generation mechanism. To exhibit a growth mode or habit, the compound must grow free of physical constraints. Examples of such conditions include the growth of crystals from the vapor phase and from solution by precipitation or evaporation. Typically, the slower the growth rate, the larger and more perfect is the crystal's structure. A subdivision of crystal habit is crystal pseudomorphs. These are forms whose outward habit reflects an earlier molecular order of an apparent order. An excellent example of this is alumina. Pseudohexagonal gibbsite ore is actually a monoclinic crystal. When calcined to aluminum oxide, it retains it pseudohexagonal shape. The result is a hexagonal shaped particle of aluminum oxide (corundum) whose apparent hexagonal symmetry is not aligned to the actual hexagonal structure of the corundum produced by the calcining. This property and the presence of typical calcination induced patterning uniquely identify calcined gibbsite.

Cleavage is the tendency of a crystal to part along specific lattice planes in response to stress. Crushing or grinding operations typically produce cleavage fragments of materials that tend to exhibit cleavage. The extent to which a crystal will exhibit cleavage is dependent upon the relative energies or bonding forces between the ions in the lattice. If the forces are nearly the same in all directions, no cleavage fragments will form. Quartz is an example of a mineral that does not form cleavage fragments. Mica, on the other hand, has very weak bonds between hydroxyl ions along one plane in its lattice. The result is a single cleavage plane that is evident in all fragments of mica. The number of specific cleavage planes exhibited varies with the distribution of the energies through the lattice. Calcite exhibits three planes of cleavage and often one additional less perfect cleavage plane diagonal to the others. Feldspars occasionally exhibit two cleavage planes. The angles of intersection for the cleavage planes of a material are often very characteristic. When combined with other optical crystallographic properties, a rapid and unique identification can often be made. The mica minerals and the divalent metal carbonates are examples.

Biologicals

Many biological entities or fragments can be defined in strict morphological terms. The morphology of biological materials is considerably more varied and complex than that of crystals, but these materials also exhibit texture, internal structure, wall patterning, functional subunits, characteristic terminus, characteristic associate structure, and so on. Every division of biology has a unique language to describe the significant forms pertinent to that form of life. Typical biologicals encountered during an analysis of airborne particles include insect parts or fibers, herbaceous plant parts or hairs, charred herbaceous plant parts, cereal plants, wood parts, charred wood, pollens, spores, conidia, bacteria, algae (including diatoms), and animal hair. Figure 3.2 shows a charred wood fragment. There are also numerous nondescript biological masses that defy simple classification. Rotted bacterial mats form a type of light absorbing mass with no distinct morphological features. This type of material can occasionally be at least identified as an organic by its interaction with or staining of the toluene soluble mounting media often used to mount these specimens. The morphological identification of biologicals is an important part of any optical microscopic analysis of airborne particulate. Other optical characteristics that assist in the identification of biologicals are discussed later in this chapter. The contribution of these biologicals to the total airborne burden in an area can be substantial.

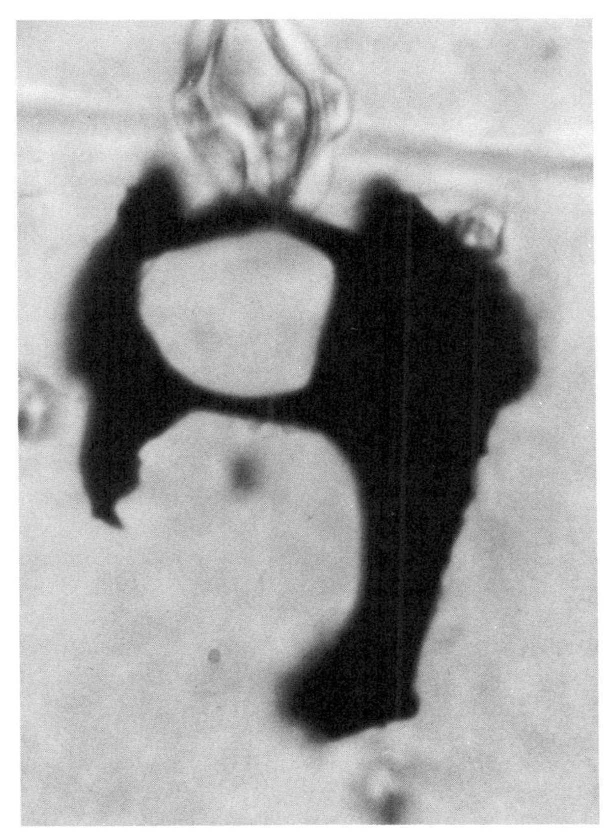

Figure 3.2. Charred wood fragment observed with slightly uncrossed polarizers (10°) at 600× magnification. Micrograph courtesy of Microlab Northest Inc.

Morphological Artifacts

Morphological artifacts is a term used here to describe all of those particles whose form reflects something specific about their source. Combustion-related particles and weathered soil particles are two typical examples, but there are many more. Charred biologicals have already been mentioned, but fossil fuels also produce many characteristic forms. Materials that burn as a liquid produce oil cenospheres. Oil cenospheres are carbonized small droplets of the original liquid fuel. They have been typically identified as oil cenospheres in the literature, though plastics, resins in coal, tars, and similar materials with melting points well below the boiling point of the material will form oil cenospheres when combusted.

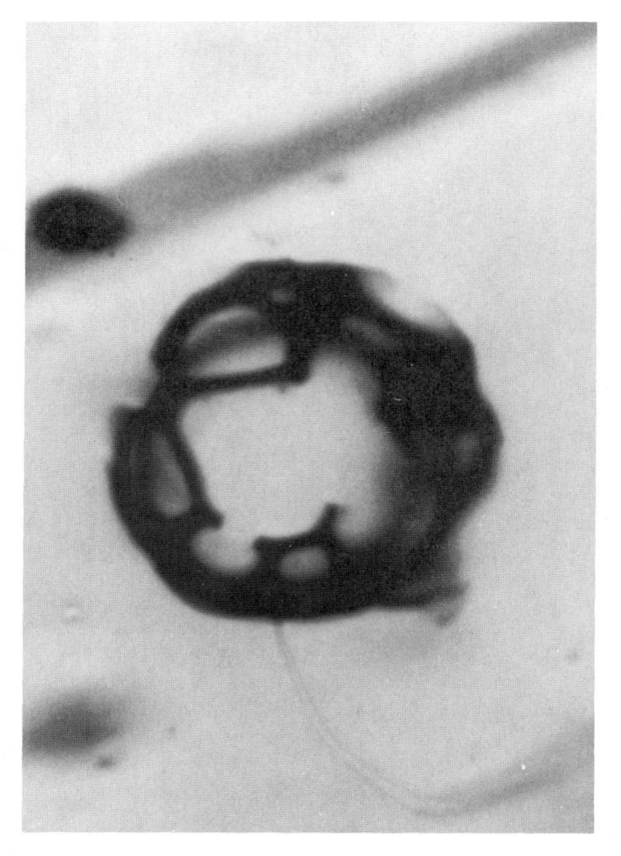

Figure 3.3. View of a cenosphere collected in Philadelphia, Pennsylvania, at 600× magnification. Micrograph courtesy of Microlab Northwest Inc.

Oil cenospheres cover a range of forms. They are all roughly spherical with a highly carbonized surface structure that may or may not be obvious. One extreme is a black solid sphere with some surface broad indentures. These particles are cenospheres that still contain a high proportion of volatile organics. These are characteristic of particles that pass quickly through the combustion chamber. If the particle remains exposed to high heat in an oxygen-deprived atmosphere, the volatile organics bake out leaving a lacy, hollow sphere, as shown in Figure 3.3. In an actual sample, the two extremes and all intermediate forms would be found. Oil cenospheres are an indication of liquid fuels and do not represent the total carbonaceous impact of the combustion of the fuels. Much, if not most, of the carbonaceous particulate produced by the combustion of liquid fuels consists of submicron "soot" particles. Figure 3.4 shows the small elemental car-

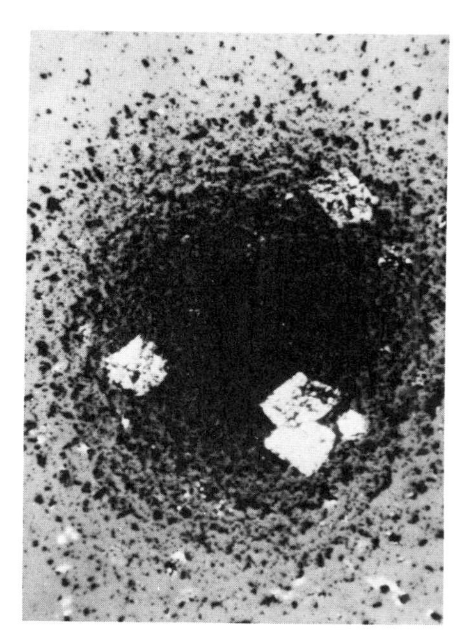

Figure 3.4. View of the impact zone of the jet in a 0.8-m cascade impactor stage from Miami, FL, showing small elemental carbon particles. The large, white particles are $(NH_4)_2SO_4$ crystals that form and grow directly on the collection medium. Micrograph taken at 25× with slightly uncrossed polarizers (20°) and is used courtesy of R. Drafts of IITRI.

bon particles on a <0.8 μm cascade impactor. The large particles are $(NH_4)_2SO_4$ that grow on the collection medium.

The combustion of solid fuels such as coal produces another set of characteristic particles. This characteristic flyash consists of the fusion products of mineral impurities in the fuel. One of these characteristic spherical particles is shown in Figure 3.5. The flyash from a pulverized coal burner reflects the mineral impurities of the coal and the conditions of the firing chamber during use. The resultant flyash can be identified as having a calcite, clay, feldspar, pyrite, or quartz origin. Calcite ($CaCO_3$) changes to $CaC + CO_2$. The carbon dioxide is trapped in the lime (CaO) matrix producing a frothy glass sphere. Clays are colored brown to yellow with a few small gas inclusions from released water. The feldspars and quartz tend to form spheres of glass without inclusions. Pyrite becomes a hemitite sphere.

Weathering produces another unique morphology which simplifies the characterization of roadside materials as distinct from fresh sources. Weathering results in siliceous gels and clays adhering or forming on the surface of clean particles. The exact mechanism of its formation is not well understood and the time required for its formation varies from area to area, but its presence indicates that the particle has been essentially incorporated into the soil. When this happens to industrial particles, it helps to distinguish between original emissions and secondary reentrainment from road surfaces or parking lots.

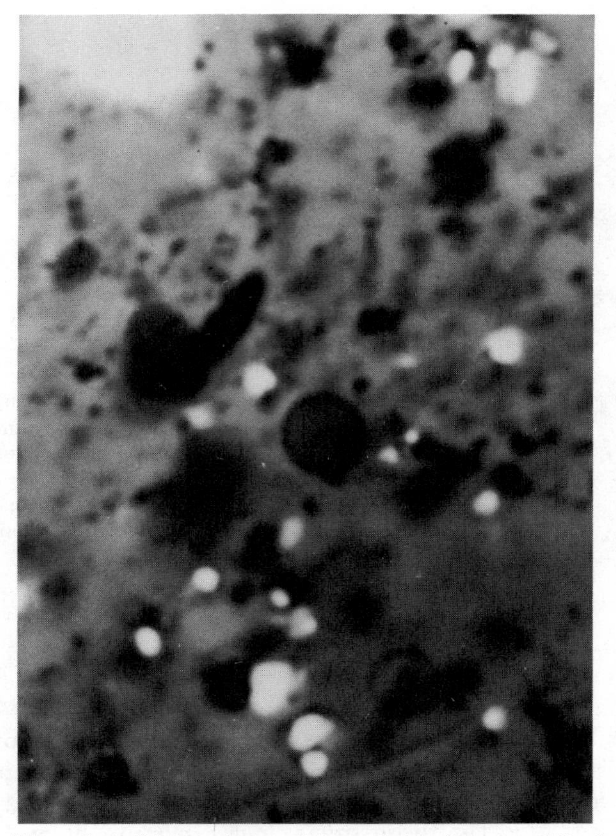

Figure 3.5. View of a solid, spherical particles probably from a coal-fired power plant at 51X magnification and slightly uncrossed polarizers (20°) observed in a sample from St. Louis, Missouri. Micrograph courtesy of R. Draftz of IITRI.

As this discussion indicates, a great deal of information can be obtained from the shape of a particle. However, it should be obvious a substantial degree of operator skill is necessary to properly interpret the information provided.

OPTICAL PROPERTIES

The second kind of data is optical, such as color that represents the interaction of the light with the electrons of the chemical constituents of the particle. Using proper conditions the optical information can often provide a highly specific classification of the particle as to its possible source. To understand this type of data, it is necessary to describe the interactions that can occur with a particle.

As light moves through a medium—gas, liquid, or solid—the propagating electromagnetic wave interacts with the material and the light wave moves at a reduced velocity. The ratio of the speed of light in a vacuum to the speed of light in a material is called the refractive index.

$$n = \frac{c}{u} \tag{3.1}$$

where n is the refractive index, c is the velocity of light in a vacuum, and u is the velocity of light in the medium. Air has a refractive index very near to 1.0. Water has a value of 1.333 while that of flint glass is 1.627. In general, the refractive index increases with average atomic number and with the density of the material.

The refractive index of any particle can be determined by successively immersing the particle in liquids of known refractive index until, in one such liquid, the particles become invisible. When this occurs, the refractive index of the invisible particle is equal to that of the liquid medium. This liquid medium is found by trial and error, but the process can be speeded up by using tests that enable the microscopist to determine whether the refractive index of the particle is greater or less than that of the liquid medium in which it is immersed.

One such test, the Becke test (Kerr, 1959), depends on the direction of movement of a bright halo (the Becke line) as the microscope is focused up and down. The Becke line is found at the boundary of the particle and the liquid medium and results from reflection and refraction of light rays at an interface between two materials of different refractive index. This halo will always move toward the higher refractive index medium as the position of focus is raised and toward the lower refractive index medium when the position of focus is lowered (McCrone and Delly, 1973a).

For particles too small to have visible Becke lines, the lens effect is very useful. If a particle has a higher refractive index than the mounting media, it behaves as a positive lens and focuses transmitted light to an intense bright spot above the particle. If it has a lower refractive index it diverges the beam, creating a dark area above the particle.

Another measure is relief. The degree of contrast, seen as a sharp dark border at the edge of the particle, is called relief. High relief, a wide black border, indicates the particle's refractive index is much higher or much lower than the media's. Low relief, a very thin black border, indicates a smaller difference in refractive indices.

Dispersion is the variation of refractive index with wavelength. The index of refraction can be expressed in a variety of ways, each adding to the understanding of this basic property of matter and electromagnetic energy. One of the simplest equations relates the dielectric constant ϵ, the magnetic permeability μ, and the refractive index n of a material

$$n = \sqrt{\epsilon\mu} \tag{3.2}$$

For the purpose of understanding the optical properties of matter in terms of the refractive index, the following equation is useful.

$$n = 1 + \frac{4Nq^2}{M} \frac{1}{(f_0^2 - f^2)} \tag{3.3}$$

where N = number of outer electrons per unit volume (resonances per molecule/cm^3)
q = charge on the electron
M = mass of the electron
f_0 = electron harmonic or natural resonance frequency
f = frequency of light used

This equation is a simplification based on the assumption that the frequency of light used is far from any important resonance frequency. From this equation it can be seen that any increase in the number of outer shell electrons, assuming the number of molecules per cubic centimeter is constant, will increase the refractive index. As the atomic number of the elements in a compound increases, so does the refractive index for the compound.

Another relationship that can be seen from the refractive index equation is the relationship between the wavelength of light used and the refractive index. Most colorless objects have the resonance frequencies in the ultraviolet (1×10^{15} Hz) end of the spectrum. If the blue light (6.6×10^{14} Hz) is used to determine the refractive index, the term $(f_0^2 - f^2)$ is smaller than if red light (5.0×10^{14} Hz) is used.

If the $(f_0^2 - f^2)$ term is made smaller, the refractive index becomes larger. This property is known as dispersion. All colorless compounds have higher refractive indices in the blue end of the spectrum than in the red end. All materials exhibit dispersion, but to see its effects using the microscope, the particle must be mounted in a media with a refractive index that matches that of the particle at some wavelength near or in the visible range. It is seen as two colored Becke lines, one for the wavelengths for which the particle has the higher refractive index and one for the wavelengths for which the liquid has the higher refractive index.

Dispersion effects can be enhanced to become a very powerful analytical technique. The methods used to enhance these effects are all referred to as dispersion staining techniques. They are all subject to the limitation of the particle being in a media with a matching refractive index near or in the visible range. When white light passes through two different materials, a particle and the media it is mounted in, even if they share the same refractive index for some wavelength, a relative phase difference is created in the different beams. Using phase contrast microscopy, this phase shift can cause color effects that can provide relative refractive index information over a range of plus or minus 0.1 refractive index units.

The refractive index is a complex number consisting of both a refractive coefficient (a real term) and an absorption coefficient (an imaginary term). Both can be measured using optical microscopy but generally only the real term is quantified. The variability of these terms with wavelength produce colored particles if the absorption frequency is in the visible band of light. Color is therefore a special topic under refractive index.

As a final note, the terms N and f_0 in the above equation are not the same in all directions through most materials. These differences are the origin of anisotropy. For more information on this subject, consult a text on optical crystallography.

During the qualitative part of an analysis, a particle's refractive index can be determined very accurately because the mounting liquids can be changed until a match is found. During the quantitative part of the analysis, the mounting media cannot be changed and the degree to which the refractive index can be deteter-mined is a function of particle size and the amount of difference between the refractive index of the particle and the mounting media. The larger the particle and the greater the difference in refractive index, the less precise the determination.

The selection of a mounting media for the quantitative part of the analysis becomes very important when refractive index information is important to the analysis. The technique with the greatest range is phase contrast. With a refractive index of 1.65 for the mounting media, minerals from quartz to corundum (1.65 to 1.76) fall within the useful range. This includes most slag minerals, many industrial minerals, and most natural minerals. When filter "clearing" techniques are used, the mounting media must match the filter, with a refractive index of about 1.51. This very significantly reduces the number of industry relevant minerals that refractive index data can characterize. However, if the examination of particles is to be made while on the filter, clearing the filter by using a mounting media of matching refractive index is essential to be able to observe the particles clearly. Thus, the problem becomes whether or not to remove the particles from the filter before proceeding with a detailed analysis. This problem is a common one in microscopy and will be raised several times in this and the next chapter.

Another consideration is that a mounting material puts a liquid in contact with the particle, raising the possibility of dissolving part of the sample. Whether clearing the filter or simply examining separated particles on a slide, the presence of the mounting media does raise the possibility of changing the sample. Figure 3.6 shows an asphalt particle that can be seen to be partially dissolving in the

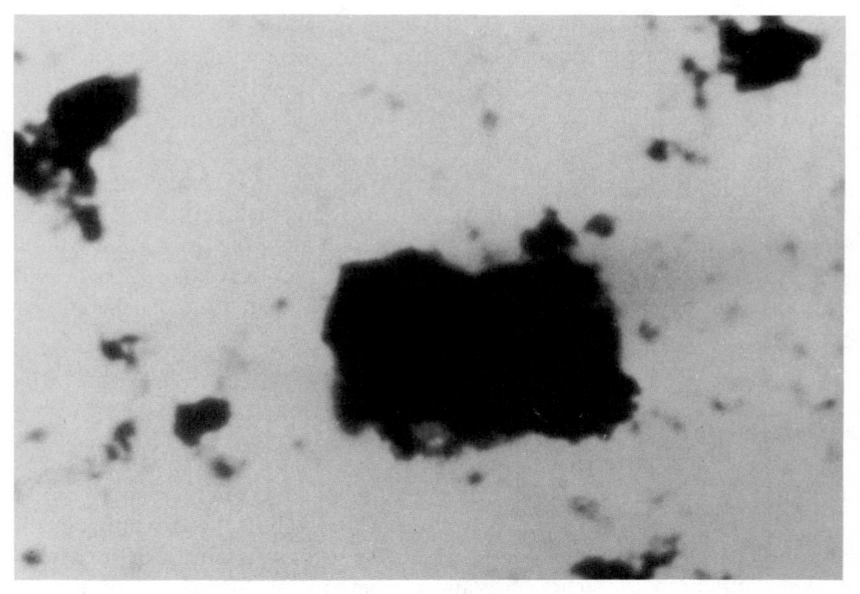

Figure 3.6. Asphalt particle showing some dissolution in the mounting media. Note the smear of dissolving material to the right of the particle. Micrograph courtesy of Microlab Northwest Inc.

mounting media. Note the smear to the right of the particle. It is one of the unavoidable disadvantages of optical microscopy.

A schematic of a polarizing optical microscopy is shown in Figure 3.7. White radiation from a lamp is passed through a polarizer that is usually made of a pleochroic compound that exhibits very strong absorption in one vibrational direction and very weak absorption in the perpendicular vibration direction. Therefore, the transmitted white light is polarized in only one vibrational direction. Amorphous particles and crystals of the cubic system exhibit only one refractive index and are called isotropic. When the second polarizer, known as the analyzer, is inserted above the sample, the microscope can be used to observe particles that have more than one refractive index (anisotropic particles). If the vibrational direction of the light transmitted by the polarizer is perpendicular to that of the light transmitted by the analyzer, the sample should be viewed between crossed polars. In this case, the background and all isotropic particles will appear dark, since all the polarized light that transmits through the sample and immersing oil is absorbed by the analyzer.

When plane-polarized light enters an anisotropic crystal, the light is doubly refracted into two components, vibrating in two perpendicular planes. The two components follow the two principal vibration directions of the crystal and have

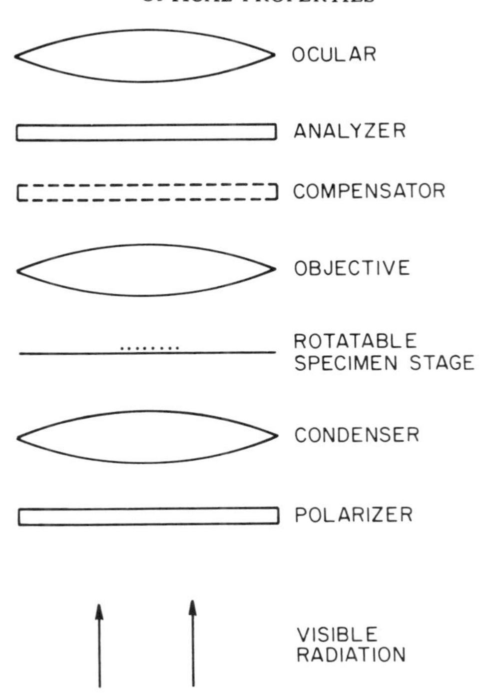

Figure 3.7. Schematic diagram of an optical microscope for polarized light microscopy.

different refractive indices, consequently, they move through the crystal at different velocities. The two components emerge from the sample with one retarded by a specific amount that depends on the difference between the two refractive indices and on the thickness of the sample. The components will recombine in the vibrational plane of the analyzer and since one component is retarded, interferences of some wavelengths will cause the image to appear colored. Destructive interference occurs for some wavelengths but constructive interference will result in the interference color observed for the image of the anisotropic particle. The interference colors are a measure of the retardation that is related to the birefringence:

$$r = 1000t(n_2 - n_1)$$

where n_1, n_2 = the two refractive indices of the anisotropic crystal
$n_2 - n_1$ = birefringence
t = thickness (μm)
r = retardation (μm)

Knowledge of the sample thickness (related to particle diameter) and the retardation of the anisotropic particle enables comparison with mineralogical tables and identification of the mineral composition of the particle. Figure 3.8 shows two views of a calcium oxalate monohydrate particle with a single polarizer set at two different orientations 90° from one another.

Birefringence can occasionally be useful for identifying materials other than minerals. There are a few biological materials that show characteristic patterns in polarized light. Figure 3.9 shows a corn starch particle that is distinguished by the Maltese cross pattern.

The observed birefringent colors can be enhanced with the use of another anisotropic crystal known as a compensator. When two anisotropic substances are superimposed, addition or subtraction of their individual retardations may occur (see Figure 3.10). The slower component (higher refractive index) is indicated by the short arrow and the faster component is indicated by the long arrow. If both substances are placed 45° away from the vibrational directions of the polarizer and analyzer, the retardations can be added when their slower components are parallel and subtracted when their components are perpendicular.

The interference color resulting from the addition (or subtraction) is of a higher (lower) order than that of either substance and is equivalent to the algebraic sum (difference) of the retardation. The vibration directions of the compensator are usually marked and fixed by inserting the compensator in a slot in the microscope at an angle 45° to the vibrational directions of the polarizers. The specimen stage can be rotated to align the vibration directions of the anisotropic particles with those of the compensator. The first-order red compensator consists of a layer of quartz of the proper thickness to produce retardation equivalent to about 530 nm. It is especially useful for observing weakly birefringent materials (gray or white when observed without the compensator).

The spectral transmission of an object is the result of the angle of incidence for the illuminating beam of light, the angle of observation, the wavelength-dependent absorption coefficient for the object, and the object's crystallographic orientation. The perceived color is dependent on the background intensity, the wavelength-dependent sensitivity of the sensor, the intensity of the light from the object toward the observer, and all of the other conditions mentioned for the spectral transmission of an object. Color can be a very sensitive criteria, but it can be equally misleading if misused. As an example silicon carbide, a common industrial abrasive, has six forms with a color variation from colorless to blue to yellow to brown to black. These colors correspond to slight changes in electron vibration frequencies with different crystal phases in pure silicon carbide and not to the presence of impurities. Potassium bromide is another case in point. It if is radiated with X rays it turns from colorless to pale blue. This change in color is the result of single electrons occupying sites normally occupied by bromine ions.

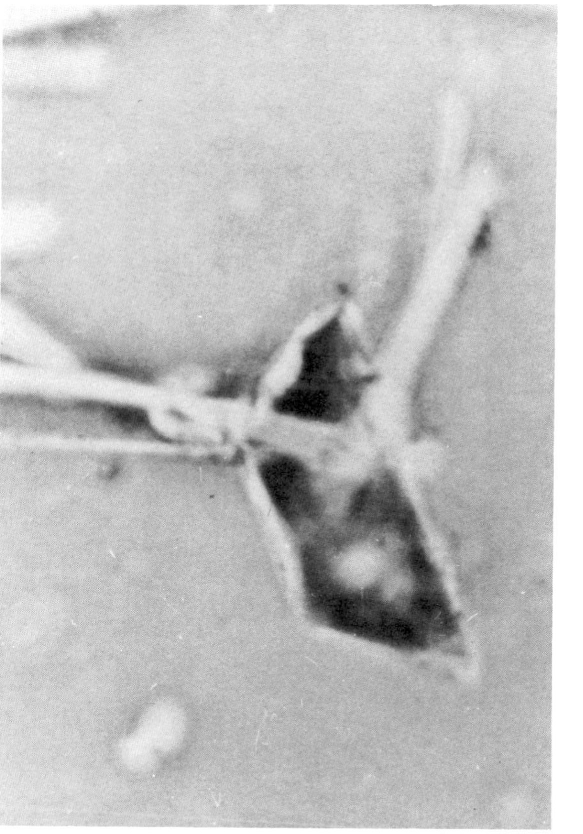

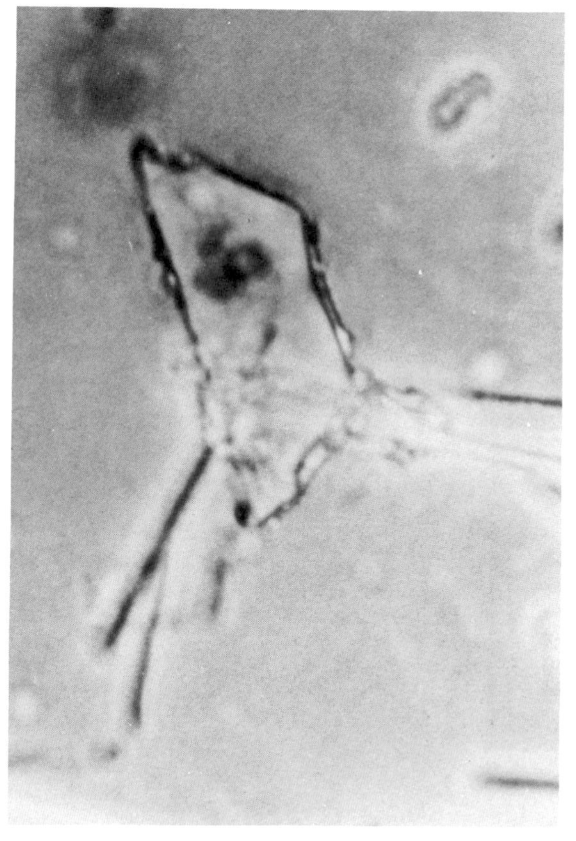

Figure 3.8. Two views of a calcium oxalate monohydrate particle with the polarizer set at two positions 90° from one another and no analyzer. Micrograph courtesy of Microlab Northwest Inc.

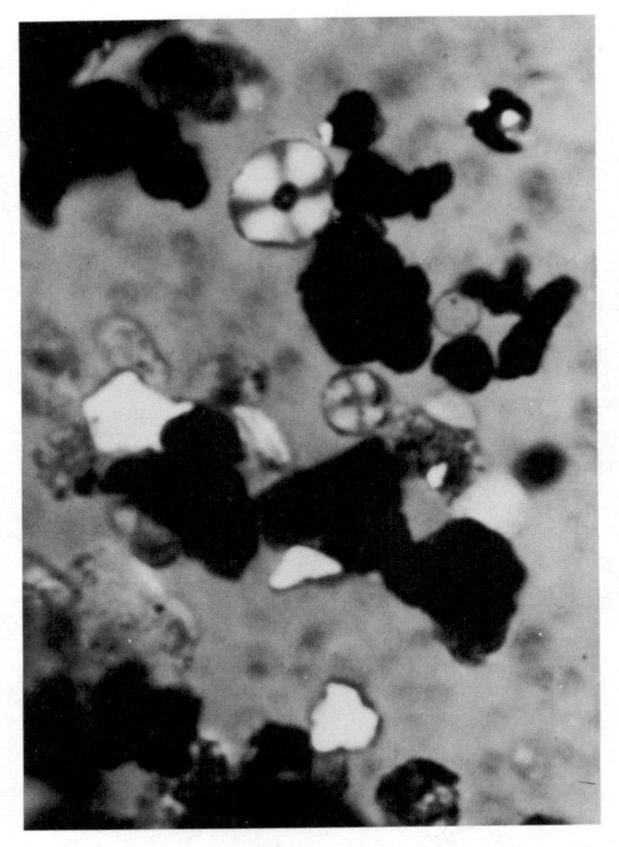

Figure 3.9. Corn starch particles with slightly uncrossed polarizers (20°) showing a distinctive Maltese cross pattern. Micrograph courtesy of R. Draftz of IITRI.

The resultant low-energy electron is a strong absorber in the red end of the spectrum. Different colors do not always indicate different compounds.

For the more common airborne particles, color can be a useful indicator of the source of a material. Fe_2O_3 is a good example. This material has two basic etiologies, oxidation through burning or oxidation through corrosion (hydration). The Fe_2O_3 from burning has made the transition from iron to FeO to Fe_3O_4 to Fe_2O_3. This transition is from a very high absorption coefficient to a final moderate absorption coefficient with very high absorption in the blue-green. The resultant particle is deep blood-red. For example, high-temperature hematite particles will appear as bright red, spherical particles. When produced by corrosion the transition is from iron to the whole Fe-O-OH-H_2O phase diagram ending in Fe_2O_3. Iron corrosion products vary in color from green-

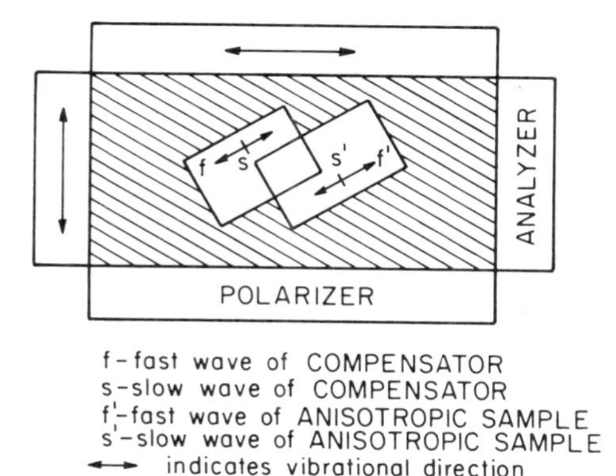

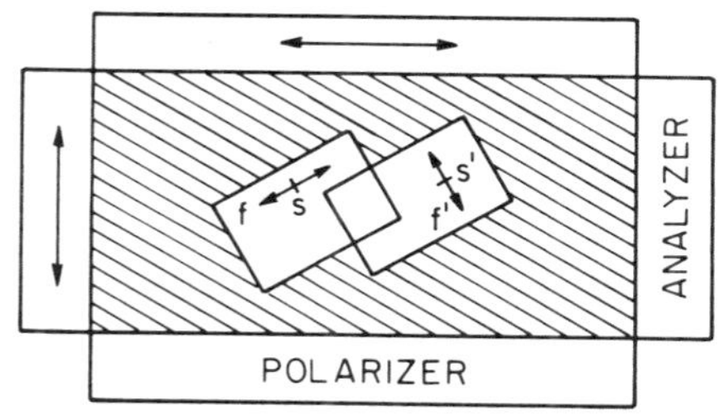

Figure 3.10. Use of a compensator with anisotropic samples viewed with crossed polarizers. Top: Vibrational components of compensator and sample parallel. Bottom: Vibrational components of compensator and sample perpendicular.

yellow to orange to red and black. They typically exhibit more than one phase in this series and have a morphology that is also different. The colors in the corrosion products are characteristic of their progression through this series of reactions.

A short glossary of terms associated with these properties is given at the end of this chapter. This is only a partial listing of the extensive vocabulary of optical microscopy. It is clear that many materials exhibit a number of these characteristics and in most cases, permit unique identification of the particle nature.

The amount of information that can be gathered for any given particle is considerable but often not relevant to the purpose of source attribution and may be too time consuming to be part of a quantitative analysis of a complex sample. The quantitative application of the light microscope is not an extended qualitative analysis. There are special sample restrictions and statistical requirements. These impose restriction on the optical properties that can be used to classify the particle types present in the sample. The manner of presenting the data and error analysis is also an important consideration for fully assessing the source impacts. The greatest single failing of most so-called quantitative light microscopical analyses is in the transition from qualitative analysis to quantitative analysis. Two major differences are ignored, statistical considerations and the optical restrictions that result.

STATISTICAL CONSIDERATIONS AND OPTICAL RESTRICTIONS

In a quantitative analysis of a particulate sample, a number of conditions should be included. First, the sample should be one that reflects the original, ambient population in such a way that an additional sample would have the same properties as the first one. To analyze individual particles, these particles must be free of other particles. The particles must also be fixed in position so that they do not move while being analyzed. A sample population should be large enough to minimize the impact of statistically unreliable events. How do these conditions impact an analysis of particles collected on a high-volume filter?

The subpopulation selected for analysis must be collected in a standardized fashion designed to minimize bias. The use of a cleared filter passes this requirement; tapping the back side of the filter to collect the dust does not. Ultrasonic collection techniques are good because the efficiency of the removal mechanism is not size dependent. However, a liquid carrier is required and this presents possible solubility problems, similar to those involved in filter clearing. The ultrasonic technique also collects particles from both sides of the filter, another advantage over approaches involving the collection of particles from only one surface. Whatever approach is used, the inherent variability of the population across the filter surface must be considered.

The next restriction is that the particles must be free from one another. The probability of particles being isolated optically decreases rapidly as the area covered by particles increases about 10%. Twenty-four-hour high-volume filters generally fail this requirement with loading equivalent to ambient conditions of 50 $\mu g/m^3$. Cleared fiber filters in fact never meet the requirement of free particles because even if the particles are sufficiently separated, the underlying fibers will affect the observations of the particle. However, the particle in a cleared glass fiber filter are often at different depth of fields and may be distinguished at differing focal points.

During a quantitative analysis, individual particles are examined and recorded. If these particles are capable of moving in the liquid mounting media, the possibility of analyzing the same particle twice becomes a very real consideration. To prevent this, the mounting media must fix the particles in position; oils are not satisfactory. Aroclor 5442, Caromount, or other permanent mounting media will meet this requirement. The need to fix the orientation of the particle precludes changing the mounting media.

The next requirement is that the sample size be large enough to minimize the impact of statistically rare events. A quantitative analysis involves the measurement of the particle, assigning the particle to a class of particles, multiplying the volume of the particle by the density-assigned particles in that class, and finally calculating the percent of the total mass represented by that particle class. A single very large particle out of a sample population of more than a thousand may represent as much as 20% of the total mass. This is an example of a statistically rare or unreliable event. There are a number of ways to deal with this problem. One approach is to count a sufficient number of particles so that no single particle accounts for more than a small percentage of the total mass for that category. Other segregated counting schemes can be employed. However, it is critical that the question of single large particles be explicitly considered in any quantitative analysis of a complex sample (Crutcher and Nishimura, 1978).

PARTICLE CLASSIFICATION

Typically, 30 to 40 different particle classes are needed for a given analysis. Each of these classes may contain more than one type of particle. For example, the category of melilite, a common slag mineral, will contain all of the different minerals in this family, provided their elemental composition does not modify their optical properties beyond the classing criteria used. The pollen category may contain 15 different types of pollen as well as three or four different spores and a few conidia. There are two very practical reasons for combining these individual particles into classes. First, the addition of subcategories adds time and cost to the analysis. Second, the addition of irrelevant categories clouds the relevant data. If all anisotropic particles are classed together, none of the sources in the area will be resolved by the data.

An essential part of any analysis is that the method be described sufficiently well to be repeatable by another skilled analyst. This means that the classing criteria must be well defined. The term "slag" is not enough. There are hundreds of different slags and many are mineralized, meaning that they contain glasses and different minerals in a complex matrix. It would be nearly impossible to write a single optical criteria that would unambiguously identify such a mineralized slag and not include many materials from other sources. The solution is to establish individual classes of particles, each well defined by their own optical and morphological properties, which can be evaluted as to their source after the quantitative particle analysis has been completed. The definition of these classes should indicate how a particle is placed into this group, such as high relief, and not just a text value for the refractive index.

SOURCE IDENTIFICATION

The purpose of the analysis is twofold; to identify emission sources and to quantify their impact on the environment at a specific site. It is often not necessary to perform a complete quantitative analysis to determine the major sources impacting a site. An estimate of the contribution from the dominant sources may suffice for days of abnormally high loading. Often these types of samples represent an out-of-control condition at a local industry. A source inventory and the location of the sample site under these conditions will be sufficient to produce a very reasonable estimate of the source contribution.

For example, examination of samples from Milwaukee, Wisconsin showed high loadings from very specific sources. In one case, yellow, reticulated spherical paint particles were observed in perfusion. These particles were the results of painting a nearby bridge. In another case, blue leather fibrils were observed. A leather working factory nearby was considered to be the likely source of this one-time occurrence (R. Draftz, private communication). The ability to distinguish these unusual events is one of the strengths of optical microscopy. The danger of an estimate is that it represents the opinion of the analyst. The value of that opinion rests in the skill of the analyst, something very difficult to evaluate. The advantage of an estimate is that it is relatively quick and easy to perform.

The quantification of the amount of material from a specific source impacting a specific site can only be based on quantitative analytical results. The transition from the analytical results to the final source apportionment requires the use of a model. Specific particles can often be attributed to a specific source but they are not the only emission from that source. Similarly many particle types have more than one source criteria. A number of limiting factors acting at the site will add scatter to the data. An incomplete source inventory can make correct source attribution impossible and lead to erroneous conclusions. Choice of sampler location is often important since incorrect location may indicate a problem where none actually exists. There may also be unusual local events that can impact a site with unsuspected particles types. An example would be a landfill that periodically received material from a building demolition. A sample site along the path to the landfill could be very significantly impacted by such an event. Aside from these considerations, determining the effect an emission has at a given site involves the use of source inventories, assemblage analysis, industrial tracers, and source samples. Together these are the foundation of receptor modeling with microscopic analytical data.

SOURCE INVENTORIES

Source inventories are available in all monitored areas though they are not always complete or up to date. This information along with the location of the sample site can be very beneficial in evaluating the impact of different sources at the site. Many sources emit similar materials. Fused glass spheres from the melting of small quartz, calcite, or clay particles may be from an accidental fire, an incineratoı, a coal-fired power plant, or a hog fuel boiler. There is nothing unique about a fused mineral that would differentiate these sources. A source inventory showing the presence or absence of these possible sources would simplify the identification of the source.

A knowledge of the types of sources in an area also helps to establish the optical categories that are least likely to overlap. The absence of a source inventory can and has significantly limited results when it was discovered that two or more sources that could have been resolved, had been lumped into one category because the source types were not known. Source inventories are occasionally incomplete with regard to identifying the source types. When only company names are available, the analyst is often not aware of the nature of that company's business. Additional information would be needed in this instance.

ASSEMBLAGE ANALYSIS

Another property of most sources is that along with a major particle type, a variety of particles are emitted. Often the presence of these other types of particles will permit a simple type of mass balance to help distribute the contents of a category with a number of different sources. Quartz would be an example. In an environment that contains cement plants, glass manufacturing, rock crushing, and roads, the apportionment of the quartz category can be very complex. Assemblage analysis can be very beneficial in such an area. The cement industry would contribute calcite, quartz, and blast furnace slag as a raw product assemblage, clinker and lime spheres from roasting, and alite, belilite, celite, and crushed clinker from the finishing, grading, packaging, or shipping facility. Each of these industries has its own typical assemblage for different parts of its processes. By recognizing these assemblages, part of the quartz category could be distributed among those sources for which assemblages were present.

Assemblage analysis can also indicate unknown sources and indicate how particles from those sources are reaching the sampling site. The shipping of landfill materials past a sampling site can produce an assemblage characteristic of a source not in the local area. The size distribution and mix of particles in the assemblage would indicate bulk transport into the area. Assemblage analysis is a very useful technique, but it may require more information about the sources and more source samples than are often available.

INDUSTRIAL TRACERS

Many industries have a particle type or assemblage that is unique to that industry or for that industry in a given area. These particles become a tracer for that industry or source. Prehydrated cement minerals, zinc tetrads, lead oxide prisms, gypsum crystals, and so on, are just a few examples. There are many such tracers and they all share two attributes: they are characteristic of one source and they represent only a fraction of the emissions from that source.

SOURCE SAMPLING

Source sampling for light microscopy is somewhat different than that for any other type of analysis. Dust collected with cellophane tape from flat surfaces around an industry can be very useful in terms of characterizing fugitive emissions and assemblages. Samples from baghouses or surfaces within the plant can indicate specific assemblages characteristic of specific operations. Soil samples indicate both local geology and the amount of modification from deposited industrial particles. Tape lifts from the surface of the ground indicate the composition of the top layer of dust, that which may be lifted by the wind on a windy day. Particles filtered from the air during periods of specific wind direction indicate typical air transport assemblages.

The collection of source samples for the light microscope is very simple in terms of sampling equipment required. A roll of tape is often sufficient. The ability of the microscopic analyst to screen types of particles that will not become airborne from the sample being examined will result in useful information from what would be useless for an elemental analysis approach. These samples from the source also indicate the variety of emissions from the source. The most critical part of source sample collection for the light microscope is explicit record keeping.

APPLICATIONS

There have been numerous applications of optical microscopy to the identification of the sources of particles and the attribution of particulate mass to those sources. For example, aerosol source characterizations for TSP have been performed for Miami, Florida (Draftz, 1979), St. Louis, Missouri (Draftz and Severin, 1980), and Phoenix, Arizona (Graf, Snow and Draftz, 1977). In these studies, the impacts of the various components have been estimated for both a high-volume sampler and a low-volume Andersen multistage impactor. The Andersen impactor separates the particles into size categories of 16.4 μm (stage 0), 9.3-16.4 μm (stage 1), 5.35-9.3 μm (stage 2), 2.95-5.35 μm (stage 3), 1.53-2.95 μm (stage 4), 0.95-1.53 μm (stage 5), 0.54-0.95 μm (stage 6), 0.38-0.54 μm (stage 7), and a backup filter to collect the <0.38-μm particles. Tables 3.1 and 3.2 illustrate the results of such estimations. These results are semiquantitative and

Table 3.1. Aerosol Composition and Size Range of High-Volume Filter Particles Municipal Court Site[a] July 23, 1975

	Estimated wt%	Geometric Size, Average	μm Range
Minerals	75–90		
Quartz feldspar	*M*[b]	15	<1–40
Carbonates	*P*	6	<1–32
Micas	*t*	12	5–60
Clays	*t*	<1	<1
Iron oxides	*t*	7	<1–10
Vehicle emissions	5–10		
Tailpipe emissions	*m*	<1	<1
Rubber tire fragments	*m*	30	5–100
Combustion products	5–15		
Fly ash	*t*	5	<1–12
Coal and coke fragments	*m–M*	6	<1–50
Partially combusted plant parts	*t*	35	—>100
Oil soot	*t*	30	—>50
Ammonium sulfate	*m*	<1	<1
Biologicals	⩽1		
Pollen, spores, conidia	*t*	30	6–110
Plant fragments	*t*	40	5–100
Starches	*t*	12	8–16
Insect parts	*t*		
Miscellaneous	<1		
Titanium dioxide	*t*	1	1
Hydrated iron oxides	*t*	1	1
Magnetic fragments	*t*	10	7–16
Total suspended particles (μg/m^3)		109.1	

[a] Draftz and Severin (1980).
[b] *P* = primary (>25%), *M* = major (5–25%), *m* = minor (0.5–5%), *t* = trace (<0.5%).

provide a good indication of the major sources of particles. In some cases, a very significant impact from a single source can be readily identified. In Baltimore, Maryland, one study found that as much as 50% of the observed TSP loading was due to cornstarch that could be uniquely identified by its distinctive Maltese cross pattern under polarized light (Koch, Schakenbach, and Severin, 1979). Figure 3.11 gives a view of a portion of a filter showing the large number of spherical particles displaying the characteristic Maltese cross. Similar studies of high-volume TSP samples from various cities in Illinois have also been con-

Table 3.2. Aerosol Composition by Impaction Stage

Municipal Court Site[a] July 23, 1975

	0	1	2	3	4	5	6	7	8
Minerals	80–95[b]	90–95	90–95	85–95	85–95	65–80	20–35	20–30	1–5
Quartz, feldspars	*M*	*M*	*M*	*m–M*	*m*	*m*	*m*	*m*	*m*
Carbonates	*P*	*P*	*P*	*P*	*P*	*P*	*M–P*	*M–P*	*m*
Micas	*m*	*m*	*m*	*t*					
Clays	*t*	*t*	*t*						
Iron oxides	*t*	*t*	*t*	*t*	*t*	*t*	*t*	*t*	
Vehicle emissions	1–5	1	1	1	1–5	5–10	10–20	1–20	85–95
Tailpipe emissions	*t*	*t*	*t*	*m*	*m*	*M*	*M*	*M*	*P*
Rubber tire fragments	*t–m*	*t*	*t*	*t*					*t*
Combustion products	1–5	1–5	1–5	1–5	5–10	15–25	55–70	60–75	1–5
Fly ash	*t*	*t*	*t*	*t–m*	*t*				
Coal and coke products	*m*	*m*	*m*	*m*	*m–M*	*M*	*m*	*m*	*m*
Partially combusted plant parts	*t*	*t*	*t*	*t*	*t*				
Oil soot	*t*	*t*	*t*	*t*	*t*		*P*	*P*	*m*
Ammonium sulfate									
Biologicals	<1	<1	<1	<1–2	<1				
Pollen, spores, conidia	*t*	*t*	*t*	*t–m*	*t*				
Plant fragments	*t*	*t*	*t*						
Starches	*t*	*t*	*t*	*t*					
Insect parts			*t*	*t*					
Miscellaneous	<1–2	1	<1–5	1–5	1–5	1–5			
Titanium dioxide	*t*	*t*	*t*	*t*					
Hydrated iron oxides	*t–m*	*t*	*m*	*m*	*m*	*m*			
Magnetic fragments	*t*								

[a] Draftz and Severin (1980).

[b] Percentage of mass on that impactor stage.

[c] *P* = primary (>25%), *M* = major (5–25%), *m* = minor (0.5–5%), *t* = trace (<0.5%).

Figure 3.11. View of a portion of a filter from Baltimore, MD, showing a high loading of corn starch particles. Micrograph courtesy of R. Draftz of IITRI.

ducted (Arnold and Draftz, 1979). In other instances carbonaceous materials, particularly natural biological materials, can be an important component of TSP. These materials can be easily distinguished as described previously. Confirmation of the carbonaceous aerosol contribution can be obtained by analysis before and after low-temperature ashing in an oxygen plasma to remove oxidizable carbon (Draftz, 1982).

In many studies, optical microscopic investigations of a few samples are used as part of a check on the quality of source apportionment by other methods. This approach has been used in the study of source in Houston, Texas (Dzubay and Stevens, 1983) and in El Paso, Texas (Janocko et al., 1983).

Optical microscopy is a powerful tool. However, accurate identification and mass apportionment require a considerable degree of expertise and must be done by a well-trained microscopist. Qualitative and semiquantitative results are quite readily achieved by these trained personnel. However, full quantitative analysis is very time consuming, and to examine enough particles to be certain of having a representative sample, is therefore, rather costly. In chapter 4, image analyzing techniques, which have permitted automation of electron microscopy, are discussed. Although similar image analysis has been described for optical microscopy (McCrone and Delly, 1973a), it has yet to be applied to receptor modeling problems of the type described here. The use of such techniques may be an approach to expand the use of this powerful analytical method.

GLOSSARY OF MICROSCOPY TERMS

The following are only a limited set of terms associated with optical microscopy.

Anistropy: An anisotropic material is bright when viewed between cross polars. It indicates that the material has more than one refractive index.

Biabsorption: The difference between the maximum and minimum absorption exhibited by this material is any possible orientation. This is a fundamental property of crystalline materials. Pleochromism is a term used to identify absorption differences in the visible wavelengths that produce different color effects in the crystal when viewed while rotating a single polarizing filter.

Biaxial Crystal: A crystal that has two optic axes is biaxial. The orthorhombic, monoclinic, and triclinic crystal systems are biaxial.

Birefringence: The difference between the maximum and minimum refractive indices exhibited by this material in any possible orientation. This is a fundamental property of crytalline materials. The term "birefringent" is synonymous with "anisotropic."

Cross Polarized Light: The term cross polarized light used here will refer to the use of a linear polarized filter below the subject and another above the subject with the angle between the polarization direction for the two filters at 90° to one another. The result is that no light can pass through both filters except where modified by an anisotropic material.

Extinction Position: The position in which the particle disappears as it is rotated against the black field of view under cross polarized light. Extinction occurs every 90° of rotation for anisotropic crystalline materials. Three types of extinction positions exist for particles exhibiting a characteristic axis or angle, parallel, symmetric, and oblique.

Interference Figure: This is the pattern at the back focal plane of the objective used to determine the orientation of the crystal on the stage and the optical crystal class, uniaxial or biaxial, to which it belongs.

Optic Axis: The optic axis of an anisotropic material is an imaginary line through the material normal to which all optical properties are the same.

Viewed down this axis with a narrow, parallel light beam, an anisotropic material would not appear to be anisotropic.

Pleochroism: The property of certain crystals to display different colors when viewed from difference directions under transmitted light.

Polarizing Filter: A polarizing filter only allows light polarized parallel to one line to pass through. A polarizing light microscope contains one such filter below the stage, called the polarizer, and one above the stage called the analyzer.

Polarized Light: The term polarized light used here will refer to linear polarized light. Light is polarized if all of its field vectors are aligned parallel to one another.

Retardation Color: An interference color effect seen in anisotropic materials viewed between crossed polars, produced by a combination of the thickness of the particle and the difference between the two refractive indices measurable in that specific orientation of the material. The change of colors with increases in retardation follow a modified Newtonian interference color scale known as a Michel–Levy chart sequence.

Sign of Elongation: Materials that have an obvious long axis can be easily tested to determine if their high refractive index is oriented more closely to that axis. If it is, the material has a positive sign of elongation, if not it is negative (do not confuse this with the optical sign; the optical sign is outside the scope of this book).

Uniaxial Crystal: A crystal that has only one optic axis is uniaxial. The hexagonal, tetragonal, and trigonal crystal systems are uniaxial.

CHAPTER

4

SCANNING ELECTRON MICROSCOPY

Ever since the first scanning electron microscope (SEM) became available commercially in 1965, the SEM has gained increasing acceptance as a powerful research tool. With the development of the energy-dispersive detector for X-ray analysis, the SEM has moved from an instrument that could produce micrographs to an extremely versatile analytical technique. The SEM equipped with X-ray detection capabilities (SEM/XRF system) has been used for research in biology, medicine, metallurgy, nuclear science, geology, agriculture, electronics, and air pollution.

The ability of the SEM/XRF system to perform elemental analysis of extremely small volumes of materials has great significance in the field of air pollution. Individual particles can be analyzed for elemental constituents and both qualitative and semiqualitative data can be obtained. With improvements in computers, the scanning and image analysis can be done in real time under computer control so that size, shape, and elemental composition can be obtained on a large number of particles in a reasonable length of time. Thus, scanning electron microscopy represents a powerful tool for the identification and classification of particles. For example, Butler, Crossley, and Colwill (1981) used scanning electron microscopy to examine an urban aerosol fractionated with an Andersen high-volume impactor. They were able to obtain both physical and chemical information on particles in various size ranges and to explore the relationships between size and composition.

SCANNING ELECTRON MICROSCOPE (SEM)

Figure 4.1 shows a schematic diagram of an SEM. The source of free electrons is obtained by thermionic emission from the heated top of the tungsten filament in an evacuated chamber. This beam of energetic electrons is focused onto the sample surface by two electromagnetic lenses. The lenses are used to demagnify the electron source (~4–6 mils in diameter) to a diameter of about 100 Å. The electron beam can be scanned over the specimen in a rectangular raster.

One type of emission resulting from the complex beam-matter interaction is secondary electrons. These are low-energy electrons ejected from the sample by

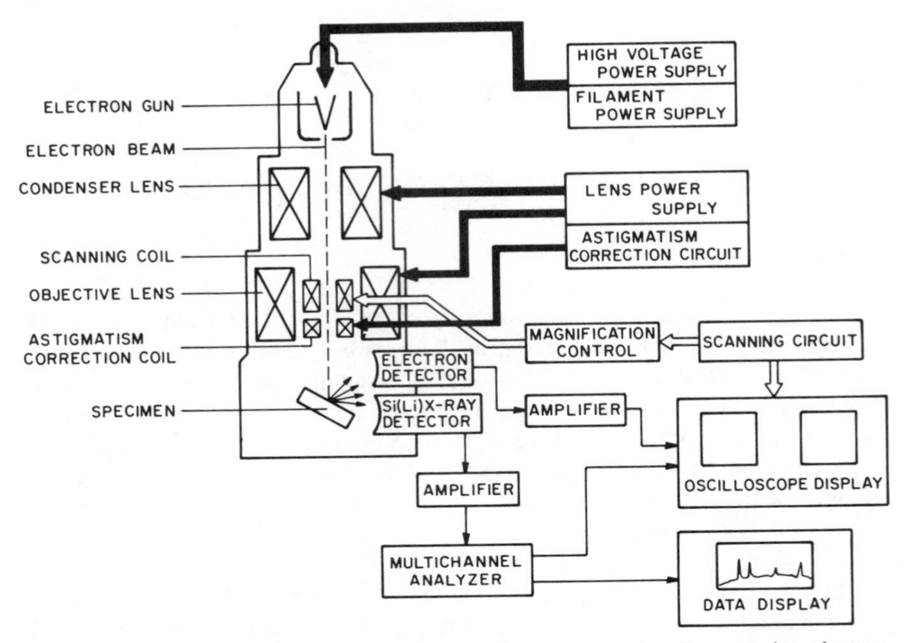

Figure 4.1. Schematic diagram showing the major components of a scanning electron microscope.

the primary electron beam and are collected by a scintillation counter. The output of the scintillation counter is connected to a cathode ray display. The deflection of the electron probe over the specimen and the deflection in the cathode ray displays are synchronized by connecting the two sets of scanning coils to the same X–Y generator. In this way, an image of the specimen appears on the display. Image contrast arises from the differences in the efficiency of producing secondary electrons from different regions of the specimen. This efficiency is dependent on the topography of the sample so that the morphology of the specimen is displayed. In a similar manner, the electrons from the microscope beam can be scattered from the particle and used for particle imaging. The energy of the backscattered electrons are sensitive to the average atomic number that also helps to distinguish particles from the background or to distinguish among particles. The current use of computer-controlled microscopy tends to prefer the use of the backscattered mode to obtain particle size and shape information.

The incident electron beam can also cause emission of X rays that are generated from a sample volume of about 1-3 μm^3. These X rays can be collected and analyzed by an X-ray spectrometer system, thereby providing a means of determining the elemental composition of a microscopic region. The X-ray detector consists of a lithium-drifted silicon diode located between two electrodes across which a bias voltage is applied. An X-ray photon creates electron-hole pairs and the energy of the X-ray is given by the number of pairs created (3.8 eV for each electron-hole pair in silicon at 77°K). The electron-hole pairs migrate to the electrodes, creating a current pulse. This pulse is amplified and converted to a voltage proportional to the collected charge from the detector. This voltage is passed into a multichannel analyzer (MCA) where the peak voltage is converted to a digital number that is used as an address for storing a single count in a computer. The detector is now ready to accept another pulse, the whole process taking about 10 μsec. A beryllium window separates the silicon detector from the electron column because the silicon detector is light sensitive and so shields any stray light or cathodoluminescence. The beryllium also absorbs back-scattered electrons and prevents any X-ray generation by high-energy back-scattered electrons.

It is generally difficult to obtain a precise quantitative analysis of the elemental composition of the particle. However, it is possible to obtain relative concentrations of elements present at $\geqslant$0.1 at% in the particle. Janossy, Kovacs, and Toth (1979) describe a method for obtaining accurate atomic weight ratios for ultrathin samples. The elements that can typically be observed include Na, Mg, Al, Si, P, S, Cl, K, Ca, Ti, V, Cr, Mn, Fe, Cu, Ni, Zn, Pb, and Br. It is possible to observe the lighter elements, Na and Mg particularly, because of the fact that the sample is in a vacuum. A particle that is observed but for which there are not observable X rays is often assigned to be a carbonaceous particle. Carbonaceous particles can be imaged and information can be obtained on the possible particle sources from the particle morphologies (Griffin and Goldberg, 1979). More work needs to be done to improve the information obtained from carbonaceous particles since they represent a major component in typical urban aerosol samples. The precision with which elemental analyses can typically be made is given in Table 4.1.

It is possible to utilize the concentration measurements made on individual particles to estimate the average bulk concentration values for the collection

Table 4.1. Precision of CCSEM Elemental Analytical Results

Elemental Concentration (wt%)	Average Relative Error (%)	95% Confidence Interval (%)
<1.0	35	0.65–1.35
2.5	32	1.94–3.06
5.0	16	4.36–5.64
10.0	8	8.80–11.20
15.0	5	14.27–15.75

of particles. Casuccio et al. (1983) and Energy Technology Consultants (1983) have described a detailed intercomparison study that compares the results of scanning electron microscopy with bulk chemical analysis for several high-volume sampler filters taken in El Paso, Texas. The chemical methods include X-ray fluorescence, both photon- and proton-induced, atomic absorption spectrometry, and ion chromatography. There were two colocated samplers, one with a glass fiber filter and one with a cellulose filter. The two filters for a single-time interval were subjected to these analyses and the results are presented in Table 4.2.

The results in Table 4.2 generally show good agreement between methods, particularly when it is recognized that there was a significant mass difference between the two filters. The glass fiber filter indicated 156 $\mu g/m^3$ whereas the cellulose fiber filter measurement was 111 $\mu g/m^3$, so the elemental values presented in this table are in percentage of TSP mass. The XRF analysis for the glass fiber filter was only performed for the heavier elements and a replicate analysis was not performed for the cellulose fiber filter. The atomic absorption analysis was only performed for the glass fiber filter while PIXE, CCSEM, and ion chromatography were performed for all of the samples. PIXE analysis of silicon on glass fiber filters shows the problems that would be anticipated in analyzing for the element that is the principal constituent of the filter. CCSEM could provide silicon analyses because the particles were separated from the filter and any filter fibers removed with the particles could be easily determined through their aspect (length to width) ratio. The CCSEM did show lower sulfur values than either PIXE or ion chromatography. These results may reflect the known catalysis of SO_2 to SO_4^{2-} on the alkaline filter media. The in situ formed sulfate would not represent readily removed particles and so may not show up in CCSEM. It should be noted that there is better agreement for sulfur on the cellulose filter where sulfate should not be formed. The lead values were in reasonable agreement although the CCSEM value for the glass fiber filter is low, possibly because of the inability to remove deeply penetrating, fine particles. In general, these results from Casuccio et al. (1983) do show that a reasonable representation of the bulk concentration values can be obtained from the microscopic results.

Another comparison has been made by Johnson et al. (1981) for particles collected on Teflon filters using a dichotomous sampler. These samples are a better set for analysis by both SEM and XRF. Table 4.3 shows the ratio of elemental values for two series of samples from Raleigh, North Carolina and Philadelphia, Pennsylvania. There are problems for some of the low abundance elements such as phosphorus and copper. It can be seen that chlorine and bromine are lower in the electron microscopy results, possibly because of a loss of these volatile elements in the microscope vacuum. The disagreements for S, Fe, and Ti in these studies are not well understood. It may be that the recent im-

Table 4.2. Comparison of Elemental Concentrations from Analyses of Collocated and Replicate Filter (%TSP)[a]

	Glass Fiber Filter					Cellulose Filter					Replicate Cellulose				
	AA	XRF	PIXE	CCSEM	IC	AA	XRF	PIXE	CCSEM	IC	AA	XRF	PIXE	CCSEM	IC
Si	*c*	*b*	0.0	10.5	*c*	*c*	6.9	11.8	10.1	*c*	*c*	*c*	10.3	9.6	*c*
S	*c*	*b*	3.1	1.7	3.3	*c*	4.2	3.2	2.8	2.1	*c*	*c*	2.9	2.1	2.0
Ca	*c*	*b*	10.3	12.5	*c*	*c*	11.9	10.7	10.8	*c*	*c*	*c*	9.6	10.8	*c*
Fe	*c*	*b*	2.1	3.3	*c*	*c*	2.3	2.6	3.0	*c*	*c*	*c*	2.4	2.3	*c*
Pb	2.1	2.2	2.3	1.3	*c*	*c*	2.6	2.6	2.5	*c*	*c*	*c*	2.6	2.7	*c*
Br	*c*	0.3	0.3	0.7	0.7	*c*	1.2	0.9	0.6	0.4	*c*	*c*	0.8	1.1	0.4
Zn	1.1	*b*	1.0	0.8	*c*	*c*	1.0	1.2	0.8	*c*	*c*	*c*	1.1	1.0	*c*

[a] Taken from Casuccio et al. (1983).
[b] Not reported.
[c] Not analyzed.

Table 4.3. Ratio of SEM to XRF Elemental Weight Analysis (SEM/XRF ± 1 Standard Deviation)[a]

Element	Raleigh Samples (n = 6)	Philadelphia Samples (n = 8)
Al	0.94 ± 0.14	0.88 ± 0.23
Si	0.97 ± 0.10	1.05 ± 0.21
P	0.36 ± 0.19	0.47 ± 0.21
S	0.78 ± 0.32	0.17 ± 0.06
Cl	0.47 ± 0.19	0.64 ± 0.32
K	0.85 ± 0.14	0.99 ± 0.26
Ca	0.92 ± 0.15	1.26 ± 0.19
Ti	1.70 ± 0.46	1.17 ± 0.62
Fe	1.23 ± 0.22	2.33 ± 0.32
Pb	0.88 ± 0.49	0.95 ± 0.57
Mn	0.79 ± 0.38	1.45 ± 0.59
Cu	0.02 (n = 1)	2.32 ± 2.4
Zn	0.83 ± 0.54	1.09 ± 1.0
Br	0.55 ± 0.39	0.43 ± 0.20

[a]Taken from Johnson et al. (1981).

provements in the scanning algorithms and statistics of counting particles may resolve these problems. It is clear that quite reasonable composition determinations can be obtained from scanning electron microscopy.

IMAGE DISPLAY

To illustrate the qualitative information that is available from the imaging process, four single particles are displayed in Figure 4.2 to 4.5. These figures show the secondary electron image, the backscattered image superimposed with the image analysis lines that are described in the next section, and the X-ray spectrum taken with the electron beam focused on the particle center as defined by the intersection of the analysis lines. It can be seen that both shape and composition can be useful in identifying particle types. Figure 4.2 shows an angularly shaped Ca-amphibole particle. It has silicon, calcium, and iron as its major constituent elements and has a distinctive crystalline shape and texture. Figure 4.3 shows a relatively spherical quartz particle that can be distinguished from the cenosphere (flyash) in Figure 4.4 and the spherical iron particle in Figure 4.5 by their differing chemical compositions and surface texture. The quartz particle is visibly not as smooth as the other two and predominantly shows the presence of silicon. The cenosphere shows an alumino-silicate composition with some

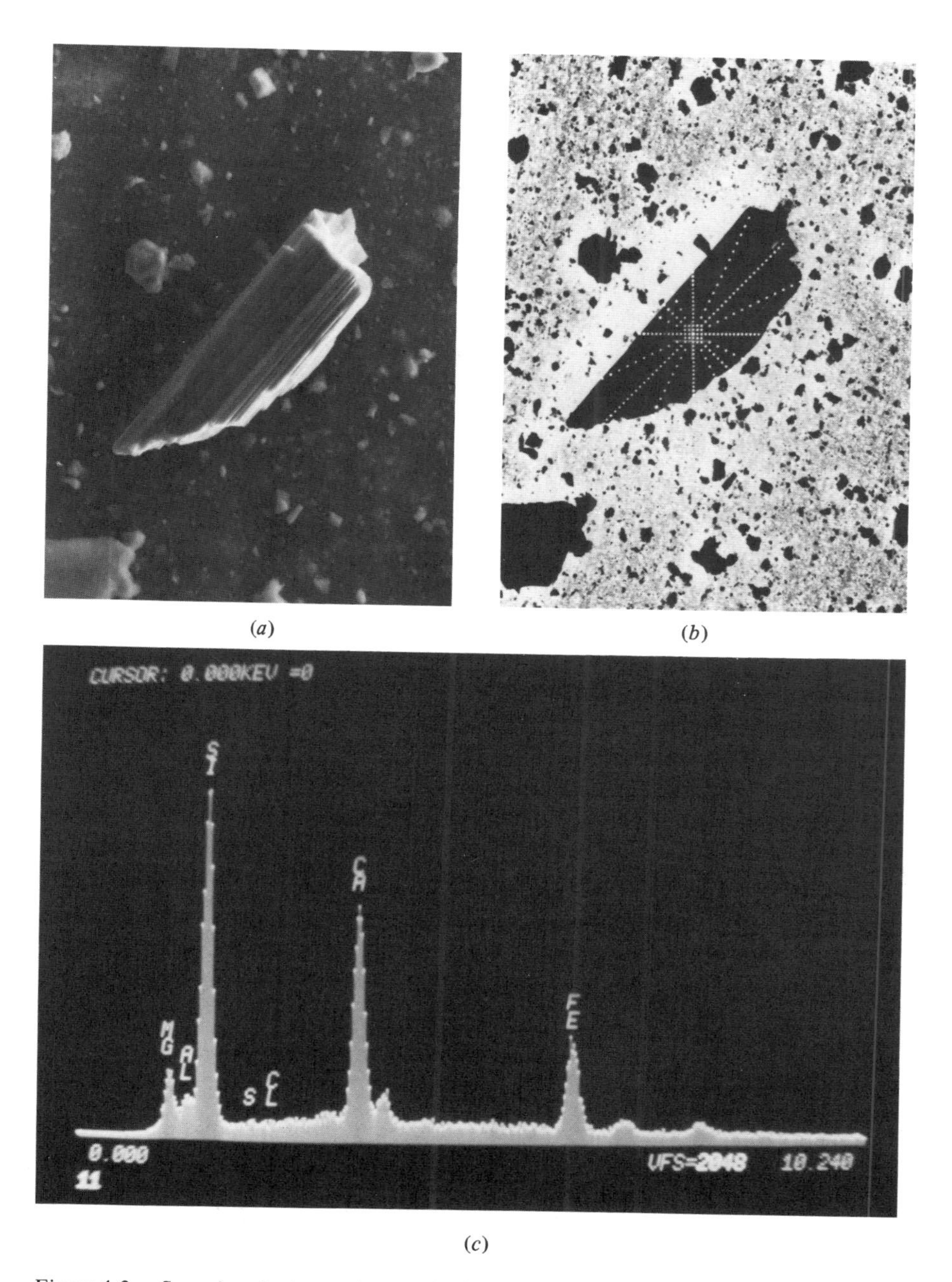

Figure 4.2. Scanning electron micrograph of a Ca-amphibole particle using secondary electron image (*a*) and backscattered electron image (*b*) showing the computer generated image analysis lines for particle sizing. (*c*) The X-ray spectrum with the electron beam focused at the central spot is shown at the bottom. Figure courtesy of Energy Technology Consultants.

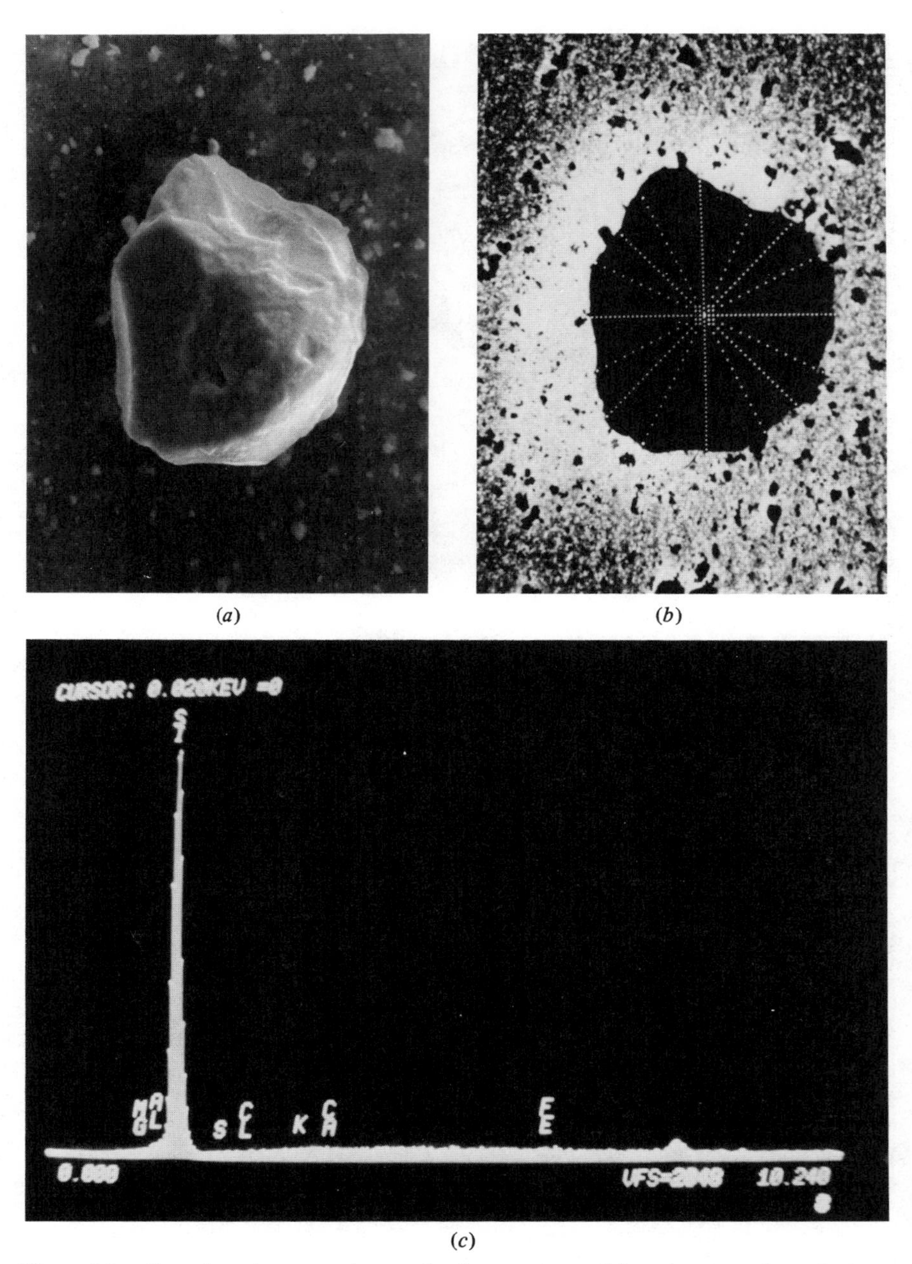

Figure 4.3. Scanning electron micrograph of a quartz particle using secondary electron image (*a*) and backscattered electron image (*b*) showing the computer-generated image analysis lines for particle sizing. (*c*) The X-ray spectrum with the electron beam focused at the central spot is shown at the bottom. Figure courtesy of Energy Technology Consultants.

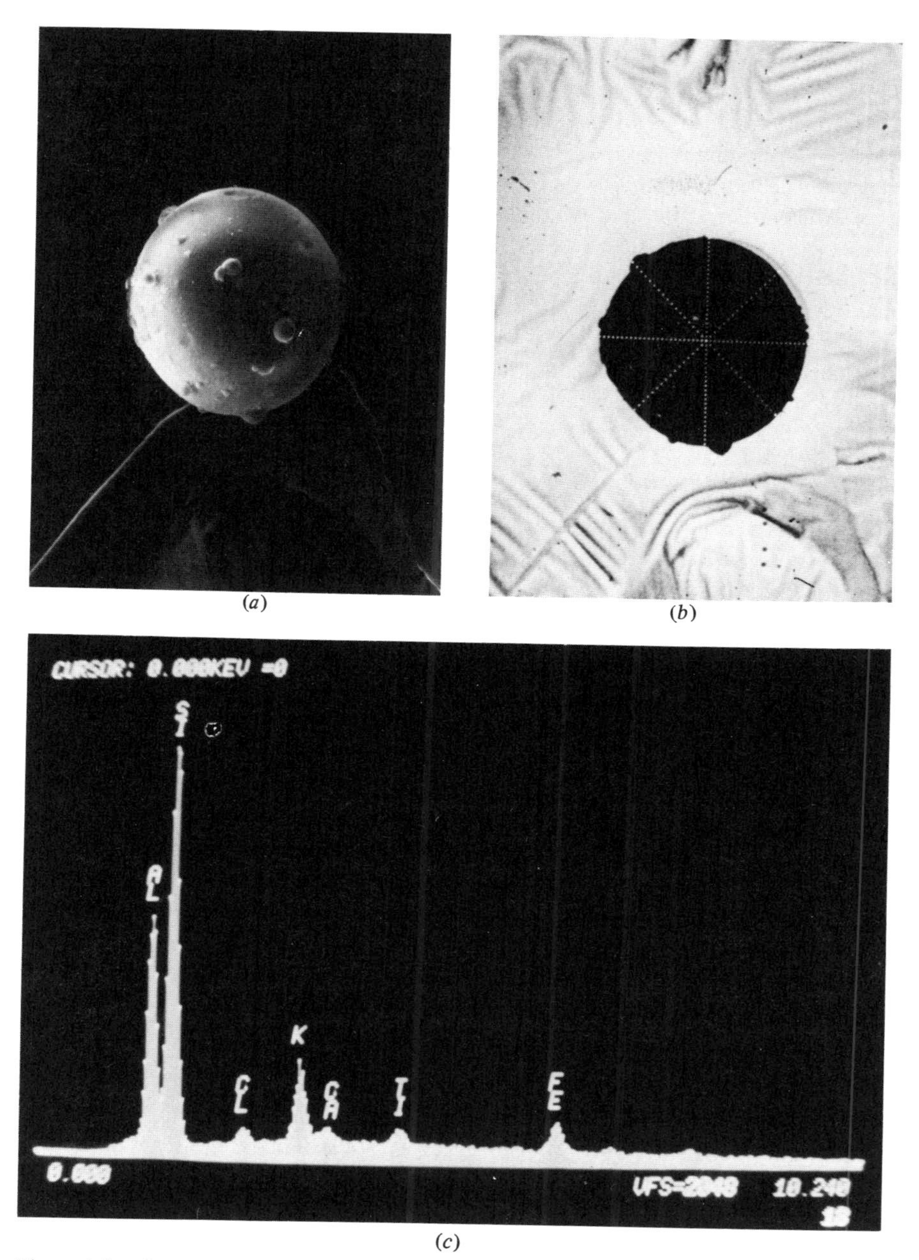

Figure 4.4. Scanning electron micrograph of a cenosphere (flyash) particle using secondary electron image (*a*) and backscattered electron image (*b*) showing the computer generated image analysis lines for particle sizing. (*c*) The X-ray spectrum with the electron beam focused at the central spot is shown at the bottom. Figure courtesy of Energy Technology Consultants.

other minor constituents. The iron particle is clearly different in composition although quite similar to the cenosphere in shape and texture.

Particles can also be distinguished by size and composition. In Figure 4.6, a particle containing only lead and bromine can be seen to have a diameter of the order of 0.33 μm. Figure 4.7 shows a larger (~1 μm) particle containing lead and zinc. The lead/bromine particle is likely to have come from automotive emission while the lead/zinc particle's probable source was a nearby lead smelter. It can be seen that a great deal of information can be obtained from the SEM image and the corresponding elemental analysis.

One way to take advantage of the elemental data is to use the intensity of a particular X-ray energy to control the image brightness. In this way the particles can be imaged with the fluoresced X rays and strong elemental associations can be rapidly detected. Linton et al. (1980) have examined the elemental relationships in a series of physically fractionated, settled urban dusts. These dust samples have been sequentially separated on the basis of physical size, magnetic susceptibility, and density in order to separate particle fractions with high lead concentrations. The procedure was outlined in Figure 2.3. The samples were then examined with an SEM in the secondary electron mode, and the X-ray intensities were mapped for key elements associated with lead sources such as automobiles or lead-based paint. Figure 4.8 shows a set of views of a sample of separated dust. Several particles show strong correlations between Fe, Pb, and Br. Such particles where mixed lead halides condense on steel exhaust particles have been confirmed in other studies (Linton, 1979). These particles can be contrasted with those in Figure 4.9 which show particles with Pb associated with Ti. These particles are from near a building and are probably lead containing paint chips. Thus, the combination of secondary electron image and elemental associations gives inferential information about possible particles sources. An excellent collection of electron micrographs and the corresponding X-ray spectra for a wide variety of particles is found in Volumes 3 and 6 of *The Particle Atlas* (McCrone and Delly, 1973c; McCrone, Brown, and Stewart, 1980).

COMPUTER-CONTROLLED SCANNING ELECTRON MICROSCOPY

The major development that has opened the use of scanning electron microscopy to quantitative receptor modeling is the coupling of computers to the SEM (Kelly, Lee and Lentz, 1980; Lee and Kelly, 1980). The computer is used both to control the electron beam and automatically process the image produced by the sample. The computer directs a digital beam control for a real-time point-by-point analysis with up to 4096 by 4096 points. The time sequence of the process of automated image analysis is shown in Figure 4.10, taken from the report of Lee and Kelly (1980).

The beam-control system moves the electron beam in a stepwise fashion

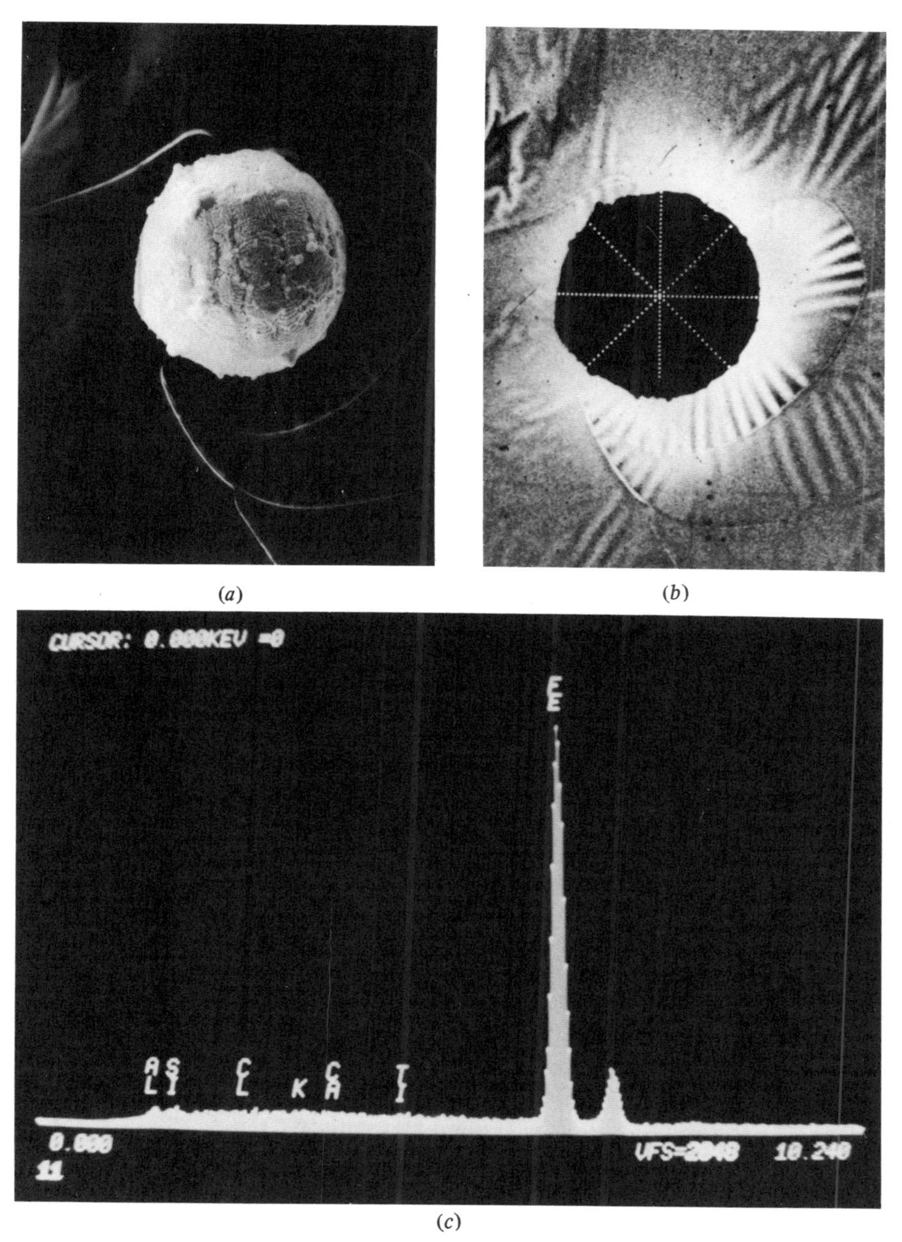

Figure 4.5. Scanning electron micrograph of a spherical iron particle using secondary electron image (*a*) and backscattered electron image (*b*) showing the computer generated image analysis lines for particulate sizing. (*c*) The X-ray spectrum with the electron beam focused at the central spot is shown at the bottom. Figure courtesy of Energy Technology Consultants.

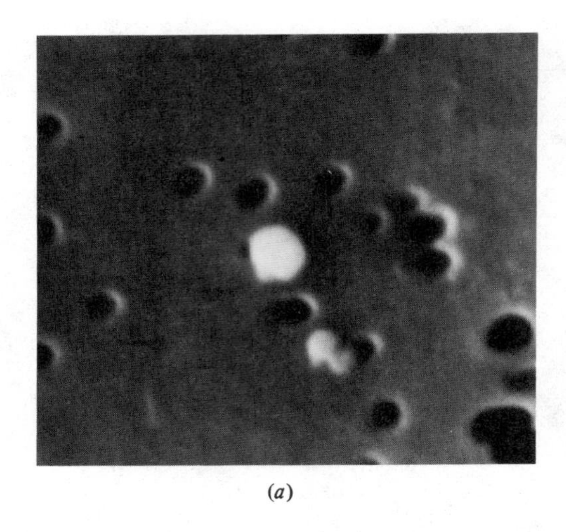

(*a*)

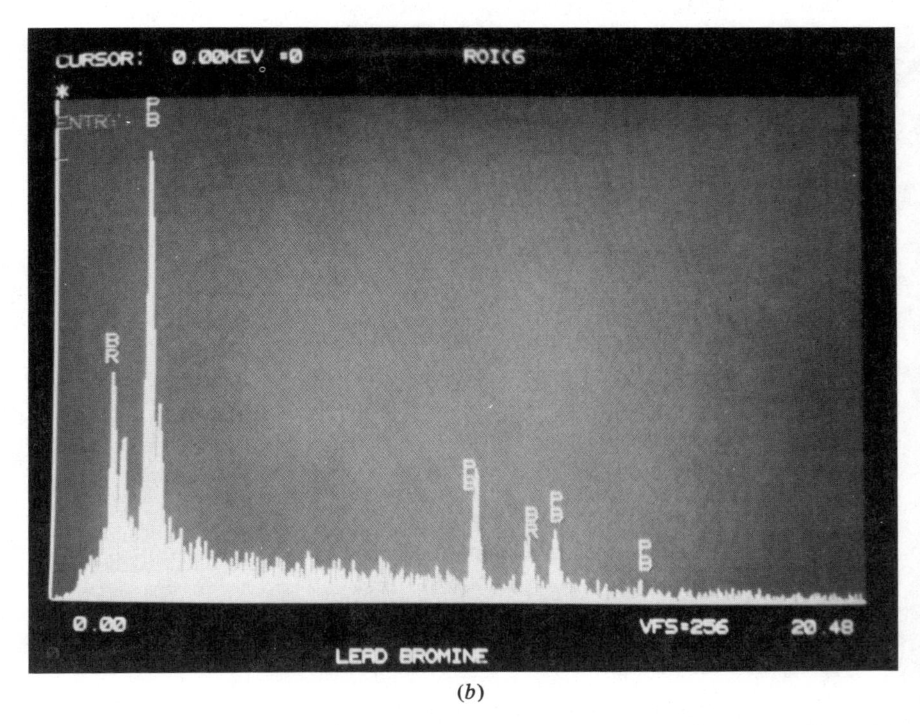

(*b*)

Figure 4.6. Secondary electron micrograph of a submicron sized lead/bromine particle and its corresponding X-ray spectrum. Figure courtesy of Energy Technology Consultants.

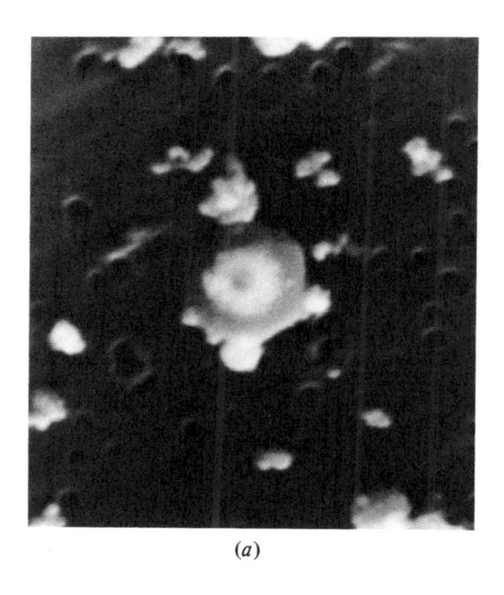

(a)

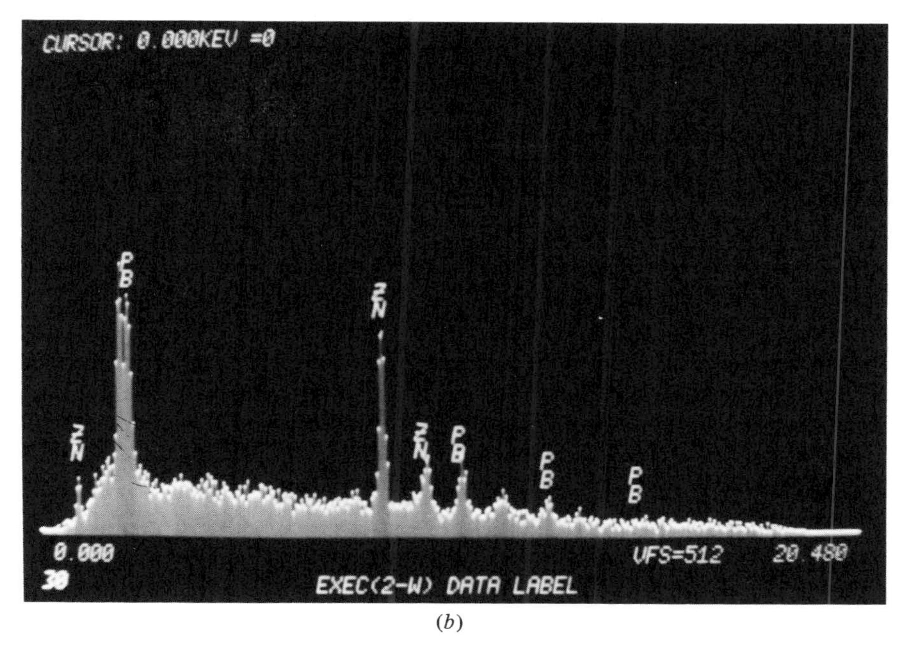

(b)

Figure 4.7. Secondary electron micrograph of a lead/zinc particle and its corresponding X-ray spectrum. Figure courtesy of Energy Technology Consultants.

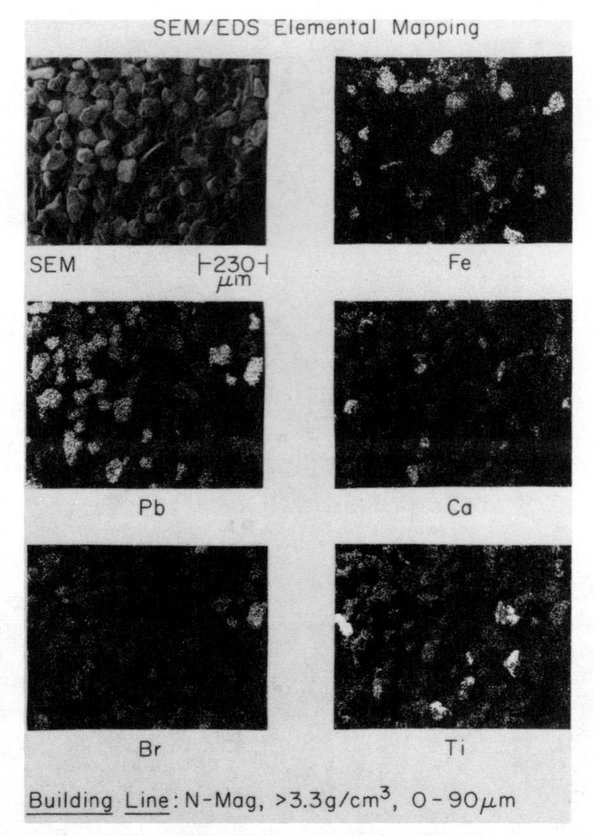

Figure 4.8. Images of physically fractionated particles obtained using secondary electrons and the intensities of X-ray peaks for various major elements. Particles with greater concentrations of the particular element will appear brighter. Figure courtesy of Dr. R. Linton, University of North Carolina at Chapel Hill.

across the sample. The spacing between the points examined determines the minimum particle size that will be observed with certainty. For the system typically used, this minimum particle size is 0.2 μm. The intensity of the signal, backscattered or secondary electron, is compared to a threshold that is established for the particular sample backing material. If the signal level is below the threshold, it assumes no particle is present at that location. If the signal is above the threshold, then a secondary beam control system drives the beam in a preset pattern to determine the size and shape of the entity or feature that is causing the increased reflectance.

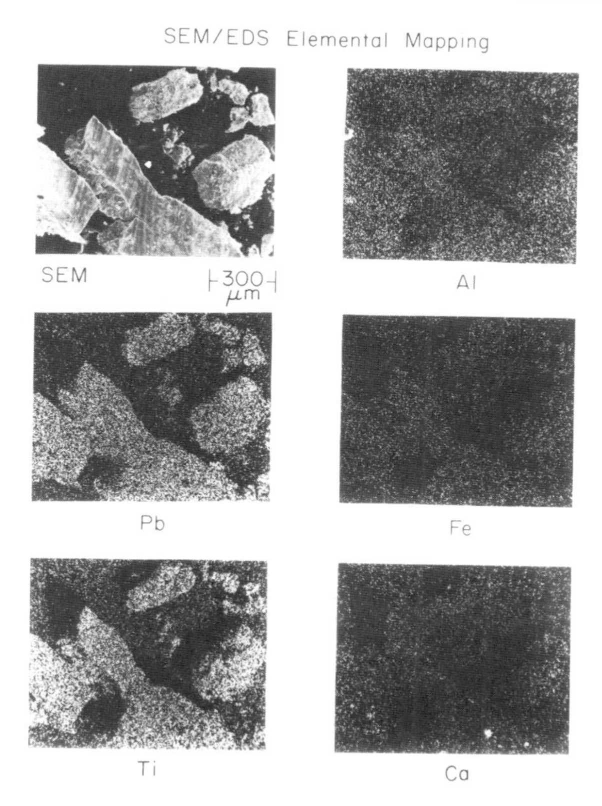

Figure 4.9. Images of paint chip particles obtained using secondary electrons and the intensities of X-ray peaks for various major elements. Particles with greater concentrations of the particular element will appear brighter. Figure courtesy of Dr. R. Linton, University of North Carolina at Chapel Hill.

The preset pattern consists of pairs of diagonals across the particle where the lengths are determined by the points where the signal drops below the threshold values. The pattern is repeated twice. The first pattern is used to locate the feature and the second provides the lengths of the diagonals through the centroid. The maximum, minimum, and average diameters are calculated and stored as well as the centroid of each particle. The measurement of the particle dimensions is made more accurate by using a closer spaced point array once the particle is located. The stored centroid locations are compared to the currently examined

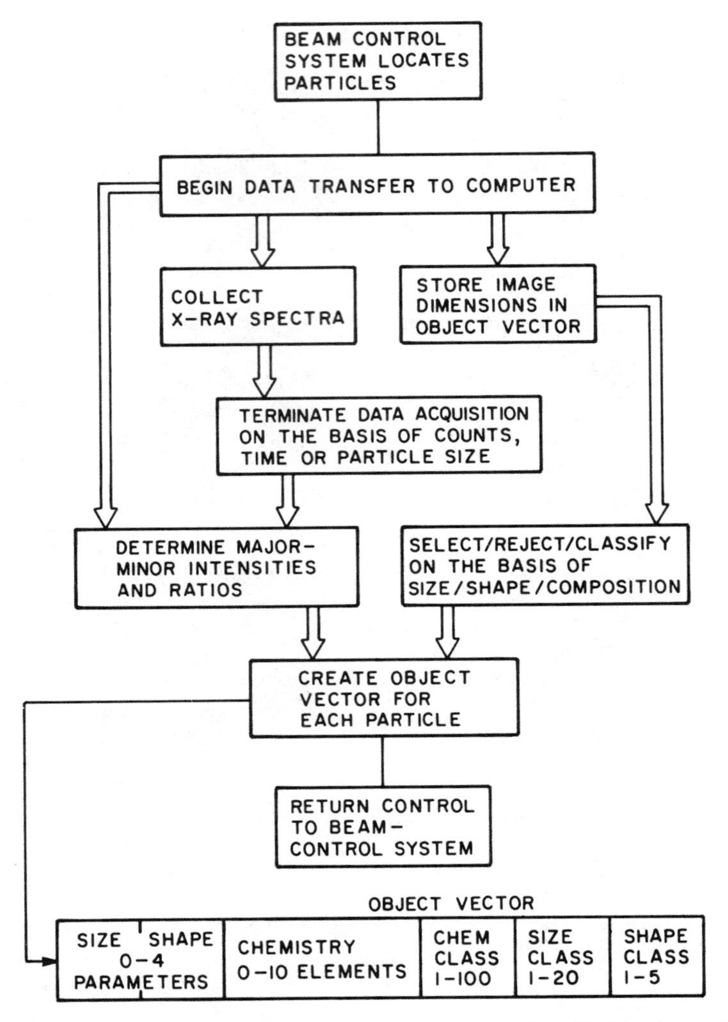

Figure 4.10. Sequence of the automated imaging process as described by Lee and Kelly (1980). Copyright by SEM, Inc. and used with permission.

particle to ensure that particles are not counted twice. Once the parameters have been stored, the beam is returned to the centroid of the particle and the X-ray spectrum is recorded in a multichannel analyzer. A density can be estimated on the basis of the chemical composition and the mass calculated assuming that the particle is a spheroid of revolution. The process takes approximately 1.5 sec/particle. The system then returns to the search mode to find additional particles.

Thus, it is possible to measure the characteristics of several thousand particles in an hour.

SAMPLE PREPARATION

The nature of the sample preparation method depends on the type of sample to be analyzed. Ideally the particles should be uniformly distributed as a monolayer with particles separated from one another on a relatively smooth background with a uniform atomic number.

The primary emphasis of SEM receptor studies has been on airborne particulate matter collected on filter media. If the particles have been collected on a fibrous material such as glass fiber, quartz fiber, or paper filters, the particles are distributed within the volume of the filter and cannot be directly analyzed. Under these conditions, it is necessary to remove the particles from the filter and redeposit them on an appropriate material. As discussed in the previous chapter, there are serious problems associated with any method to transfer particles. It is difficult to ensure representative removal of particles since there will be differential penetration of particles into the fibrous filter pad, depending on particle size. Thus, more of the large particles may be removed from the surface layer than smaller particles from the deeper regions of the filter If the primary interest is in particle mass, this may not be a problem. In addition, the solvent used to remove the particles from the filter may dissolve some particles or leach materials from others.

The results of a recent study of particles in El Paso, Texas (Janocko et al., 1983) indicated that the CCSEM results of the average aerosol composition were similar to those of other methods, indicating that it may be possible to transfer a representative sample of TSP samples. A punch of material from the glass fiber filter is sonicated with acetone forced through the filter. The acetone flush may be performed in each direction using clean solvent for each. The process is repeated six times and the particulate matter is then redeposited on a 0.2-μm pore Nucleopore filter. For a paper filter, a less rigorous procedure has been used. In this case, 50 ml and a punch are ultrasonically agitated for 5 min. The piece of filter is removed from the solvent and washed with a stream of filtered acetone. The material is again redeposited on a 0.2-μm pore Nucleopore filter. A comparison of the reproducibility of removal and CCSEM elemental analysis for both glass fiber and paper filters is presented in Table 4.4. As can be observed, reasonable precision is obtained and it appears that a reproducible sample can be removed from the original filter. The SEM results in El Paso, Texas were also compared with optical microscopy with generally reasonable agreement (Energy Technology Consultants, 1983). In both of these analyses, particles were removed

Table 4.4. Comparison of CCEM Elemental Results for Two Blind Replicate Samples ($\mu g/m^3$)[a]

	Glass Fiber		Paper	
Element	Original Sample	Blind Replicate	Original Sample	Blind Replicate
Na	1.6	1.1	0.7	0.6
Mg	1.1	1.4	1.6	1.6
Al	3.6	4.2	2.2	2.7
Si	11.1	11.3	11.2	11.1
P	0.4	0.5	0.3	0.2
S	3.6	2.3	1.8	2.3
Cl	0.6	0.8	0.6	0.2
K	1.4	1.8	1.3	1.3
Ca	5.2	6.3	14.6	14.9
Ti	1.0	1.3	0.5	0.6
V	0.6	0.9	0.4	0.4
Cr	1.5	1.0	0.7	0.2
Mn	1.3	1.4	0.4	0.4
Fe	4.6	3.6	1.7	1.7
Ni	0.4	0.6	0.3	0.2
Cu	Trace	Trace	0.5	Trace
Zn	0.6	0.7	0.4	0.3
Br	0.4	0.3	0.5	Trace
Pb	1.0	0.5	0.2	0.8

[a]Taken from Energy Technology Consultants (1983).

from the original filter medium and it appears that representative samples can be obtained for such analyses.

Collection on membrane-type filters can directly provide an acceptable sample if the particle loading is not too high. In this case, it is fairly simple to obtain a sample. Often optical microscopic examination is made to find locations for taking samples. Small sections are then removed from the filter and mounted on an aluminum stub attached with a dispersion of amorphous graphite in butanol. To prevent the accumulation of charge on the insulating particles that would distort the image, the sample needs to be coated with a conducting material. For the best imaging, gold has been used as the conducting material. However, gold makes the X-ray fluorescence analysis impossible. The usual method is to deposit a thin coating of carbon on the sample by evaporation. This coating process does subject the sample to a moderate vacuum, potentially resulting in evaporative loss of particles. The sample is then available for analysis.

SIZE DISTRIBUTIONS

One of the advantages of SEM is the direct determination of the physical sizes of the particles in the sample. Thus, more detailed information can be obtained on particle sizes than would be available from the aerodynamic sorting provided by a sampling device. Even a cascade impactor only separates particles into a few categories. As discussed in Chapter 2, there are often problems of particle bounce in cascade impactors or fine particles in the coarse particle mode sample in dichotomous samplers. Thus there are uncertainties in the actual size distribution measured by indirect means. Automated SEM makes the direct measurement of the size distribution feasible because of the ability to examine a sufficiently large number of particles.

To illustrate the physical size distributions that can be obtained, Figure 4.11 shows the mean physical particle diameter distribution for the fine and coarse mode filter samples from a dichotomous sampler located in Philadelphia, Pennsylvania (Casuccio and Janocko, 1983). The nominal 50% efficiency cut point is designed to be 2.5 μm. Although there are some larger diameter particles in the fine fraction, this figure shows that dichotomous sampler is basically separating

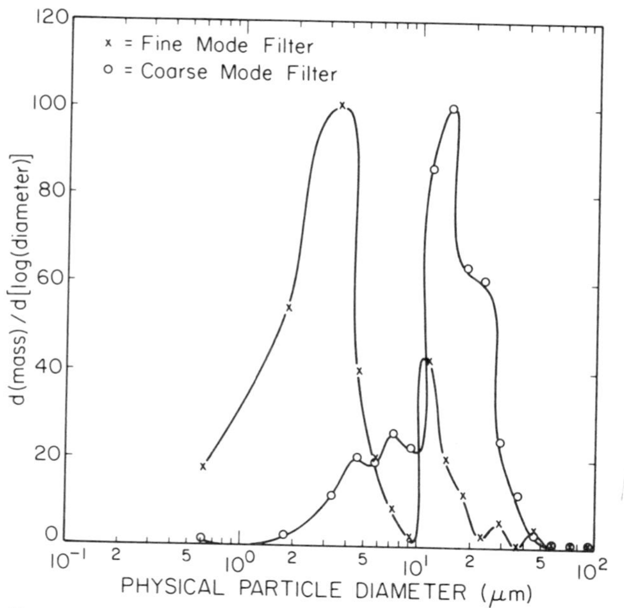

Figure 4.11. Particle size distributions for fine and coarse dichotomous sampler samples taken in Philadelphia, Pennsylvania. Physical size measurements were made using automated scanning electron microscopy. Figure courtesy of Energy Technology Consultants.

particles as designed. It should be noted that there are particles in the coarse mode greater than 20 μm where the nominal design specification would suggest that there should not be as many large particles.

It is possible to estimate the aerodynamic size distribution from the physical size and estimated density (Casuccio et al., 1983). It is not possible to precisely convert a measured physical diameter to an equivalent aerodynamic value for nonspherical particles. The approximate aerodynamic diameter is given by

$$D_a = \chi D_p (\rho)^{1/2} \tag{4.1}$$

where D_a is the aerodynamic diameter, D_p is the measured physical diameter, χ is a shape factor, and ρ is the density as estimated from the chemical composition. The shape factor is included to correct for the nonspherical shape of the particles. Values of χ are obtained from Dallavalle (1948). The aerodynamic size distributions corresponding to the physical size distributions are shown in Figure 4.12. The factor of the square root of the density further distorts the distributions to larger diameters, and there is appreciable fine mode mass in particles with >10 μm diameters and coarse mode mass in particles with >20 μm diameters. Generally the sampler is separating the particles, but there appear to be some problems with the precision of classification. Scanning electron microscopy

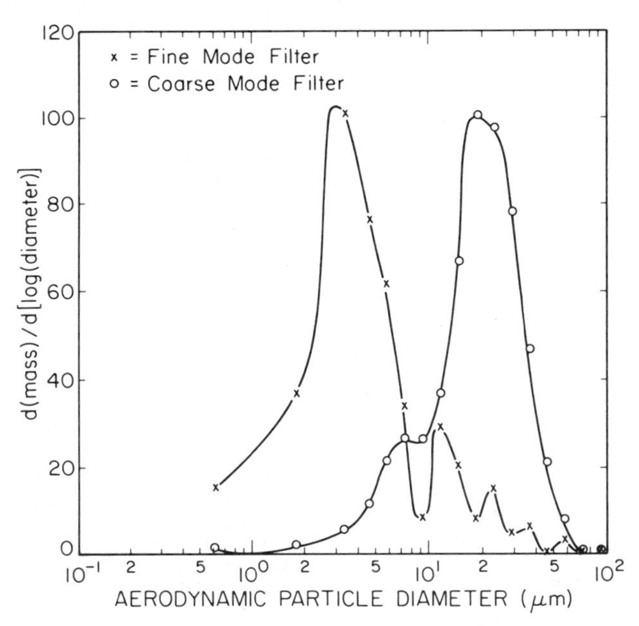

Figure 4.12. Aerodynamic size distributions derived from the physical size distributions given in Figure 4.11. Figure courtesy of Energy Technology Consultants.

is the only way that the complete physical size distribution of the actual sample can be determined.

DATA ANALYSIS

The process described above provides a substantial amount of information on each individual particle. The question then arises as to how to utilize these data to provide an understanding of the system under study. Several approaches have been applied to utilize these data in receptor models. Since the discussion of the mathematical models begins in the next chapter, the use of such models will be analyzed after the models have been introduced. However, some further details of the conceptual framework of these analyses need to be presented here.

The general approach that has been employed in the intepretation of the microscopic analytical data has been to assign each particle to one of a number of empirically defined particle categories (Johnson et al., 1981; Casuccio et al., 1982; Lee and Fisher, 1980). A listing of particle category descriptions and densities is given in Table 4.5. For example, a particle that is rich in iron must be placed in one of the particle-type categories containing significant iron concentrations. These types include iron-rich, spherical iron, chlorite, pyrite, calcium ferrites, and so on. If there are no other major elements observed besides the iron, the particle is assigned as iron-rich unless the aspect ratio (length to width) is less than 1.33:1. For such a situation, the particle should be classified as spherical iron. For a particle with approximately twice the sulfur as iron, the classification would be pyrite. The criteria for these category assignments have been developed and verified by analyzing a large number of reference materials with known characteristics (Lee and Fisher, 1980).

The particles that do not fit into any of the predefined categories are assigned to an "unknown" class. These particles can be examined to determine if there are enough similar characteristic particles to permit the definition of a new particle class. It is possible to determine the number of particles of a given size range within a given particle characteristic class. These numbers of particles in each of the particle size/chemistry categories become the variables that are used in the further analysis. These new particle characteristic category variables can also be employed in a mass balance as given by equation (1.1).

Typically, authentic source material is collected and the source material is apportioned into the particle characteristic categories. The ambient samples can then be considered to be a linear sum of contribution of particles from each source in each category to the total particles assigned to that category. Further details of this analysis are presented in Chapter 6.

If authentic source materials are unavailable, it is necessary to group particles from the same source together by some alternative means. Some initial efforts

Table 4.5. **Description and Density of CCSEM Particle Types**[a]

Particle Type	Description	Density
Na-rich	Predominantly sodium-rich	1.0
Mg-rich	Predominantly magnesium-rich	3.6
Al-rich	Predominantly aluminum-rich	4.0
Si-rich	Silicon-rich, may include minerals other than quartz, such as feldspars	2.6
S-rich	Predominantly sulfur-rich	2.0
Cl-rich	Predominantly chlorine-rich	3.2
K-rich	Predominantly potassium-rich	2.3
Ca-rich	Predominantly calcium-rich	2.8
Ti-rich	Predominantly titanium-rich	4.2
Fe-rich	Predominantly iron-rich, nonspherical	5.3
Mn-rich	Predominantly manganese-rich	5.0
Zn-rich	Predominantly zinc-rich	5.6
Cr-rich	Predominantly chromium-rich	7.2
Pb-rich	Predominantly lead-rich	9.1
V-rich	Predominantly vanadium-rich	3.4
Cu-rich	Predominantly copper-rich	7.4
P-rich	Predominantly phosphorous	2.3
Br-rich	Predominantly bromine-rich	3.0
Ni-rich	Predominantly nickel-rich	8.9
R. Mix Clay	Spherical particle of any clay mineral	2.3
Na/K–S	Sodium and/or potassium associated with sulfur	2.6
R. Fe	Round iron, predominantly spherical iron rich particles	5.3
Ca–S	Predominantly calcium and sulfur in approximately equivalent preparations	2.6
Fe–Ca	Predominantly iron and calcium	4.0
Mg–Si	Predominantly silicon-rich, associated with magnesium	3.2
Mg–Ca	Predominantly calcium with magnesium, such as dolomite	2.9
Halide	Any combination of elements associated with substantial chlorine	2.2
Fe–Cr	Predominantly iron-rich with chromium	7.6
Fe–Mn–Cr	Predominantly iron-rich with manganese and chromium	7.5
Fe–Mn	Predominantly iron-rich with manganese	5.2
Fe–Zn	Predomianntly iron-rich, associated with zinc	5.4
Fe–Si	Predominantly iron with smaller amounts of silicon	3.9
Mix–Zn	Any combination of elements associated with zinc	5.6
Mix–Clay	Any clay mineral	2.6

Table 4.5. (*Continued*)

Particle Type	Description	Density
Clay–S	Any clay mineral, associated with an elevanted sulfur level	2.5
S-Bearing	Any element, associated with an elevated sulfur level	1.8
Pb-Bearing	Any combination of elements associated with lead, includes lead-rich	5.0
Ca–Si	Predominantly silicon and calcium	2.7
Si–S–Ca	Predominantly silicon and calcium with elevated sulfur	2.5
Si–Ca–Fe	Predominantly silicon and calcium, associated with with iron	2.8
Carbon	Low levels or no detectable elements; elements lighter than sodium are not detected	2.0
Unknown	Elemental composition outside of those defined by the other types	2.3

[a]Taken from Energy Technology Consultants (1983).

have been made to apply multivariate statistics to the problem of source apportionment using scanning electron microscopy results. These methods and results are discussed in Chapter 7.

The SEM under computer control and with automated image analysis represents a powerful method for obtaining a great deal of information on individual particles. It is possible to combine these individual particle results to obtain a reasonable approximation to the bulk concentration of the collection of particles. The advantages and disadvantages of this methodology have been summarized by Casuccio et al. (1983) as follows:

Advantages

1. Physically measures particles within a broad size range, from 0.2 to 300 μm.
2. Elemental chemistry is obtained from every particle.
3. Each particle is classified by size and composition.
4. Distributions are obtained for each particle class as a function of size.
5. Both geometric (optical) and aerodynamic size distributions are calculated.
6. Size and X-ray data are stored for future retrieval.

7. Analysis time averages less than 1.5 sec/particle, including sizing and chemical analysis.
8. Data acquisition is designed so that size and weight distributions have an absolute uncertainty, independent of the size range selected.
9. The analysis is compatible with most sampling methods.
10. Effects of operator bias, fatigue, and subjectivity, inherent in manual microscopic techniques, have been minimized.
11. Results are reproducible.

Disadvantages

1. All samples must be prepared for the SEM. Most samples must be redeposited.
2. The nature of dichotomous fine fraction samples presents special sample preparation problems.
3. Species with an average atomic number close to that of the substrate are difficult to detect.
4. Particle volume is inferred from projected area.
5. Calculated particle mass assumes that the density is known for each particle type.
6. Chemical inhomogeneities within a particle may not be recognized.
7. The software sorting algorithm must be modified if a significant fraction of undefined particles is encountered.

These advantages and disadvantages must be given strong consideration when formulating a receptor modeling study. As with other analytical methods, CCSEM may not, by itself, provide all the desired information. However, it seems likely that there will be an increasing role for such a methodology in future receptor modeling studies. The use of CCSEM in specific applications will be presented as part of the discussion of regression-type mass balance models in Chapter 6.

CHAPTER

5

INTRODUCTION TO MATHEMATICAL RECEPTOR MODELS

The objectives of mathematical receptor models are to utilize information about the pollutants at a particular site to identify their sources and to develop a plan of effective air quality management. In order to accomplish this objective, it is necessary to have a framework in which to develop models of the atmospheric processes that permit us to make the connection between source and receptor. In a dispersion model, the concentration of material at a specific site impacted by multiple sources is given by

$$\bar{x}_{ij} = \sum E_{ijk} D_{kj} \tag{5.1}$$

where $\bar{x}_{ij}$ is the average concentration of the ith component during the jth time interval, and E_{ijk} is the emission rate of the ith component for the kth source during the jth interval. In order to use such a model to predict concentrations at a specific location, it is necessary to know the rate and chemical nature of the emissions from each of the p sources as well as all of the factors involved in the transport and transformation of the material from the sources to the collection site. There have been very substantial efforts to develop dispersion models to include the complexities of actual situations including complex terrain, buildings, heat island effects, and so on. There are still, however, substantial difficulties in accurately predicting pollutant concentrations because of uncertainties in both the emission and dispersion terms. In addition, there should be additional terms added to equation (5.1) to account for the changes in concentrations that occur from interactions between pollutants in the transport period, such as the formation of SO_4^{2-} droplets or particles from SO_2 in transit. An attempt to account for TSP at a site without inclusion of secondary aerosol will be doomed to failure.

The development of receptor models follows the development of new or improved elemental analytical methods. During the 1960s substantial improvements were made in the ability to determine many elements with excellent sensitivity using energy-dispersive X-ray fluorescence and instrumental neutron activation. Atomic absorption spectrometry also became widely available so that it has become possible to accurately determine a number of elements in the small sized samples of airborne particulate matter typically collected. Once this kind of data began to become available, it was necessary to find new approaches to interpret the data. A number of methods have been developed to manipulate atmospheric chemical data in order to understand the nature of the atmosphere. The basic premise in all of the mathematical models is that we can assume that the measured property of the atmosphere is a linear sum of independent contributing sources. For example, consider the quantity of airborne particulate lead. We can assume that the lead collected on a filter over a given time interval is the result of collecting the lead-carrying particles from the variety of sources contributing to the air being sampled, so that the calculated average airborne particulate lead concentration for the sampling period can be expressed as

$$(\text{Pb})_{\text{total}} = (\text{Pb})_{\text{auto}} + (\text{Pb})_{\text{smelter}} + (\text{Pb})_{\text{incinerator}} + \cdots \tag{5.2}$$

However, particle sources such as automobiles emit particles that contain elements other than lead. Therefore, the lead concentration (ng/m^3) attributed to automobiles, $(\text{Pb})_{\text{auto}}$, can be considered to be the product of two terms; the mass concentration of lead in automobile particles (ng/mg) and the mass of automotive particles per unit volume of air sampled (mg/m^3):

$$(\text{Pb})_{\text{auto}} = a_{\text{Pb, auto}} f_{\text{auto}} \tag{5.3}$$

where $a_{\text{Pb, auto}}$ is the concentration of lead in automotive particles at the receptor site and f_{auto} is the airborne mass loading of automotive particles. Clearly if we can measure other properties of the collected particles, a series of such equations can be developed. This mass balance approach is not restricted to particulate matter. It can be applied to any linearly additive property of the air. Generalizing equation (5.2) yields

$$x_{ij} = \sum a_{ik} f_{kj} \tag{5.4}$$

where x_{ij} is the volume concentration (ng/m^3) of the ith species in the jth sample, a_{ik} is the mass concentration (ng/mg) of the ith species in material from the kth source, and f_{kj} is the volume concentration (mg/m^3) of material from the kth source collected in the jth sample. The mass balance on given species as given in equation (5.4) is subject to the constraint that the total mass of particles sampled is equal to the sum of contributions from each source type:

$$m_j = \sum_{k=1}^{p} f_{kj} \tag{5.5}$$

where m_j is the measured total mass concentration (mg/m^3) of airborne paticulate matter.

Various approaches are taken to utilize equation (5.4) to provide an understanding of the properties of an airshed. These methodologies include chemical mass balance (chemical element balance), factor analysis, and regression methods. These techniques are discussed at length in subsequent chapters. Other mathematical approaches to the interpretation of air quality data, such as time series analysis, are also presented. In order to present this variety of mathematical approaches to analyzing air quality data, some of the basic mathematics that are common to many of these techniques will be reviewed.

MATHEMATICAL BACKGROUND

The following is a brief introduction to a variety of mathematical concepts and operations that will be used in developing the mathematical calculational approaches to receptor modeling. If more details of these mathematical procedures are needed, there are a number of good textbooks that can be used as references, such as Horst (1963), Cooley and Lohnes (1971), and Lawson and Hanson (1974).

Definitions

A *matrix* is defined as a rectangular array of numbers arranged in rows and columns. For example the 4 × 5 matrix (4 rows by 5 columns), A, can be written as

$$A = \begin{bmatrix} a_{11} & a_{12} & a_{13} & a_{14} & a_{15} \\ a_{21} & a_{22} & a_{23} & a_{24} & a_{25} \\ a_{31} & a_{32} & a_{33} & a_{34} & a_{35} \\ a_{41} & a_{42} & a_{43} & a_{44} & a_{45} \end{bmatrix} \tag{5.6}$$

The *transpose* of matrix A, denoted by A', is simply a 5 × 4 matrix obtained by interchanging all of the rows and columns. For example,

$$A' = \begin{bmatrix} a_{11} & a_{21} & a_{31} & a_{41} \\ a_{12} & a_{22} & a_{32} & a_{42} \\ a_{13} & a_{23} & a_{33} & a_{43} \\ a_{14} & a_{24} & a_{34} & a_{44} \\ a_{15} & a_{25} & a_{35} & a_{45} \end{bmatrix} \tag{5.7}$$

A *square matrix* is one in which the number of columns equals the number of rows.

A *symmetric matrix* is a square matrix in which the corresponding elements on either side of the main diagonal are equal, that is,

$$a_{12} = a_{21}, a_{13} = a_{31}, \text{etc.}$$

A *vector* is directed line segment in space and can be represented by an $n \times 1$ (column vector) or $1 \times n$ (row vector) matrix:

$$\mathbf{x} = \begin{bmatrix} x_1 \\ x_2 \\ x_3 \end{bmatrix}, \qquad \mathbf{x}' = [x_1, x_2\ x_3] \tag{5.8}$$

A *scalar* is a single number and could be considered to be a matrix with one column and one row.

A *sample* is a collection of material obtained from the environment by some kind of sampling device.

The *variables* in the system are the properties of the sample as determined by a variety of possible processes. Variables include parameters such as elemental composition, average size, the wind speed or direction during the sampling interval, and so on. The collection of values for the variable describing a single sample can be considered to be a column vector in the overall data matrix. The collection of values for a single variable over all of the samples in a data set would be a row vector.

Elementary Vector Operations

A variety of properties of vectors are useful in the discussion of the various mathematical receptor models. We can multiply a vector by a scalar resulting in the increase in each element in the vector by that scalar

$$\mathbf{y} = a\mathbf{x} = \begin{bmatrix} ax_1 \\ ax_2 \\ ax_3 \end{bmatrix} \tag{5.9}$$

There are several ways that vectors can be multiplied together to make a vector product. The first is normally referred to as the scalar product or dot product but is sometimes referred to as the minor product (Horst, 1963). It is formed by multiplying a row vector by a column vector

$$\mathbf{x}'\mathbf{y} = \sum x_k y_k \tag{5.10}$$

This product is only defined for vectors of the same order. The order of multiplication can be reversed:

$$\mathbf{y}'\mathbf{x} = \mathbf{x}'\mathbf{y}$$

A second vector product, the cross product, is obtained if a column vector is multiplied into a row vector. The resulting matrix is sometimes referred to as the major product

$$\mathbf{x}\mathbf{y}' = \begin{bmatrix} x_1y_1 & x_1y_2 & \cdots & x_1y_p \\ x_2y_1 & x_2y_2 & \cdots & x_2y_p \\ \vdots & & & \\ x_py_1 & x_py_2 & \cdots & x_py_p \end{bmatrix} \tag{5.11}$$

For each of these vector products, a product moment can be calculated. The minor product of a column vector with its transpose is the sum of the squares of the vector elements:

$$\mathbf{x}'\mathbf{x} = \sum x_j^2 \tag{5.12}$$

The major product moment is a matrix obtained by postmultiplying a column vector by its transpose. Several geometric properties of vectors will be of interest. The length of a vector will be the square root of its minor product. As illustrated in Figure 5.1, the angle between two vectors may be useful to show how close in

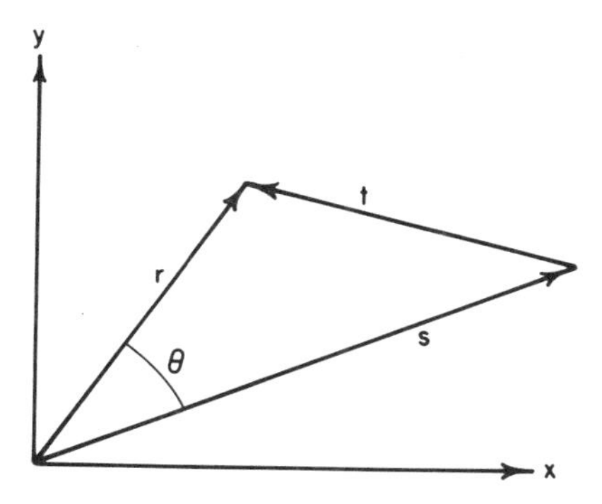

Figure 5.1. Illustration of the angle between two vectors as an indicator of their similarity to one another.

space to one another the vectors lie. The cosine of the angle between the vectors can be expressed as

$$\cos\theta = \frac{\mathbf{r}'\mathbf{s}}{(\mathbf{r}'\mathbf{r})^{1/2}\,(\mathbf{s}'\mathbf{s})^{1/2}} \tag{5.13}$$

A vector is defined as a normalized vector if it has unit length. Any vector can be normalized by dividing each element by a scalar equal to its length. The length of a vector is given by $(\mathbf{x}'\mathbf{x})^{1/2}$ so that a normalized vector would be $\mathbf{x}/(\mathbf{x}'\mathbf{x})^{1/2}$.

Matrix Arithmetic

In the descriptions of the various methods, there are a number of instances where one matrix is multiplied by another.

$$C = AB \tag{5.14}$$

To accomplish this multiplication process, the number of columns in the first matrix A must equal the number of rows in the second matrix B. The resultant matrix C will be of a size determined by the number of rows in A by the number of columns in B. The values of the individual elements in C are given by

$$c_{ij} = \mathbf{a}_i'\mathbf{b}_j = \sum_{k=1}^{m} a_{ik}b_{kj} \tag{5.15}$$

where c_{ij} is the element in the ith row and jth column. Because of the nature of this process, it is generally the case that AB is different from BA. For example if X is an $m \times n$ matrix, $X'X$ is an $n \times n$ matrix while XX' is an $m \times m$ matrix. Thus, it is important to be careful as to the order of multiplication of matrices.

Several other matrices of importance include the identity matrix I, defined by

$$I = \begin{bmatrix} 1 & 0 & 0 & \cdots & 0 \\ 0 & 1 & 0 & \cdots & 0 \\ 0 & 0 & 0 & \cdots & 0 \\ . & . & . & \cdots & . \\ 0 & 0 & 0 & \cdots & 1 \end{bmatrix} \tag{5.16}$$

$$IA = AI \tag{5.17}$$

A diagonal matrix is one with elements only on the principal diagonal of the matrix and zeros everywhere else:

$$\begin{bmatrix} \lambda_1 & 0 & 0 & \cdots & 0 \\ 0 & \lambda_2 & 0 & \cdots & 0 \\ 0 & 0 & \lambda_3 & \cdots & 0 \\ . & . & . & \cdots & . \\ 0 & 0 & 0 & \cdots & \lambda_n \end{bmatrix} \tag{5.18}$$

where $\lambda_1, \cdots, \lambda_n$ are numbers.

Several other properties of matrices are of interest to this discussion. The product of a matrix times its transpose will be used in several contexts. As discussed above, different dimension results will be obtained, depending on the order of multiplication. A nomenclature similar to the vector products can be defined where XX' is the major product moment and $X'X$ is the minor product moment. The elements of the major product moment matrix represent the cross-product terms

$$c_{ik} = \sum_{j=1}^{n} x_{ij} x_{kj} \tag{5.19}$$

with the diagonal elements being the sums of square of the values of a variable over the row. The elements of the minor product moment matrix represent the sums of squares for a sample over the column of variables:

$$e_{jk} = \sum_{i=1}^{m} x_{ij} x_{ik} \tag{5.20}$$

There are several properties of square matrices and particularly square symmetric matrices that need to be mentioned. The inverse of a square matrix, A^{-1}, exists if $A^{-1} A = I$. Only square matrices can have inverses but not every square matrix will have an inverse. The method of calculating the inverse of any given square matrix is beyond the scope of this discussion. See Horst (1963) for the details of this procedure.

Square matrices can also have the property of being orthogonal. The concept of orthogonality can best be understood beginning by discussing two vectors. Two vectors are orthogonal if their minor product is zero:

$$\mathbf{x}'\mathbf{y} = 0$$

If the vectors are in a two- or three-dimensional space, we would say they are perpendicular to one another. In higher order spaces, they are orthogonal. We can say a matrix is orthogonal if

$$X'X = D \tag{5.21}$$

where D is a diagonal matrix, implying that all of the possible pairs of column vectors are orthogonal. Since D is a diagonal matrix and all elements off the main diagonal are zero, it is its own transpose

$$D = D' \tag{5.22}$$

Therefore,

$$X'X = XX' \tag{5.23}$$

If each of the columns of the matrix are also normalized, then

$$D = I$$

and the matrix is said to be orthonormal.

Rotation of Coordinate Systems

In several applications to be discussed later and as part of the introduction to the discussion of determining the structure of a matrix, it is useful to describe how one set of coordinate system axes can be transformed into a new set. Suppose that we wish to change from the coordinate system x_1, x_2 to the system of y_1, y_2 by rotating through angle ϕ as shown in Figure 5.2. For this two-dimensional

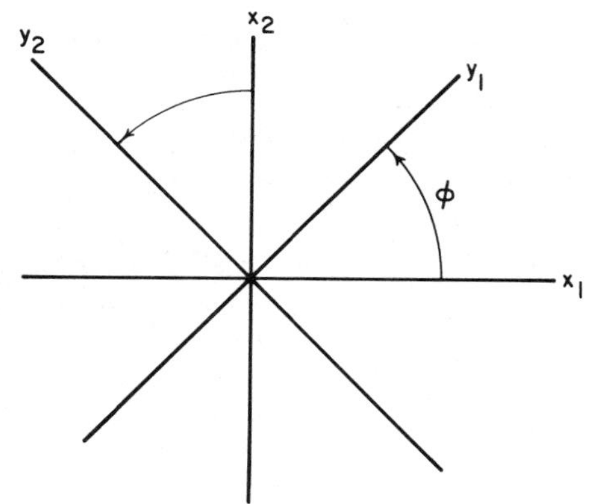

Figure 5.2. Illustration of the rotation of a coordinate system (x_1, x_2) to a new system (y_1, y_2) by an angle ϕ.

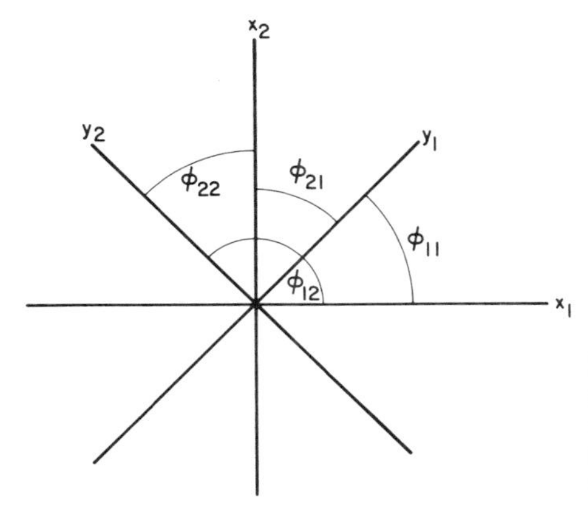

Figure 5.3. Illustration of the rotation of a coordinate system ($\mathbf{x}_1$, $\mathbf{x}_2$) to a new coordinate system ($\mathbf{y}_1$, $\mathbf{y}_2$) showing the definitions of the full set of angles used to described the rotation.

system, it is easy to see that

$$
\begin{aligned}
y_1 &= \cos \phi x_1 + \sin \phi x_2 \\
y_2 &= \sin \phi x_1 + \cos \phi x_2
\end{aligned}
\tag{5.24}
$$

In matrix form, this equation could be written as

$$\mathbf{y}' = \mathbf{x}' T \tag{5.25}$$

In order to obtain a more general case, it is useful to define the angles of rotation in the manner shown in Figure 5.3. Now the angle ϕ_{ij} is the angle between the ith original reference axis and the jth new axis. Assuming that this is a rotation that maintains orthogonal axes (rigid rotation), then, for two dimensions,

$$
\begin{aligned}
\phi_{12} &= \phi_{11} + 90^\circ \\
\phi_{21} &= \phi_{11} - 90^\circ \\
\phi_{22} &= \phi_{11}
\end{aligned}
\tag{5.26}
$$

There are also trigonometric relationships that exist,

$$
\begin{aligned}
\sin \phi_{11} &= \sin (\phi_{21} + 90) = \cos \phi_{21} \\
-\sin \phi_{11} &= -\sin (\phi_{12} - 90^\circ) = \sin (90 - \phi_{12}) = \cos \phi_{12}
\end{aligned}
\tag{5.27}
$$

so that equation (5.27) can be rewritten as

$$
\begin{aligned}
y_1 &= \cos \phi_{11} x_1 + \cos \phi_{21} x_2 \\
y_2 &= \cos \phi_{12} x_1 + \cos \phi_{22} x_2
\end{aligned}
\tag{5.28}
$$

This set of equations can then be easily expanded to n orthogonal axes yielding

$$\begin{aligned} y_1 &= \cos\phi_{11}x_1 + \cos\phi_{21}x_2 + \cdots + \cos\phi_{n1}x_n \\ y_2 &= \cos\phi_{12}x_1 + \cos\phi_{22}x_2 + \cdots + \cos\phi_{21}x_n \\ &\cdots \\ y_n &= \cos\phi_{1n}x_1 + \cos\phi_{2n}x_2 + \cdots + \cos\phi_{nn}x_n \end{aligned} \tag{5.29}$$

A transformation matrix T can then be defined such that its elements are

$$t_{ij} = \cos\phi_{ij} \tag{5.30}$$

Then, for a collection of N row vectors in a matrix X with n columns,

$$Y = XT \tag{5.31}$$

and Y has the coordinates for all N row vectors in terms of the n rotated axes. For the rotation to be rigid, T must be an orthogonal matrix. Note that the column vectors in Y can be thought of as new variables made up by linear combinatins of the variables in X with the elements of T being the coefficient of those combinations. Also a row vector of X gives the properties of a sample in terms of the original variables while a row vector of Y gives the properties of a sample in terms of the transformed variables.

Structure of a Matrix

As a part of the development of the receptor models, we will want to represent a data matrix as a product of two other matrices containing specific information regarding the sources of the material present in the samples. It can be shown (Horst, 1963) that any matrix can be expressed as a product of two matrices

$$X_{nm} = A_{np}F_{pm} \tag{5.32}$$

where the subscripts denote the dimensions of the respective matrices. There will be an infinite number of different A and F matrices that satisfy this equation.

To divide the matrix into two cofactor matrices as in equation (5.32), a question is raised about the minimum value p can have and still yield a solution. This value is the "rank" of matrix X (Horst, 1963, p. 335). The rank clearly cannot be greater than a matrix's smaller dimension and the rank of a product moment matrix cannot be greater than the smaller of the number of columns or the number of rows. The rank of the product moment matrix must be the same as that of the matrix from which it was formed.

Associated with the idea of matrix rank is the concept of linear independence of variables. We can look at the interrelationships between columns (or rows) of a matrix and determine if columns (or rows) are linearly independent of one another. To understand linear independence let us examine the relationship between the two vectors in Figure 5.1. We can find the vector **t** such that

$$\mathbf{r} = \mathbf{s} - \mathbf{t} \tag{5.33}$$

The vector **t** can then be generalized to be the resultant of the sum of **r** and **s** with coefficients a_r and a_s:

$$\mathbf{t} = a_r\mathbf{r} + a_s\mathbf{s} \tag{5.34}$$

If $\mathbf{t} = 0$, then **r** and **s** are said to be colinear or linearly dependent vectors. Thus, a vector **y** is *linearly dependent* on a set of vectors, $\mathbf{v}_1, \mathbf{v}_2, \ldots, \mathbf{v}_m$ if

$$\mathbf{y} = a_1\mathbf{v}_1 + a_2\mathbf{v}_2 + \cdots + a_m\mathbf{v}_m \tag{5.35}$$

and at least one of the coefficients, a_i, is nonzero. If all of the a_i values in equation (5.35) are zero, then **y** is linearly independent of the set of vectors, $\mathbf{v}_i$. The number of linearly independent column vectors in a matrix defines the minimum number of dimensions needed to contain all of the vectors. The idea of the rank or true dimensionality of a data matrix is an important concept in receptor modeling as it defines the number of separately identifiable, independent sources contributing to the system under study. Thus finding the rank of a data matrix will be an important task. In addition, the ability to resolve sources of material with similar properties or the resolution of various receptor models needs to be carefully examined. To examine these question, several additional mathematical concepts need to be discussed.

A given data matrix can be reproduced by one of an infinite number of sets of independent column vectors or *basis* vectors that will describe the axes of the reduced dimensionality space. We can define the rank of the matrix and develop a set of linearly independent basis vectors by the use of an eigenvalue analysis. In this discussion we will only consider the analysis of real, symmetric matrices such as those we obtain as the minor or major product of a data matrix. Suppose we have a real, symmetric matrix R that we wish to analyze for its rank, remembering that the rank of a product moment matrix is the same as the data matrix from which it is formed.

An eigenvector of R is a vector **u** such that

$$R\mathbf{u} = \mathbf{u}\lambda \tag{5.36}$$

where λ is an unknown scalar. The problem then is to find a vector so that the vector $R\mathbf{u}$ is proportional to $\mathbf{u}$. This equation can be rewritten as

$$R\mathbf{u} - \mathbf{u}\lambda = 0 \tag{5.37}$$

or

$$(R - \lambda I)\,\mathbf{u} = 0$$

implying that $\mathbf{u}$ is a vector that is orthogonal to all of the row vectors of $(R - \lambda I)$. This vector equation can be considered as a set of p equations where p is the order of R:

$$\begin{aligned}
u_1(1-\lambda) + u_2 r_{12} + u_3 r_{13} + \cdots + u_p r_{1p} &= 0\\
u_1 r_{21} + u_2(1-\lambda) + u_3 r_{23} + \cdots + u_p r_{2p} &= 0\\
u_1 r_{31} + u_2 r_{32} + u_3(1-\lambda) + \cdots + u_p r_{3p} &= 0\\
\cdot \quad \cdot \quad \cdot \quad \cdots \quad \cdot \quad &\cdot\\
u_1 r_{p1} + u_2 r_{p2} + u_3 r_{p3} + \cdots + u_p(1-\lambda) &= 0
\end{aligned} \tag{5.38}$$

Unless $\mathbf{u}$ is null vector, equation 5.37 can only hold if

$$R - \lambda I = 0 \tag{5.39}$$

There is a solution to this set of equations only if the determinant of the left side of the equation is zero:

$$|R - \lambda I| = 0 \tag{5.40}$$

This yields a polynomial in λ of degree p. It is then necessary to obtain the p roots of this equation, λ_i, $i = 1, p$. For each λ_i there is an associated vector $\mathbf{u}_i$ such that

$$R\mathbf{u}_1 - \mathbf{u}_1\lambda_1 = 0 \tag{5.41}$$

If these λ_i values are placed as the elements of a diagonal matrix Λ, and the eigenvectors can be collected as columns of the matrix U, then we can express (5.37)

$$RU = U\Lambda \tag{5.42}$$

The matrix U is a square orthonormal so that

$$U'U = UU' = 1$$

Postmultiplying (5.42) by U' yields

$$R = UU'\Lambda$$

Thus any symmetrix matrix R may be represented in terms of its eigenvalues and eigenvectors:

$$R = \lambda u_1 u_1' + \lambda_2 u_2 u_2' + \cdots + \lambda_p u_p u_p' \tag{5.43}$$

so that R is weighted sum of matrices $\mathbf{u}_i\mathbf{u}_i'$, of order p by p and of rank 1. Each term is orthogonal to all other terms so that for $i \neq j$

$$\mathbf{u}_i'\mathbf{u}_j = 0$$

and

$$\mathbf{u}_i\mathbf{u}_i'\mathbf{u}_j'\mathbf{u}_j = 0$$

Premultiplying (5.42) by $\mathbf{u}'$ yields

$$U'RU = \Lambda \tag{5.44}$$

so that U is a matrix that reduces R to a diagonal form. The eigenvalues have a number of useful properties (Joreskog, Klovan, and Reyment, 1976).

1. Trace Λ = trace R; the sum of the eigenvalues equals the sum of the elements in the principal diagonal of the matrix.
2. $\Pi_{i=1}^{p} \lambda_i = |R|$; the product of the eigenvalues equals the determinant of the matrix. If one or more of the eigenvalues is zero, then the determinant is zero and the matrix R is called a singular matrix. A singular matrix cannot be inverted.
3. The number of nonzero eigenvalues equals the rank of R.

Therefore, if for a matrix R of order p, there are m zero eigenvalues, the *rank* of R is $(p - m)$. The $(p - m)$ eigenvectors corresponding to those nonzero eigenvalues form a set of orthogonal basis vectors that span the space of the $(p - m)$ linearly independent vectors of R.

Another approach can be taken to examine the basic structure of a matrix. This method is called a singular value decomposition of an arbitrary rectangular matrix. A very detailed discussion of this process is given by Lawson and Hanson (1974). According to the singular value decomposition theorem, any matrix can be uniquely written as

$$R = UDV' \tag{5.45}$$

where R is an n by m data matrix, u is an n by n orthogonal matrix, V is an m by m orthogonal matrix, and D is an n by m diagonal matrix. All of the diagonal elements are non-negative and exactly k of them are strictly positive. The col-values of $A'A$. These elements are called the singular values of R. By making appropriate choices of A and F in equation (5.32), the singular value decomposition is one method to partition any matrix. The singular value decomposition is also a key diagnostic tool in examining colinearity problems in regression analysis (Belsley, Kuh, and Welsch, 1980). The application of the singular value decomposition to regression diagnostics will be discussed in Chapter 6.

STATISTICAL BACKGROUND

Moments of the Distribution of Values

The data that are to be analyzed are distributed in some initially unknown manner. It is typical to begin the examination of a variable by calculation of its mean value,

$$\bar{x} = \frac{1}{n} \sum_{j=1}^{n} x_j \tag{5.46}$$

and its standard deviation,

$$s_j = \left[\frac{\sum_{j=1}^{n} (x_j - \bar{x})^2}{n} \right]^{1/2} \tag{5.47}$$

A point to be raised about these two very commonly reported parameters is that in many environmental systems, the variable will not be normally distributed. In fact, they are more likely to be distributed lognormally. It may, therefore, be useful to examine the geometric mean $\bar{x}_g$ and standard deviations σ_g:

$$\ln \bar{x}_g = \frac{1}{n} \sum_{j=1}^{n} \ln x_j \tag{5.48}$$

$$\sigma_g = \left[\frac{1}{n} \sum_{j=1}^{n} (\ln x_j - \ln \bar{x})^2 \right]^{1/2} \tag{5.49}$$

Although these parameters are rarely calculated and reported, it may be useful to keep in mind that data points are not normally distributed if the standard deviation is greater than the mean value. The calculation of higher moments of the distribution can also provide information about the shape of the distribution function. The two more commonly used values are called skewness and kurtosis. The skewness is defined by

$$sk = \frac{1}{n}\sum_{j=1}^{n}(x_j - \bar{x})^3 \tag{5.50}$$

where a positive value indicates a distribution that is skewed toward the high values and a negative value indicates the distribution that is skewed toward the lower values. Kurtosis is given by

$$k = \frac{1}{n}\sum_{j=1}^{n}(x_j - \bar{x})^4 \tag{5.51}$$

Kurtosis gives the relationship between the spread in the central part of the distribution to the spread in the tail of the distribution. The sign and magnitude of these quantities help to further define the shape of the distribution of values of the particular parameter. In fact, we can define a generalized moment of the distribution as

$$\mu_v = \frac{1}{n}\sum_{j=1}^{n}(x_j - \bar{x})^v \tag{5.52}$$

These distribution parameters are important in deciding what are the statistical properties of the data set and what statistical inference can be made from the results of various analyses that can be made.

For some statistical procedures, it is necessary to remove the effects of using different metrics in describing the variables and so the variables are put in standard form. First, the deviation is calculated by subtracting the mean value from each sample value:

$$d_{ji} = x_{ji} - \bar{x}_j \tag{5.53}$$

The standardized variable z_j can be calculated by dividing the deviation by the standard deviation

$$z_{ji} = \frac{d_{ji}}{s_j} = \frac{x_{ji} - x_j}{s_j} \tag{5.54}$$

The standardized variable then has a mean value of zero and a standard deviation of unity, and thus, all standardized variables have the same metric.

The initial step in the analysis of the data generally requires the calculation of a function that can indicate the degree of interrelationships that exists within the data. Functions exist that can provide this measure between the two variables when calculated over all of the samples or between the samples when calculated over all of the variables. The most well-known of these functions is the product-moment correlation coefficient. To be more precise, this function should be referred to as the correlation about the mean. The "correlation coefficient" between two variables, x_j and x_k, over all n samples is given by

$$r_{jk} = \frac{\sum_{i=1}^{n} (x_{ij} - \bar{x}_j)(x_{ki} - \bar{x}_k)}{\left(\sum_{i=1}^{n} (x_{ji} - \bar{x}_j)^2\right)^{1/2} \left(\sum_{i=1}^{n} (x_{ki} - \bar{x}_k)^2\right)^{1/2}} \tag{5.55}$$

By utilizing the standardized variables, equation (5.55) can be simplified to

$$r_{jk} = \frac{1}{N} \sum_{i=1}^{n} z_{ji} z_{ki} \tag{5.56}$$

There are several other measures that can also be utilized. These measures include covariance about the mean

$$c_{jk} = \sum_{i=1}^{n} d_{ji} d_{ki} \tag{5.57}$$

the covariance about the origin

$$c_{jk}^{0} = \sum_{i=1}^{n} x_{ji} x_{ki} \tag{5.58}$$

and the correlation about the origin

$$r_{jk}^{0} = \frac{\sum_{i=1}^{n} x_{ji} x_{ki}}{\left(\sum_{i=1}^{n} x_{ji}^2 \sum_{i=1}^{n} x_{ki}^2\right)^{1/2}} \tag{5.59}$$

The matrix of either the correlations or covariances, called the dispersion matrix, can be obtained from the original or transformed data matrices. The data matrix contains that data for the m variables measured over the n samples. The correlation about the mean is given by

$$R_m = ZZ' \tag{5.60}$$

where Z' is the transpose of the standardized data matrix Z. The correlation about the origin

$$R_0 = Z^0 Z^{0\,\prime} = (XV)'(XV) \tag{5.61}$$

where

$$z_{ji}^0 = \frac{x_{ji}}{\left(\sum_{i=1}^{n} x_{ji}^2\right)^{1/2}}$$

which is a normalized variable still referenced to the original variable origin and V is a diagonal matrix whose elements are defined by

$$v_{ij} = \delta_{ij}\left(\sum_{i=1}^{n} x_{ji}^2\right)^{-1/2} \tag{5.62}$$

The covariance about the mean is given as

$$C_m = DD' \tag{5.63}$$

where D' is the transpose of the matrix of deviations from the mean calculated using equation (5.53). The covariance about the origin is

$$C_0 = XX' \tag{5.64}$$

the simple product of the data matrix by its transpose. As written, these product matrices would be of dimension m by m and would represent the pairwise interrelationships between variables. If the order of the multiplication is reversed, the resulting n by n dispersion matrices contain the interrelationships between samples.

The relative merits of these functions to reflect the total information content contained in the data have been discussed in the literature (Rozett and Petersen,

1975; Duewer, Kowalski, and Fasching, 1976). Rozett and Petersen (1975) argue that since many types of physical and chemical variables have a real zero, the information regarding the location of the true origin is lost by using the correlation and covariance about the mean that include only differences from the variable mean. The normalization made in calculating the correlations from the covariances causes each variable to have an identical weight in the subsequent analysis. In mass spectrometry where the variables consist of the ion intensities at the various m/e values observed for the fragments of a molecule, the normalization represents a loss of information because the variable metric is the same for all of the m/e values. In environmental studies where measured species concentrations range from the trace level (sub part per million) to major constituents at the percent level, the use of covariance may weight the major constituents too heavily in the subsequent analyses. The choice of dispersion function depends heavily on the nature of the parameters being measured.

Besides having a convenient measure of the interrrelationship between two variables, it is also useful to develop procedures to describe the relationships between samples so that subsequently the samples can be grouped according to how similar or dissimilar they are to one another. One set of possible functions to describe the relationship between samples are the correlation and covariance function defined in equations (5.56)-(5.59) with the meaning of the indexes changed so that j and k refer to different samples and the summations are taken over the m variables in the system.

An alternative approach is to define one of several geometrical measures. Consider an m dimensional space. Each axis represents one of the variables in the system so that the values of all of the measured variables for a single sample can be represented by a point. The distance between two points would be an indication of the dissimilarity between two samples since the larger the value of distance, the more dissimilar the two samples are from one another. One measure of distance is merely the extension of the simple Euclidean distance to an m dimensional space. The Euclidian distance (ED) is given by

$$\mathrm{ED}_{jk} = \sqrt{\sum_{i=1}^{m} (x_{ji} - x_{ki})^2} \tag{5.65}$$

Other functions have also been used, including squared Euclidian distance (SED):

$$\mathrm{SED}_{ik} = \sum_{i=1}^{m} (x_{ji} - x_{ki})^2 \tag{5.66}$$

mean character difference (MCD):

$$\mathrm{MCD}_{jk} = \sum_{i=1}^{m} (x_{ji} - x_{ki}) \tag{5.67}$$

mean Euclidian distance (MED):

$$\mathrm{MED}_{ik} = \left[\frac{1}{m} \sum_{i=1}^{m} (x_{ii} - x_{ki})^2 \right]^{1/2} \tag{5.68}$$

and the mean square Euclidian distance (MSED):

$$\mathrm{MSED}_{jk} = \frac{1}{m} \sum_{i=1}^{m} (x_{ji} - x_{ki})^2 \tag{5.69}$$

An alternative approach is to consider a vector drawn from the origin to each of the n points. A measure of the similarity between two samples could be the cosine of the angle between their respective vectors (Imbrie and Van Andel, 1964). The cosine is given by

$$\cos \theta_{jk} = \frac{\sum_{i=1}^{m} (x_{ji})(x_{ki})}{\left(\sum_{i=1}^{m} x_{ji}^2 \sum_{i=1}^{m} x_{ki}^2 \right)^{1/2}} \tag{5.70}$$

With only positive valued data, this measure has a range of 0 to 1. A value of 0.0 signifies there is nothing in common between the two samples, 1.0 shows identical samples, and 0.7071 ($\cos 45°$) shows that the two vectors are about as similar as columns of random digits. The $\cos \theta$ is clearly just the correlation about the origin between the jth and kth samples which can be given this geometrical interpretation. Now that the various interrelationship parameters have been defined, the mechanisms of the mathematical methods can be discussed.

CHAPTER

6

CHEMICAL MASS BALANCE

The first formal statement of equation (5.4) was given by Miller, Friedlander, and Hidy (1972) and independently by Winchester and Nifong (1971). The name originally given by Miller, Friedlander, and Hidy to their method is chemical element balance (CEB). A number of workers have adopted this terminology. More recently, Cooper and Watson (1980) have suggested that chemical mass balance (CMB) is a more appropriate name for the methodology. This terminology is being incorporated in the series of guidelines being developed by the U.S. Environmental Protection Agency (O.A.Q.P.S., 1981b) and will, therefore, be used here.

The premise of the chemical mass balance is that through an initial study of the airshed the sources of material to be sampled can be identified and their characteristic properties can be individually determined. Thus, a key step in the use of the chemical mass balance model is the development of the inventory of sources and the determination of the composition of material emitted by a variety of air pollution sources.

LEAST-SQUARES FITTING

The idea of the mass balance approach to receptor models is to know the number and composition of sources that contribute material to the measured values. The problem then is to obtain the best estimates of the mass values in equation (5.1). For the purposes of this discussion of the mathematics of the fitting process, we will assume the equation takes the form

$$x_i = b_0 + b_1 Z_1 + b_2 Z_2 + \ldots + b_j Z_j + \ldots + b_p Z_p \qquad (6.1)$$

where the x_i are functions of independent variables, Z_j. In this case, they are the measured concentrations of the emissions from the p sources indentified as the ones impacting the receptor site where the ambient measurements were made. It will generally be assumed for our purposes that b_0 should be zero. The other p b_j values are the contribution of each source to the total mass of material being sampled. It is assumed that this is, in fact, the correct form of the equation for this system; that, for example, there are no higher order terms or cross products. It is further assumed that the x values are typical and statistically uncorrelated. Each observed x_i value can be thought of as being the sum of a true part $x(Z)$, and a random error e,

$$x_i = x(Z) + e \tag{6.2}$$

The independence assumption can then be restated as

$$E(e_j e_j') = 0 \tag{6.3}$$

where e_j and e_j' are any two of the random components in the system and $E(e_j e_j')$ is the population average or expectation value of the product $e_j e_j'$.

Three additional minor assumptions are also made in performing a linear least-squares fit. The first assumption is that all of the observations of x have the same variance, although that variance is unknown. Second, it is assumed that the independent variable values are known without error. We examine the problem of including our knowledge of the uncertainties in the Z values in another section. Finally, it is assumed that the errors, e_j's, are normally distributed about a mean value of zero.

Thus, equation (6.1) can be expressed as a sum over p terms for each of the n species determined in the ambient sample. Using the notation from Chapter 5, the mass equation becomes

$$x_i(a) = \sum_{j=1}^{p} f_j a_{ji}, \qquad i = 1, n \tag{6.4}$$

From this expression, we can define a value for chi-squared as

$$\chi^2 = \left\{ \frac{1}{\sigma_i^2} \left[x_i - x_i(a_i)\right]^2 \right\} \tag{6.5}$$

where σ_i is the uncertainty in the measured value of x_i. The approach is to minimize the value of χ^2 with respect to each of the $p + 1$ coefficients, yielding a set of $p + 1$ simultaneous equations to be solved for the a_j values. This approach is a conventional multiple regression analysis and computer programs to perform this calculation are readily available. The standard least-squares equations can then be expressed in matrix form as

$$F = (A'A)^{-1} A'X \tag{6.6}$$

Another assumption commonly employed in multiple regression analysis is that there is no uncertainty in the values of the source compositions, a_{ji}'s in equation (6.4). The first workers to recognize the need to consider the errors in all of the observed quantities on the calculated mass contributions were Hammerle and Pierson (1975). In their approach, they assume that there are equal uncertainties in all of the measurements, or homoscedasticity. It is unusual for such a condition to hold in typical ambient monitoring situations. It is then necessary to look at methods that allow realistic inclusion of all of the known errors in the least-squares fitting process.

In 1979, Watson and Dunker independently suggested an approach, generally called effective variance weighting, that utilizes the known uncertainties in both the ambient concentrations and the source compositions. Since neither of these reports is currently in the jounal literature, a description of the mathematical basis for the approach will be presented. The derivation will follow that given by Watson (1979), that he developed from Britt and Luecke (1973).

Assuming that the errors are in both the measured ambient concentrations and source profiles, the probabilities of obtaining specific values in the range between x_i and $x_i + dx_i$ and between a_{ji} and $a_{ji} + da_{ji}$ are given by

$$p(x_i)\, dx_{ji} = \frac{1}{\sqrt{2\pi}\, \sigma_{x_i}} \exp\left\{-\frac{1}{2}\, \frac{[x_i - x_i(a)]^2}{\sigma^2_{x_i}}\right\} dx_i, \qquad i = 1, n$$

$$p(a_{ji})\, da_{ji} = \frac{1}{\sqrt{2\pi}\, \sigma_{x_i}} \exp\left\{-\frac{1}{2}\, \frac{(a_{ji} - a'_{ji})^2}{\sigma^2_{a_{ji}}}\right\} da_{ji}, \qquad \begin{array}{l} i = 1, n \\ j = i, p \end{array} \tag{6.7}$$

where a'_{ji} is the true composition value of the ith species emitted by the jth source. Since we have assumed that these errors are uncorrelated, the joint probability of simultaneously observing ambient concentration and source composition values in the indicated intervals for all of the measured species is simply the product of all of the probabilities given by equation (6.7):

$$\begin{aligned} P(x_1, x_2, \ldots, x_n, a_{11}, a_{12}, \ldots, a_{pn})\, dx_1 \ldots dx_n da_{11} \ldots da_{pn} \\ = [(2\pi)^{n(1+p)/2}\, \sigma_{x_1} \ldots \sigma_{x_n} \sigma_{a_{11}} \ldots \sigma_{a_{pn}}]^{-1} \\ \times \exp\left\{\frac{-1}{2}\left[\sum_{i=1}^{n} \frac{[x_i - x_i(a)]^2}{\sigma^2_{x_i}} + \sum_{i=1}^{n} \sum_{j=1}^{p} \frac{(a_{ji} - a'_{ji})^2}{\sigma^2_{a_{ji}}}\right]\right\} \\ dx_1 \ldots dx_n da_{11} \ldots da_{pn} \end{aligned} \tag{6.8}$$

The unknown true values of $x_i(a)$ and a'_{ji} are subject to equation (6.4) and are chosen to maximize this probability of observation. This probability will be maximized when x^2 as now defined by

$$\chi^2 = \sum_{i=1}^{n} \frac{[x_i - x_i(a)]^2}{\sigma_{x_i}^2} + \sum_{i=1}^{n} \sum_{j=1}^{p} \frac{(a_{ji} - a'_{ji})^2}{\sigma_{a_{ji}}^2} \tag{6.9}$$

is a minimum. To obtain the minimum subject to the constraint of equation (6.4), it can be rewritten as

$$x_i(a) - \sum_{j=1}^{p} f_j a'_{ji} = 0 \tag{6.10}$$

Equation (6.10) can then be multiplied by an arbitrary Lagrangian multiplier λ_i, and added to the equation (6.9), yielding

$$\chi^2 = \sum_{i=1}^{n} \frac{[x_i - x_i(a)]^2}{\sigma_{x_i}^2} + \sum_{j=1}^{p} \frac{(a_{ji} - a'_{ji})^2}{\sigma_{a_{ji}}^2} + \lambda_i \left[x_i(a) - \sum_{j=1}^{p} f_j a'_{ji} \right] \tag{6.11}$$

After differentiating equation (6.11) with respect to $x_i(a)$, a'_{ji}, λ_i, and f_j, the unknown quantities in the expression, a set of p complicated equations, are obtained that are cubic with respect to a particular a_q. The particular coefficient f_q is dependent on the $p - 1$ remaining f_j values and weakly on itself. Therefore, it is necessary to perform an iterative fit. This procedure is quite cumbersome and Watson outlines a computationally simplier approach that reduces to a standard multiple linear regression fit which in matrix notation can be written

$$f = [A'(V_e)A]^{-1} A'V_e X \tag{6.12}$$

where F is the matrix of source contributions, A is the matrix concentration profiles, X is the ambient data matrix, and V_e is a diagonal matrix whose elements are given by

$$v_{e_{ii}} = \sigma_{x_i}^2 + \sum_{j=1}^{p} \sigma_{a_{ij}}^2 a_j^2 \tag{6.13}$$

so that the weights are dependent on the values in A so an iterative fit is again necessary. The algorithm is generally written to use the values of the source compositions without their uncertainties to make an initial estimate of the source contribution values, a_j's. From these an iterative fitting process can be carried on to convergence. This "effective-variance" approach has been widely adopted for solving receptor model problems. At this point, we would like to outline some of the approaches used in receptor modeling and the results of these applications.

Tracer Element Methods

The earliest mass balance approaches to the apportionment of source contributions were to try to identify particular sources with unique characteristics of their emissions. For airborne particulate matter, for example, lead and bromine were used to identify motor vehicular traffic, sodium to identify sea salt particles, vanadium to identify residual fuel combustion, and so forth. For example, Miller, Friedlander, and Hidy (1972) used four elements to apportion airborne particulate mass between four sources, motor vehicles, soil, sea salt, and fuel oil combustion. In other words, the amount of particulate mass contributed by a source is simply the airborne concentration of the tracer element divided by the concentration of that element in emissions from that particular source:

$$f_j = \frac{x_j}{a_{ji}} \tag{6.14}$$

As might be anticipated, this very simple approximation leads to rather poor agreement between the calculated and observed ambient concentrations.

Friedlander (1973) repeated the calculation adding three additional elements, Mg, K, and Ca, and adding one additional source, cement dust. Since there are now more elements that sources, the simple linear regression approach [equation (6.6)] was used. In addition, he attempted to partition the airborne carbon concentration using an emission inventory approach where the amount of automobile exhaust calculated from the airborne lead value is used as the base for calculating the amount of diesel, aircraft exhaust, tire dust, and industrial emissions. In this approach, the emission inventory results are used to estimate the mass ratio of a source for which there is no good tracer element to one for which the tracer is well defined. The mass contribution of the nontracer source is the mass of the tracer course times this ratio. Since the amount of secondary carbon particles is unknown, their mass is calculated by the difference from these identified source contributions. The results seem reasonable although unverifiable.

The results of this initial mass balance were then used in conjunction with particle size distributions of various sources to estimate the overall size distribution for the Pasadena aerosol (Heisler, Friedlander, and Husar, 1973). It is necessary to include the modifications caused by the formation of secondary aerosols and several models were used. The calculated size and volume distributions compare well with the experimental results. Calculations of the distribution of the measured chemical elements with size were also presented.

Gartrell and Friedlander (1975) then used a more sophisticated approach to include aerosol formation and dynamics into the model. Using approximations for gas-to-particle conversion rates, coagulation, and other growth processes, they were able to make a much better fit to the data. Again they used an emission inventory to further partition the impacts of various groups of sources into more specific apportionments. This approach is quite dependent on the quality of assumed rates for the processes included and on the emission inventory used for the partitioning. However, it does give more specific results than the statistical analysis alone.

This tracer element approach has been applied in a number of other airsheds. Winchester and Nifong (1971) examined the impacts of six major source types on airborne particle compositions in Chicago, northwest Indiana, and Milwaukee using a combination of particulate emission compositions and source emission estimates rather than a direct fitting approach. Bogen (1973) examined the sources of particles in Heidelberg, West Germany, Heindryckx and Dams (1974) used airborne particle compositions to apportion continental, marine, and anthropogenic contributions to the aerosol mass in Ghent, Belgium. A simple tracer element analysis was also performed for Boston, Massachusetts (Gordon et al., 1972).

Gatz (1975) has reexamined a "model" Chicago area particle composition using the same six sources as Winchester and Nifong and six elements, Pb, Ca, V, Mn, Fe, and Al. Since Al was used for both coal-burning and soil, some assumption was necessary regarding the partitioning of aluminum between the two sources and several possibilities are presented. These results were then extended to data from a network of 22 sites in the Chicago area. For the elements, Br, Co, Cr, Cu, Fe, and K, agreement within a factor of 2 was obtained between observed and calculated ambient concentrations. However, for the As, Cd, Cl, Hg, La, Mg, Na, Sc, and Zn, differences of one to three orders of magnitude were observed, indicating that either the assumed compositions of the source emissions were incorrect or that additional sources were required to account for the observed concentrations. Hammerle and Pierson (1975) also used a tracer method that they call a "single-source model" and examined the simple pairwise correlation results between Pb, Br, Fe, Zn, Ni, Mn, Ti, V, and Ca. Using Friedlander's (1973) sources and source compositions for Pasadena, they basically repeated the earlier calculation although looking in more detail at the propagation of error and data point weighting in their calculation. The tracer element approach was employed in a recent study of sources of the aerosol in Bombay, India (Kamath and Kelkar, 1981).

The tracer approach has advantage of calculational simplicity, and, for some sources, it has the ability to quantify sources quite well. For example, the use of lead as an automotive tracer works well to predict the bromine levels in cities that do not contain substantial nonferrous metal smelters. However, the tracer method clearly limits the resolution of sources since many sources do not emit unique tracer species. The ability to use multiple species for each source to make the fit therefore will result in an increased ability to separate sources.

Linear Least-Squares Fitting

By the middle 1970s, it was recognized that better source composition data were essential and that this increased data base needed to be incorporated into the fitting process. Several major studies of source concentrations were undertaken including work by the University of Maryland and the Oregan Graduate Center. Under the direction of G. E. Gordon and W. H. Zoller, a number of specific source studies were made including coal-fired power plants (Gladney, 1974; Small, 1976; Gladney, Small, Gordon, and Zoller, 1976), oil-fired power plants (Cahill, 1974; Mroz, 1976), motor vehicles (Ondov, 1974; Ondov, Zoller, and Gordon, 1982), municipal incinerators (Greenberg, 1976; Greenberg, Zoller, and Gordon, 1978; Greenberg et al., 1978), soil (Thomae, 1977), and other industrial sources (Small, 1979).

As part of the Portland Aerosol Characterization Study (PACS), Watson (1979) examined 30 different sources or source types, many in two different size ranges, so that the composition of particles from a large number of source types have now been measured. Cass and McRae (1981) have compiled many source compositions into a fairly comprehensive source inventory. In order to make this and all of the other source composition information more readily available, it has been compiled in the appendix of this volume. Utilizing this improved source composition data, there have been a number of applications of the mass balance approach to airborne particle mass apportionment. In this section we shall review those applications using ordinary least-square fitting methods and in the next, we shall treat the "effective-variance weighted" least-squares fitting results.

Mayrsohn and Crabtree (1976) presented the use of an iterative least-squares approach to resolving six sources of airborne hydrocarbon compounds. They performed the least-squares fit to hydrocarbon compounds determined using gas chromatography to determine the concentrations of eight compounds. Their ordinary least-squares source reconciliation algorithm recognized that not all sources may contribute to every sample, and, if negative contributions were obtained, a different configuration of sources was employed with certain qualifying assumptions (Mayrsohn et al., 1975). With the six sources, they reduced their consideration to only twelve source configurations because (1) automotive exhaust needed to be included to account for the acetylene observed, (2) either gasoline or gasoline vapor had to be included to account for the butanes and pentanes observed, and (3) either natural gas or liquefied petroleum gas was required to account for the propanes and ethane observed. Each possible configuration with positive coefficients was considered and the one with the lowest standard error was chosen as the optimum solution.

An additional report has been made utilizing this mass balance approach for resolving hydrocarbon sources. Nelson, Quigley, and Smith (1983) have examined the atmospheric hydrocarbons in Sydney, Australia. They have used much more extensive hydrocarbon profiles for their sources and have obtained good agreement between the mass balance approach and an emissions inventory.

In the case of the hydrocarbon analysis just described, the least-squares fitting was performed without weighting the concentrations with the uncertainties. For these compounds, their concentrations can be considered as major species with the lowest concentrations being a few percent so that all of them are of importance to the fit. When trace elemental data are fitted in the presence of major elements, the value and variation in the absolute level of the trace elements are very small compared to the major components and will not affect the fit unless weighting is employed. Kowalczyk, Choquette, and Gordon (1978) illustrated the use of a weighted least-squares fit for six source components with eight elements in the analysis of 10 samples from Washington, D.C. Table 6.1 gives the source profiles used in this study. Only Al, Na, Fe, V, Zn, Pb, Mn, and As are used in the fit. From these elemental concentration values, x_{ij}, the average mass contributions for the sources were calculated, yielding the chemical element balance presented in Table 6.2. The use of a subset of elements permits the determination of the other elements and comparison with the observed values. It is observed that the volatile elements, Cl, Br, I, Se, and Sb, are rather poorly fit. In addition there is a substantial underestimation for Ca, Mg, Cr, Ni, and Cd. The Se, Sb, and Cd may be in error because of their volatility and the fact that the measurements of the composition are made on emitted particles extracted from stacks while still at elevated temperature. New approaches (Houck, Core, and Cooper, 1982) to stack sampling which cool particles and the stack gases to ambient conditions before particle collection should relieve many of these problems. The variations in chromium and nickel probably arise from the high degree of variability in the compositions of oil burned in oil-fired power plants. The calcium and magnesium results would apparently reflect the need for an additional component.

In the analysis of an expanded data set of 130 samples of total suspended particulate matter in Washington, Kowalczyk, Gordon, and Rheingrover (1982) add a limestone source using a composition given by Mason (1966). In this work, they also increased the number of elements included in the fitting process based on the study they made of the use of varying numbers of included elements (Gordon et al., 1981). In that study they examined the fit using 9–30 marker elements and found little difference in the quality of fit as long as several key elements (Pb, Na, and V) were included. They found that the use of Br and Ba as markers even with lead gave serious underpredictions of lead concentrations and those elements were, therefore, excluded from the fit. Table 6.3 summarizes the results using 28 elements in the weighted ordinary least-squares fit. There are still a number of elements including ones used in the fitting processes that are poorly predicted by the analysis, although now all of those elements that represent significant percentages of the mass are relatively well fit. However, the overall

Table 6.1. Source Components (μg/g) Used to Resolve Sources of Washington, D.C. Suspended Particulate Matter (Kowalczyk, Choquette, and Gordon, 1978)

Element	Soil	Marine	Coal	Oil	Refuse	Motor Vehicle
Al	56,000	0.29	120,000	1,300	11,000	–
Na	3,000	310,000	1,900	37,000	61,000	–
Fe	35,000	0.29	84,000	8,400	4,900	7,000
V	73	0.0059	360	70,000	23.0	–
Zn	78	0.0029	600	4,900	90,000	2,100
Pb	11	0.000087	480	1,200	60,800	140,000
Mn	900	0.0059	360	300	550	–
As	42	0.0087	720	84	180	–
K	10,600	11,000	16,000	1,300	–	–
Mg	5,200	40,000	6,240	12,000	9,900	–
Ca	4,500	12,000	11,000	25,000	13,000	7,000
Ba	490	0.87	1,000	5,800	520	1,800
Cl	190	561,000	240	37,000	150,000	14,000
Br	6.7	1,900	480	160	1,200	53,000
I	4	1,800	480	–	–	–
Sc	10	0.00012	42	0.56	0.99	–
Ti	3,600	0.0029	7,200	77	1,800	–
Cr	55	0.00015	190	180	370	–
Co	15	0.0015	58.0	490	5.0	–
Ni	31	0.0059	240	12,000	120	–
Cu	16	0.0087	480	2,500	1,200	–
Se	0.06	0.12	180	110	28.0	–
Cd	0.07	0.00031	31.0	8.4	1,130	–
Sb	0.56	0.0015	30	21	1,580	–
La	52	0.00087	72	49	2.80	–
Ce	67	0.0012	140	42	12.0	–
Th	7.8	0.00015	24	4.9	1.40	–

TSP mass was poor since 23 of the 65 μg/m³ are unaccounted for. Similar quality results have been reported for Osaka, Japan (Mizohata and Mamuro, 1979), St. Louis, Missouri (Dzubay, 1980) and Madison, Wisconsin (Stolzenburg, Andren, and Stolzenburg, 1982).

EFFECTIVE-VARIANCE WEIGHT LEAST-SQUARES FITTING

As was previously mentioned, the concept of using the measured uncertainties in the analysis were independently reported by Watson (1979) and Dunker (1979). Dunker made a reanalysis of the results for Washington, D.C. originally presented

Table 6.2. Chemical Element Balance of Washington, D.C. Aerosols

Element	Contributions from Components (ng/m^3)						Total Concentration (ng/m^3)		
	Soil	Marine	Coal	Oil	Refuse	Motor Vehicle	Predicted	Observed[a]	Larger/ Smaller[b]
Elements Used in Fitting									
Al	1200	<0.001	720	1	14	–	1940	1680 ± 1100	1.15[c]
Na	63	280	12	27	82	–	460	470 ± 470	1.0[c]
Fe	750	<0.001	510	6.1	6.5	67	1340	1260 ± 940	1.06[b]
V	1.6	<0.001	2.2	51	0.031	–	55	54 ± 57	1.0[c]
Zn	1.7	<0.001	3.6	3.6	120	20	150	105 ± 90	1.0[c]
Pb	0.23	<0.001	2.9	0.86	81	1300	1380	1400 ± 1660	0.01[c]
Mn	19	<0.001	2.2	0.22	0.73	–	22	27 ± 16	1.2[c]
As	0.09	<0.001	4.4	0.061	0.24	–	4.8	5.7 ± 5.0	1.2[c]
Remaining Elements									
K	230	10	94	1.0	–	–	340	510 ± 350	1.5
Mg	110	37	38	8.6	13	–	210	440 ± 380	2.1
Ca	97	11	66	18	17	67	280	770 ± 500	2.8
Ba	10	0.001	6.6	4.2	0.7	17	39	27 ± 24	1.4
Cl	3.9	510	1.5	27	200	130	870	140 ± 75	6.2[c]
Br	30.14	1.8	3.0	0.12	1.6	490	500	190 ± 220	2.6[c]

I	0.08	1.6	3.0	—	—	—	4.7	9.3 ± 7.2	2.0[c]
Sc	0.22	<0.001	0.25	<0.001	0.001	—	0.47	0.63 ± 0.63	1.3
Ti	77	<0.001	44	0.056	2.0	—	120	120 ± 95	1.0
Cr	1.2	<0.001	1.2	0.13	0.50	—	3.0	11 ± 9	3.7
Co	0.32	<0.001	0.35	0.36	0.007	—	1.0	1.1 ± 1.0	1.1
Ni	0.67	<0.001	1.5	8.6	0.16	—	11	27 ± 28	2.5
Cu	0.34	<0.001	3.0	1.8	1.6	—	6.7	13 ± 12	1.9
Se	0.001	<0.001	1.1	0.076	0.037	—	1.2	3.5 ± 2.7	2.9[c]
Cd	0.002	<0.001	0.19	0.006	1.5	—	1.7	3.5 ± 2.2	2.1
Sb	0.012	<0.001	0.18	0.015	2.1	—	2.3	9.8 ± 8.5	4.2
La	1.1	<0.001	0.44	0.036	0.004	—	1.6	1.9 ± 1.7	1.2
Ce	1.4	<0.001	0.86	0.030	0.016	—	2.3	3.4 ± 3.1	1.5
Th	0.17	<0.001	0.15	0.004	0.002	—	0.33	0.32 ± 0.24	1.0
								Average	2.0

[a] Uncertainty is standard deviation of a single observation.
[b] Ratio of observed/predicted or predicted/observed, whichever is larger.
[c] Not included in calculation of average larger/smaller value.

Table 6.3. Average Result of Chemical Element Balances of 130 Samples from Washington, D.C., Area for Summer 1976

Element	Predicted Contributions[a] (ng/m^3)							Predicted	Observed[d]	Larger/ Smaller[b]	Missing Values[c]
	Soil	Limestone	Coal	Oil	Refuse	Motor Vehicle	Marine				
Na[e]	43	0.83	8.3	12	35	–	201	300	300 ± 20	1.00[f]	0
Mg[e]	74	101	27	3.8	5.4	32	26	270	440 ± 30	1.63[f]	3
Al[e]	812	9	517	0.4	6.1	–	<0.01	1340	1350 ± 110	1.16[f]	0
K[e]	154	6	67	0.4	47	13	7	295	400 ± 20	1.38[f]	0
Ca[e]	66	635	47	8.2	7.1	47	7.6	820	860 ± 40	1.09	0
Sc[e]	0.15	0.002	0.18	0.0002	0.0006	–	<0.0001	0.33	0.33 ± 0.03	1.19[f]	0
Ti[e]	52	0.83	31	0.03	1.0	–	<0.0001	85	110 ± 10	1.35[f]	0
V[e]	1.06	0.042	1.6	23	0.013	–	<0.0001	26	25 ± 2	1.07	0
Cr	0.81	0.023	0.84	0.062	0.21	–	<0.0001	2.0	14 ± 2	7.07[f]	5
Mn[e, g]	13	2.3	1.6	0.10	0.31	1.3	<0.0001	18	17 ± 2[g]	1.26[f]	0
Fe[e]	511	8.3	362	2.8	2.8	34	0.0002	920	1000 ± 60	1.19[f]	0
Co[e]	0.22	0.0002	0.25	0.16	0.003	0.05	<0.0001	0.68	0.83 ± 0.08	1.19[f]	0
Ni	0.46	0.042	1.04	4.0	0.07	0.34	<0.0001	6.0	17 ± 2	4.35[f]	1
Cu	0.23	0.009	2.2	0.89	0.71	3.0	0.0001	7.1	17 ± 2	2.68[f]	23
Zn[e]	1.14	0.04	2.6	1.6	51	7.3	<0.0001	64	85 ± 6	1.41[f]	0
Ga[e]	0.38	0.0008	0.46	0.0001	–	–	<0.0001	0.85	1.29 ± 0.17	1.89[f]	42
As[e]	0.061	0.002	3.1	0.028	0.10	–	0.0001	3.32	3.25 ± 0.2	1.58[f]	0
Se	0.0009	0.0002	0.78	0.035	0.016	0.035	0.0001	0.87	2.5 ± 0.2	4.10	0
Br	0.097	0.013	2.1	0.054	0.66	167	1.25	171	139 ± 9	1.84	0
Rb[e]	1.19	0.0064	0.60	0.0001	–	–	0.0023	1.8	2.1 ± 0.2	1.54[f]	17
Sr[e]	3.65	1.29	3.5	0.09	0.027	–	<0.0001	8.6	10 ± 1	1.40[f]	32
Ag[e]	0.0007	0.0001	–	0.006	0.23	–	<0.0001	0.24	0.20 ± 0.01	1.44[f]	26
Cd[e]	0.0011	0.0001	0.13	0.0028	0.64	1.03	<0.0001	1.80	2.4 ± 0.02	1.89[f]	0
In	0.0007	0.0001	0.0023	<0.0001	0.0024	–	0.0004	0.0059	0.020 ± 0.001	5.12[f]	3
Sb[e]	0.0081	0.0004	0.13	0.007	0.89	0.60	<0.0001	1.6	2.1 ± 0.2	1.39[f]	0
I	0.058	0.0026	2.1	–	–	–	1.14	3.3	2.0 ± 0.1	2.41[f]	12
Cs[e]	0.028	0.0011	0.039	0.0005	0.0025	–	<0.0001	0.07	0.17 ± 0.05	3.36[f]	2
Ba	7.1	0.02	4.7	2.0	0.30	6.4	0.0006	21	19 ± 2	1.42[f]	1

La[e]	0.75	0.014	0.31	0.016	0.0016	–	<0.0001	1.1	1.5 ± 0.1	1.42[f]	0
Ce[e]	0.98	0.024	0.62	0.014	0.007	–	<0.0001	1.6	2.0 ± 0.2	1.30[f]	1
Sm[e]	0.068	0.002	0.057	0.0012	0.0003	–	<0.0001	0.13	0.20 ± 0.02	1.59[f]	0
Eu[e, h]	0.015	0.0004	0.014	0.0005	0.0002	–	<0.0001	0.030	0.030 ± 0.003	1.32[f]	5
Yb	0.037	0.0011	0.030	0.0004	0.0009	–	<0.0001	0.070	0.034 ± 0.003	2.60[f]	19
Lu	0.0067	0.0004	0.0080	0.0001	0.0004	–	<0.0001	0.016	0.0056 ± 0.0006	3.34[f]	1
Hf[e]	0.031	0.0006	0.022	0.0008	0.0004	–	<0.0001	0.055	0.10 ± 0.01	1.80[f]	1
Ta[e, h]	0.041	0.0008	0.0077	0.0011	0.0018	–	<0.0001	0.052	0.036 ± 0.004	1.84[f]	39
W	0.014	0.0013	0.038	0.0004	0.0072	–	<0.0001	0.061	0.24 ± 0.02	4.76	23
Pb[e]	0.15	0.019	2.1	0.39	34	428	<0.0001	465	400 ± 20	1.28[f]	0
Th[e]	0.11	0.0036	0.10	0.0016	0.0008	–	<0.0001	0.22	0.25 ± 0.02	1.19[f]	1

[a] Contributions designed by "–" indicate that concentration of the element in particles from the source is not known.

[b] Larger/smaller is the average of the ratio of predicted/observed or vice versa, whichever is larger over all samples.

[c] Number of samples for which no value is available because peak is not strong enough in spectra for determination or concentration is not above filter blank.

[d] Uncertainty is standard deviation of mean value.

[e] Element fitted by least-squares procedure.

[f] Larger/smaller value included in average given below.

[g] For reasons described in text, all observed values reduced by factor of 0.69 prior to fitting. Actual average observed values was 26 + 3 ng/m.

[h] As noted in the text, three observed values were eliminated from fits.

by Kowalczyk, Choquette, and Gordon (1978) and discussed in the previous section. Dunker also used a logarithmic transformation of the data to ensure that all of the source composition coefficients were positive, although some of the mass contributions were found to be negative. All of the negative values were found to be very small. The results of the analysis performed by Dunker resulted in several small changes in the values of the source concentrations and a very large increase in the manganese concentration attributed to refuse incineration. Gordon and coworkers have had a consistent problem with identifying a source of fine particle manganese that they have observed in Washington and in their subsequent report Kowalczyk, Gordon, and Rheingrover (1982) arbitrarily decreased the manganese concentrations by a factor of 0.69 before performing the seven source fits they report in that paper. In this case the effective-variance results are not substantially better than the ordinary least-square fit because there is at least one important source, limestone, that was not included in the fit.

The most extensive use of effective-variance fitting has been made by Watson in his work on data from the Portland Aerosol Characterization Study (PACS) (Watson, 1979; Cooper, Watson, and Huntzicker, 1979). Since that study, a number of other applications of this approach have been made including Medford, Oregon (Cooper, 1979; DeCesar and Cooper, 1981), Philadelphia, Pennsylvania (Chow et al., 1981), and at a number of locations in the Inhalable Particulate Network (IPN) (Chow, Watson, and Shah, 1982).

The PACS program is a particularly good example of the use of the mass balance approach using 37 sources for which direct measurements were made during the time the ambient samples were being taken. Therefore, the composition and variation in composition are quite applicable to the aerosol particle mass apportionment problem being considered. Another important innovation that was introduced into the PACS study was the attempt to obtain a representative yearly average without having to sample every day of the year. In their planning they stratified the year into a number of defined meteorological regimes. When conditions and time of year were appropriate, samples were taken. However, although many samples were obtained, only enough of the samples were analyzed to provide a meaningful average value for that meteorological regime. Ninety-four days per year were sampled with a 32-day subset selected for intensive chemical analysis. In this way, annual average values that are reasonably representative can be obtained at a quite reasonable effort for both sampling and analysis.

Samples were taken of both total suspended particles and particles with aerodynamic diameter less than 2.5 μm. The average chemical composition of the downtown Portland aerosol is given in Figure 6.1. The aerosol mass source apportionment is given in Figure 6.2. Although there are a large number of sources identified in the resolution, there is a sizable "unidentified" amount as well as several nonspecific sources including volatilizable carbon and nonvolatilizable carbon. In a latter section of this chapter, there will be a closer examination of the ability of the mass balance approach to obtain such detailed resolutions.

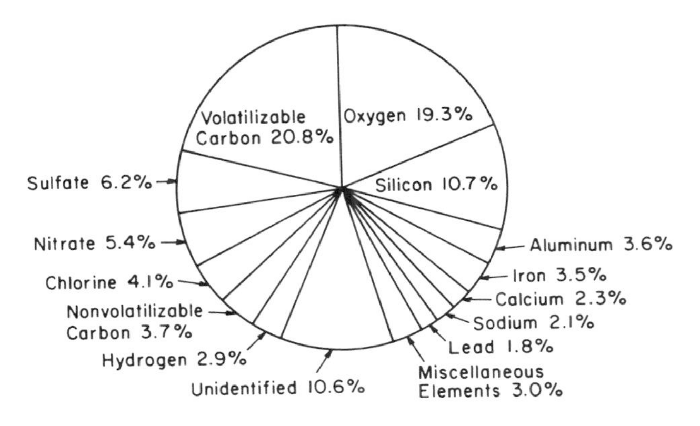

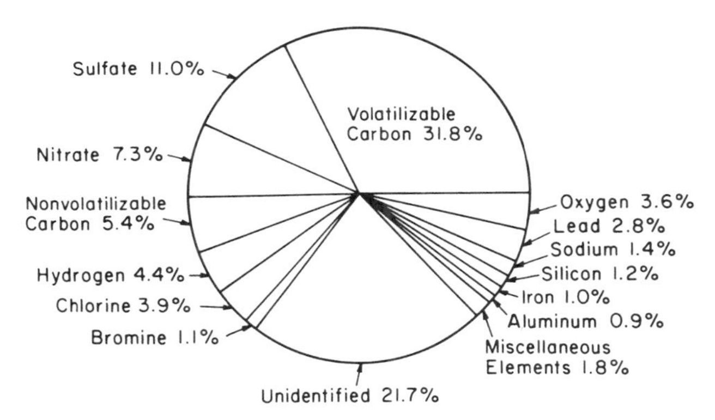

Figure 6.1. Average chemical composition of the total suspended particulate matter (top) and the fine (<2.5 μm) particulate matter (bottom) of the downtown Portland, Oregon, aerosol as presented by Core et al. (1982).

One of the important uses of the mass balance results that were obtained in the PACS program was to calibrate the dispersion models being used for making air quality management decisions (Core, Hanrahan, and Cooper, 1981; Core, et al., 1982). The Portland airshed was modeled using GRID (Fabrick and Sklarew, 1975), a conservation of mass, advection-diffusion code that has been designed to model the complex terrain of the Portland area. The area is divided into 2 by 2 km grid cells. For each of the 5000 cells, it is necessary to know the topography, the wind field flows for each of the eight meteorological regimes, and the spatially and seasonally resolved point and area source emissions as well as the stack parameters for each point source.

The results of the dispersion model can be separated into the same source

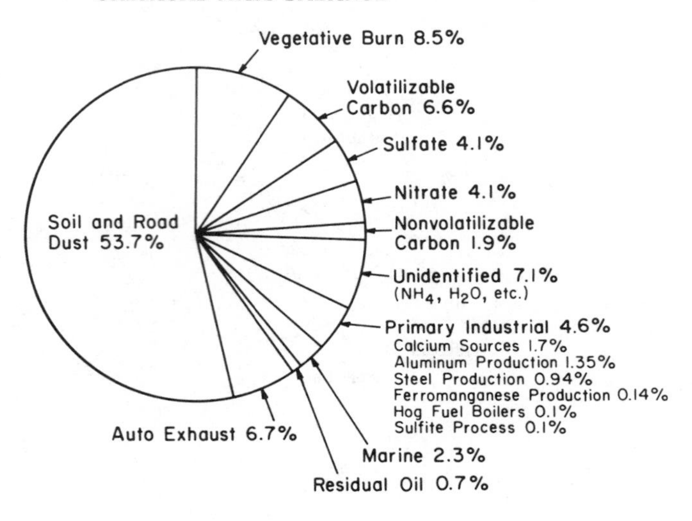

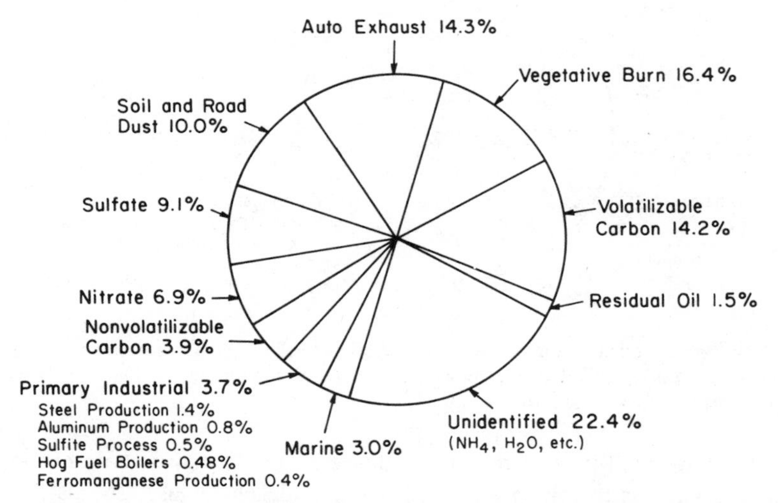

Figure 6.2. Source apportionment of the total suspended particulate matter (top) and fine (<2.5 μm) particulate matter (bottom) of the downtown Portland, Oregon, aerosol as given by Core et al. (1982).

groups as the mass balance results so that direct comparisons can be made. In Figures 6.3, 6.4, and 6.5, the results of the GRID and mass balance models are compared for road dust, auto exhaust, and residual oil combustion. Road dust and residual oil impacts are in substantial disagreement with dust substantially underpredicted by the dispersion model. For the auto exhaust, three sites agree within the uncertainty in the mass balance results. However, for one location

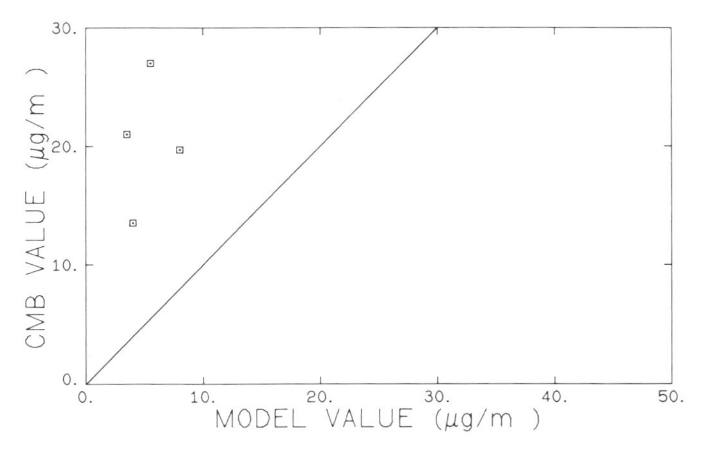

Figure 6.3. Initial comparison of Portland, Oregon, dust impacts comparing the mass balance results against the dispersion model predictions. Taken from Core et al. (1982) and used with permission.

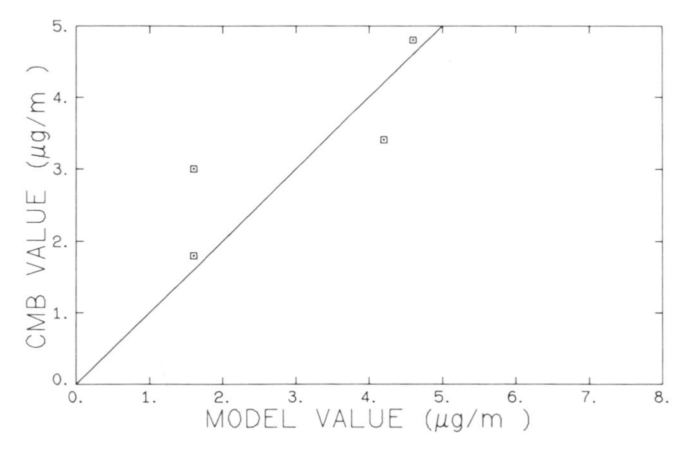

Figure 6.4. Initial comparison of Portland, Oregon, motor vehicle tailpipe emissions comparing the mass balance results against the dispersion model predictions. Taken from Core et al. (1982) and used with permission.

(site 5) there is an underprediction by the GRID model. Core, Hanrahan, and Cooper have indicated that these discrepancies can be readily understood. For the auto exhaust, a heavily traveled road was incorrectly assigned to an adjacent grid. When properly placed, the dispersion model value increases by 10%.

Several problems were found for residual oil combustion emissions. First, the emissions had been assumed to be constant over the entire year. The dispersion model was thus modified to account for the actual monthly operating schedules

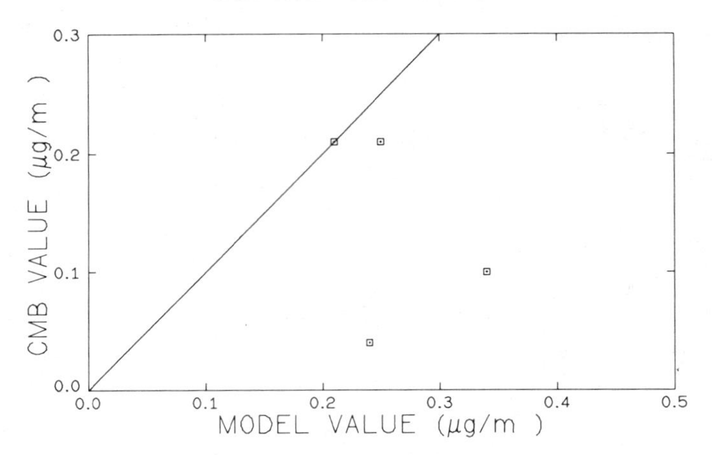

Figure 6.5. Initial comparison of Portland, Oregon, residual oil combustion impacts comparing the mass balance results against the dispersion model predictions. Taken from Core et al. (1982) and used with permission.

for each meteorological regime. Second, the topographical assignments for each cell were examined. Each grid cell is given one of five altitude variations within a cell. Small changes in plume height assignment can then substantially affect the predictions. Adjustments were made to the point source stack heights to better reflect the differences between the true stack and receptor heights. Finally, a single major source was modeled with an unrealistic operating schedule.

The road dust was underpredicted based on the EPA generalized paved road emission factor. There are few unpaved roads, so it was felt that the paved road emission factor should be increased. Since the automotive emission factor appeared to be correct because of the generally good agreement seen in Figure 6.4, the road dust emission factor was scaled by the ratio of the mass balance result for road dust to that for auto exhaust. This emission was also suggested to be larger in heavy industrial areas of Seattle, Washington (Roberts et al., 1979) compared to commercial land use areas. This adjustment substantially increased the road dust contribution to the airborne particle loading. This adjustment brings the receptor and dispersion model results into agreement for road dust.

The test of these adjustments is then to compare dispersion model predictions of total suspended particulate matter (TSP) with the measured values. Figure 6.6 shows these comparisons before and after, making the adjustments described above. Clearly, the use of the mass balance results to tune the dispersion model has made a noticeable improvement in the quality of the dispersion model predictions, and hence its value as an air quality management tool.

Another area of air quality research to which the mass balance approach is being applied is in the apportionment of acid precipitation. Cooper (1983) has

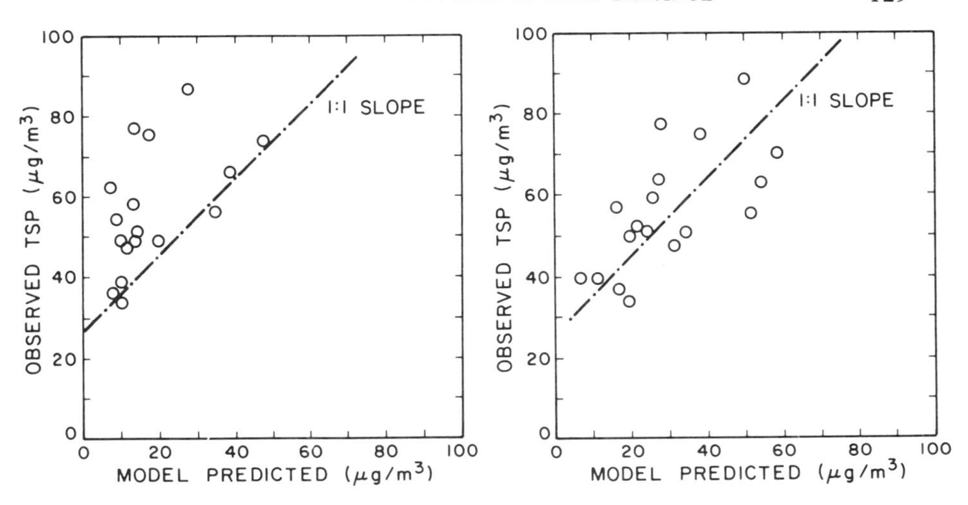

Figure 6.6. Comparison of annual dispersion model predictions with measured TSP before (left) and after (right) emission inventory improvements deduced from the mass balance results. Taken from Core et al. (1982) and used with permission.

outlined a general conceptional framework. A more specific approach has been developed and applied by Liljestrand and Morgan (1978, 1979, and 1981) for southern California and by Feeley and Liljestrand (1983) for Texas. Gatz (1983) has also used an effective-variance weighted chemical mass balance to apportion sources of species determined in central Illinois precipitation. Although these initial efforts do provide some resolution of the observed concentrations, there is only the very limited ability to relate the observed sulfate and nitrate to the observed acidity. In an effort to relate the acidic secondary species back to their primary emission as well as the gas-phase materials, Liljestrand (1983) has made an initial attempt to perform a mass balance on precipitation using theoretically deried fractionation coefficients. Encouraging initial results have been obtained.

LIMITS OF RESOLUTION IN MASS BALANCE

Several questions arise with regard to the use of these models. An intercomparison study was initiated by the U.S. Environmental Protection Agency (Stevens and Pace, 1983) using particle composition data from an intensive sampling program in Houston and several simulated composition data sets prepared for this study by the National Bureau of Standards. These studies have tested the state of the art in mass balance models and have lead to a number of conclusions.

The Houston study (Dzubay and Stevens, 1983) involved the mass balance analysis of 18 samples taken sequentially during an intensive sampling program in September 1980 using multiple dichotomous samplers with several different filter substrates. From the subsequent analyses, mass, elemental, and ionic compositions were obtained for both the fine and coarse fractions. These results were then available for the mass balance modeling. The compositions of particles from sources in Houston were not available and were not measured during this program so that source composition profiles had to be obtained from other sources such as Watson (1979) or Cass and McRae (1981).

The lack of source data immediately raises two problems in the use of the mass balance methods and comparison of results from different investigators. First, it is not always certain exactly which sources should be included in the analysis. Although emission inventories may be available for the region, it may be that the measured source composition for a coal-fired power plant in Maryland is not a particularly good representation for a plant in Texas. Even more of a problem are local sources of soil and road materials that may be quite different from locations where source measurements are available. *It is clearly essential to know the area that is to be modeled well.* Specific features such as nearby point sources or other unusual activities must be known in order to assure that a reasonable resolution can be made.

In addition, the choice of the label that investigators put on sources can be a problem. One of the major difficulties that occurred in this Houston intercomparison was in trying to develop correspondences between the different names and somewhat different compositions used for the various source types. The sources were then condensed into the eight general categories outlined in Table 6.4. The various mass balance results are presented in Table 6.5 for the fine and coarse fractions for the 12-hour sample taken beginning at 7 A.M. on September 10, 1980. The first two results are two different investigators using the same program and source profile library. It is an interactive system allowing the investigator to add or delete sources until an acceptable fit is obtained using the effective variance regression approach described previously. Using a different version of the same basic algorithm, the results obtained by a different group with a different source library are presented in the third column. Even with differences just discussed, the agreement among these results is quite good.

A problem with this kind of intercomparison is that there is no known absolute results with which the models can be compared. Therefore, it would be helpful to have a data set created to be a realistic approximation to ambient conditions so that model results could be compared directly against the known "truth." Three data sets were created using the RAM dispersion model (Gerlach, Currie, and Lewis, 1983). Fourteen different source types and two different source position maps (Figures 6.7 and 6.8) were used. The first set (Figure 6.7) used one city plan and nine active sources, 8 of which were among the 13 possible sources made known to the analysts. Forty samples with 19 elements in each were generated using the general structure described in Table 6.6. In data sets 2

Table 6.4. Source Categories and Marker Elements Used in Source Apportionment of Houston Aerosol

Marker Elements	General Source Category	Specific Source Categories
Na, Cl	Marine	Marine
Al, Si, K, Ca, Mn, Fe	Crustal	Wind erosion of soil
		Paved and unpaved roads
		Construction
		Limestone crushing and handling
		Cement
		Lime kiln
		Fly ash
		Slag
C, Br, Pb	Transportation	Noncatalyst vehicles
		Catalyst cars
		Diesels
Mn, Fe	Steel	Iron and steel production
		Steel finishing and handling
Cu, Zn, As, Sm, Sb	Nonferrous metals	
NH_4^+, SO_4^{2-}	Sulfate	Primary emissions in Houston airshed
		Conversion of SO_2 in Houston airshed
		Regional background
C	Carbonaceous	Refineries
		Botanical
		Vegetative burning
		Tire wear
NO_3^-	Nitrate	Photochemical

and 3, the second city plan (Figure 6.8) was used with all 13 known source included. The data set was expanded to include ^{14}C measurements to assist in resolving sources of contemporary carbon from fossil fuel carbon. In data sets 1 and 2, the source compositions are constant and they vary with a lognormal distribution in set 3. Only set 2 was analyzed by several groups using the mass balance procedures as part of the intercomparison exercise. The results are presented in Table 6.7 and Figure 6.9. There are three effective-variance weighted results and two sets of ridge regression results. The ridge regression procedure will be discussed shortly. It should be noted that much better results are obtained for the averages of the individual analysis than can be obtained from the analysis of the average concentrations in the data set. Information is lost in the averaging of the

Table 6.5. Intercomparison of Source Apportionment Results for the 12-h Period Beginning at 0700 on September 10, 1980[a]

Size Range	Category	ERT-A CMB(EV)	ERT-B CMB(EV)	NEA CMB(EV)
Fine	Marine	0.58	0.53	
Fine	Crustal	3.9	1.9	1.7
Fine	Transportation	3.8	2.3	3.7
Fine	Steel	1.9	1.8	0.74
Fine	Nonferrous metals			0.11
Fine	Sulfate	23.4	25.6	25.7
Fine	Carbon	6.8	8.8	7.7
Fine	Nitrate			
Fine	Total	40.5	40.9	39.8
Coarse	Marine	0.26	0.23	0.27
Coarse	Crustal	17.0	24.3	31.4
Coarse	Transportation	0.69	0.42	0.95
Coarse	Steel		0.12	0.33
Coarse	Nonferrous metals			0.04
Coarse	Sulfate	1.4	1.0	1.1
Coarse	Carbon	1.1	1.6	3.1
Coarse	Nitrate	1.3	0.57	1.0
Coarse	Total	21.6	28.2	38.3

[a]The mass concentrations measured in the fine and coarse fraction were 56.3 ± 2.8 and 38.4 ± 2.8 g/m^3, respectively.

sample composition data that is apparently not recoverable by the analysis methods. It, therefore, seems to be worth the extra effort to analyze every sample. A final point to be made regarding these results is the nature of the estimation of the precision of the mass balance results. Figure 6.10 presents the absolute errors for each of the estimations by each of the three groups who performed the analyses. It can be seen that all of the EV-1 results are biased in one direction and almost all of the ridge regression results are biased in the other. One of the effective-variance results appears to have a more reliable estimation of their uncertainties. On the whole, the mass balance approach seemed to give quite good correspondence to the actual values.

There is yet one more fundamental question of resolution to be addressed and we shall refer to this problem as the collinearity question. In the least-squares fit whether ordinary or effective-variance weighted, a key element in the fit is $(A'A)^{-1}$. This inverse square matrix represents a problem if we have two or more sources with similar composition profiles. Then if two of the columns in this matrix are almost the same or one is almost a linear combination of several others, the determinate of this matrix will be almost zero and the solution to the

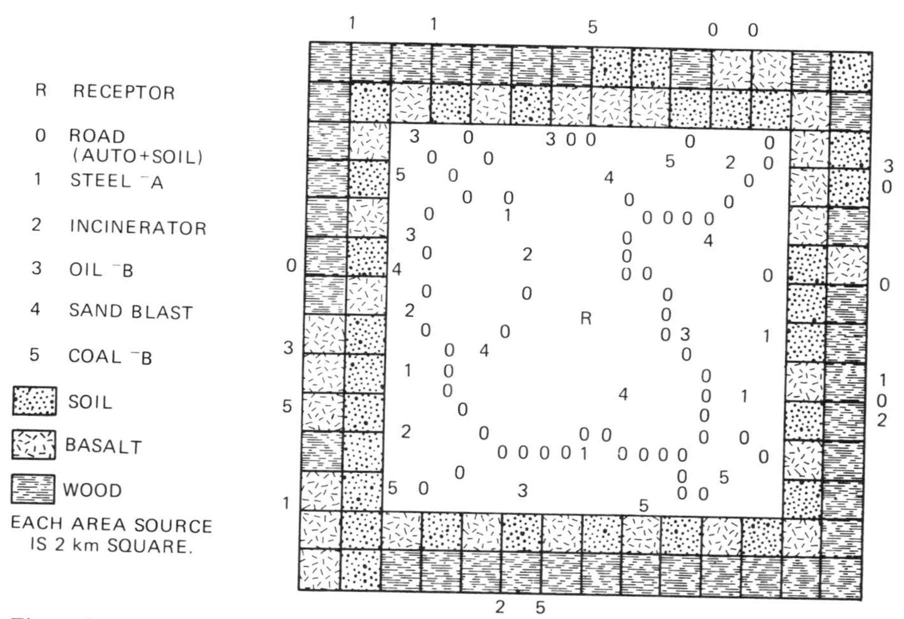

Figure 6.7. City plan used by N.B.S. to create artificial aerosol data set number 1. Taken from Gerlach, Currie, and Lewis (1983).

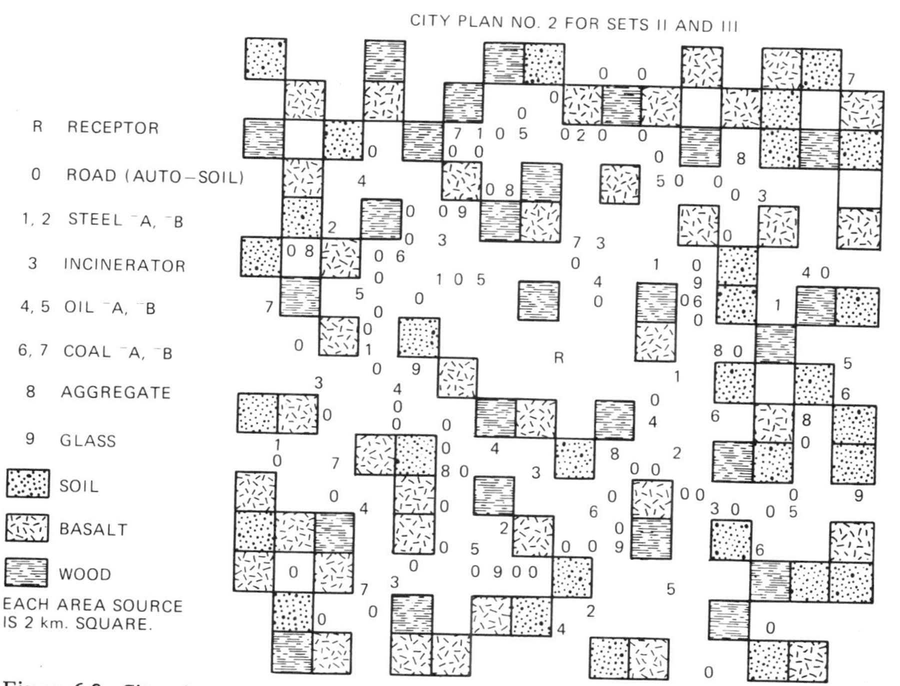

Figure 6.8. City plan used by N.B.S. to create artificial data sets number 2 and 3. Taken from Gerlach, Currie, and Lewis (1983).

Table 6.6. Structure of the Simulated Data Sets

Generating Equation

$$C_i^{(k)} = [A - e_m - e_H^{(k)}]_{ij} S_j^{(k)} + e_i^{(k)}$$

where k = sampling period $[1 \leqslant k \leqslant 40]$

$C_i^{(k)}$ = "observed" concentration of species i, period k $[1 \leqslant i \leqslant N, N \leqslant 20]$

$S_j^{(k)}$ = true intensity (at receptor) of source j $[1 \leqslant j \leqslant P, P \leqslant 13]$

A_{ij} = "observed" source profile matrix (element i, j)

e_i = random measurement errors, independent and normally distributed

e_m = systematic source profile errors, independent and normally distributed (systematic, because fixed over the 40 sampling periods)

e_H = random source profile variation errors, independent and log-normally distributed

Data Set Characteristics

[P = no. of active sources; P_0 = no. in the world list = 13]

Set I : $P = 9$ (including one unknown source)[a]; e_i, e_m; city plan 1 (Fig. 1)

Set II : $P = P_0$ (all known); errors = e_i, e_m; city plan 2 (Fig. 2)

Set III : $P = P_0$ (all known); errors = e_i, e_m, e_H; city plant 2

[a]For data set I, participants were told only that $P \leqslant P_0$.

least-squares problem will be unstable. Very small changes in the elements of this matrix result in large changes in the calculated mass values. The least-squares procedure also has a tendency with two similar sources to make one mass value large and positive and the other large and negative so the resultant is the difference between large values. However, in a mass balance, source contributions should only be positive. It is possible to use a constrained least-squares fit, but this approach has not yet been seriously explored.

An alternative approach that has been suggested to solve the collinearity problem is ridge regression (Williamson, Balfour, and Schmidt, 1983). In this approach, the instability caused by $(A'A)^{-1}$ is smoothed by adding a constant k to each of the elements along the principal diagonal of this matrix. The behavior of the regression results as a function of the value of this smoothing parameter is examined and the best value is chosen. The addition of the smoothing parameter does,

Table 6.7. **Estimated Source Impacts ($\mu g/m^3$) Compared with True Values for Set II[a]**

Steel A	N	CMB (EV-1)[b]	CMB (EV-1)[c]	CMB (RR)[c]	CMB (RR)[c, d]	CMB (EV-2)[b, d, e]	Truth
Steel A	(19)	0.58 ± 0.033	0.97 ± 0.14	0.29 ± 0.73	0.66 ± 2.4	0.55 ± 0.11	0.42
Steel B	(13)	0.24 ± 0.017	–	0.31 ± 0.49	0.26 ± 1.6	0.22 ± 0.08	0.29
Oil A	(17)	0.44 ± 0.033	0.47 ± 0.12	0.11 ± 0.66	0.45 ± 0.56	0.47 ± 0.10	0.56
Oil B	(11)	1.2 ± 0.13	1.7 ± 1.1	2.3 ± 3.3	0.11 ± 2.7	0.72 ± 0.25	0.41
Incinerator	(27)	1.5 ± 0.05	1.5 ± 0.1	1.3 ± 0.84	1.5 ± 0.29	1.5 ± 0.43	1.81
Glass	(3)	0.16 ± 0.03	–	0.15 ± 1.3	0.31 ± 1.1	0.28 ± 0.11	0.46
Coal B	(21)	3.7 ± 0.17	3.8 ± 1.1	4.3 ± 3.1	4.9 ± 2.3	3.8 ± 0.91	3.3
Coal A	(11)	4.3 ± 0.3	6.2 ± 1.3	3.0 ± 4.9	4.6 ± 4.7	5.3 ± 1.40	3.7
Aggregate	(11)	2.1 ± 0.3	–	7.3 ± 4.0	2.5 ± 6.9	2.1 ± 0.57	0.46
Basalt	(24)	8.2 ± 0.5	11.2 ± 1.7	9.0 ± 3.4	8.2 ± 6.1	9.5 ± 1.28	11.3
Soil	(33)	9.5 ± 0.4	7.8 ± 1.4	6.3 ± 2.1	8.6 ± 3.5	8.4 ± 1.26	7.9
Auto	(40)	2.5 ± 0.05	2.5 ± 0.1	2.3 ± 0.78	2.5 ± 0.2	2.5 ± 0.31	2.5
Wood	(11)	0.5 ± 0.1	0.6 ± 0.5	1.1 ± 0.38	1.0 ± 0.54	0.91 ± 0.16	0.94
Total		34.92	36.74	37.76	35.59	36.25	34.05

[a] Uncertainties represent approximately one standard division.
[b] Average of CMB results for each observation period (40 total).
[c] CMB result for the average of the observation periods.
[d] Results submitted after completion of workshop.
[e] Average intercept = 0.52 $\mu g/m^3$.

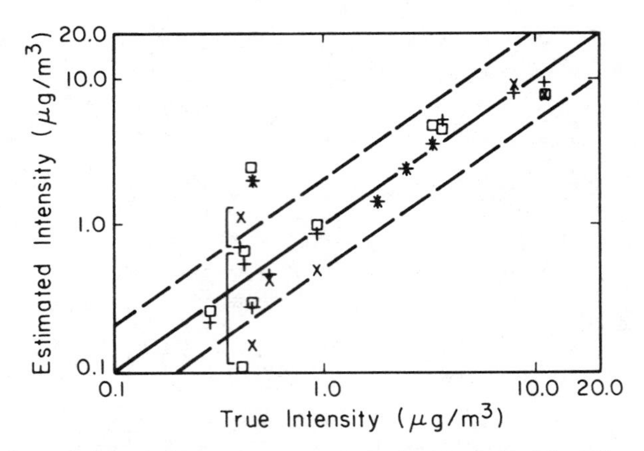

Figure 6.9. Plot of estimated average source strengths for each of the 13 sources in set II using chemical mass balance (EV-1, EV-2) and ridge regression. Dashed lines indicate region within a factor of two of the truth. [X = CMB (EV-1), += CMB (EV-2), ⊡ = CMB (RR)]. Taken from Gerlach, Currie, and Lewis (1982).

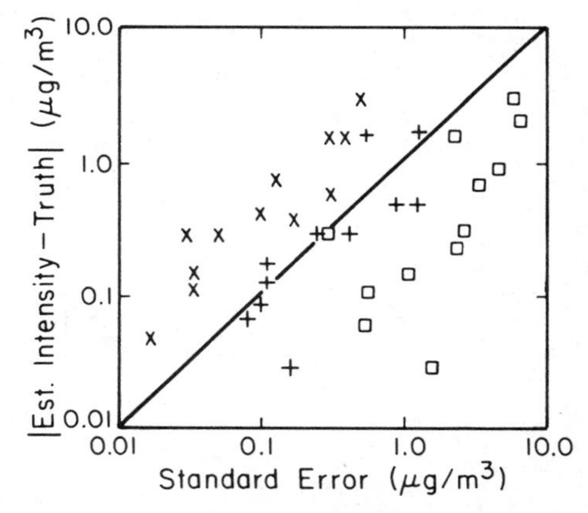

Figure 6.10. Actual (absolute) errors for each of the 13 × 3 estimates given in Figure 6.9, plotted as a function of the stated standard errors. (On the average, one would expect about two-thirds of the points for any correct method to lie below the diagonal.) Taken from Gerlach, Currie, and Lewis (1983).

of course, introduce a systematic error into the results. Several criteria are available to aid in making this choice so as to give both a small uncertainty and a small bias to the results. However, in the analysis of the simulated data set 2, the ridge regression yielded results with relatively large uncertainties and uncertainties that were large compared to the true error. It should also be noted that for the second set of ridge regression results reported in Table 6.7, the best value of the smoothing parameter proved to be zero so that the ridge regression corresponds to an effective-variance weighted least-square fit.

A more serious criticism of ridge regression has recently been presented by Henry (1983). In this work, Henry has calculated the source matrix that would give this equivalent result to the ridge solution

$$(A'A - kI)^{-1} A' = (B'B)^{-1} B' \tag{6.15}$$

where k is the smoothing parameter and I is an identity matrix of the same rank as $A'A$. He has examined the analysis of samples from Portland, Oregon using seven sources given originally by Watson (1979) and presented in Table 6.8. Using a smoothing factor of 0.1, Henry was able to obtain standard errors in the mass contributions of about 1.2 $\mu g/m^3$ or less. He was then able to obtain the set of source profiles that would give the same results with an ordinary least-squares fitting approach. These profiles are given in Table 6.9. By comparison of the original profiles in Table 6.8 with the modified profiles in Table 6.9, the changes caused by the ridge regression smoothing procedure become clear. The reason the ridge approach yields a more stable solution is that it has modified the problem to the extent that it becomes physically questionable since particle sources do not emit negative concentrations of elements. Particularly note that "urban" and "continental" dusts have quite similar compositions initially but "urban" dust has been significantly modified by the ridge regression approach. These results raise quite serious questions regarding the value of the ridge regression approach in mass balance modeling.

In the same report, Henry (1983) also presents an approach to determine the fundamental limits to source resolution using least-squares methods. The singular value decomposition method described in Chapter 5 is used to determine the number of estimable sources form the chosen matrix of source profiles by determining the true rank of the source matrix. As part of any CMB analysis, the number and nature of the sources must still be known. The point is that there are limits to how far sources can be resolved even with perfect information as to the number and composition of contributing sources.

As described above, the least-squares estimate of the mass contributions can be written as

$$F = (A'WA)^{-1} A'WX \tag{6.16}$$

The weight matrix W can be rewritten as the major product of a matrix G containing the inverse of the uncertainties (not the variances) in the elemental concentrations:

$$W = G'G \tag{6.17}$$

Table 6.8. Source Composition Matrix for Singular Value Decomposition Example.[a]

Element/Source	Marine (sea spray)	Leaded Auto Exhaust	Residual Oil Combustion	Ferro manganese Production	Vegetative Burning	Urban Dust	Continental Dust	Assumed Error ($\mu g/m^3$)
Nonvolatile C	0.0	3.8	3.1	1.5	3.5	1.85	0.59	0.5
Mg	4.8	0.0	0.0	0.0	0.0	1.3	1.76	0.05
Al	0.0	1.1	0.53	0.64	1.44	8.84	11.7	0.3
Si	0.0	0.82	0.96	0.99	0.89	22.3	25.4	0.7
Cl	40.0	3.0	0.0	0.42	5.55	0.0	0.0	0.3
K	1.4	0.072	0.28	10.5	0.6	1.03	1.0	0.15
Ca	1.4	1.25	1.56	1.3	1.07	2.44	0.93	0.10
Ti	0.0	0.0	0.11	0.046	0.0	0.66	0.76	0.02
V	0.0	0.0	3.44	0.024	0.0	0.023	0.025	0.02
Cr	0.0	0.0	0.047	0.042	0.0	0.045	0.03	0.01
Mn	0.0	0.0	0.046	17.3	0.12	0.123	0.2	0.20
Fe	0.0	2.1	2.97	2.1	0.19	6.0	6.8	0.25
Ni	0.0	0.018	5.36	0.0	0.0	0.0093	0.0002	0.025
Zn	0.0	0.35	0.4	0.58	0.0	0.11	0.041	0.01
Br	0.2	5.0	0.013	0.16	0.055	0.02	0.0	0.05
Pb	0.0	20.0	0.11	0.045	0.0	0.37	0.006	0.20

[a]Values are given as the percent mass fraction.

Table 6.9. Source Composition Matrix Equivalent to Ridge Regression Solution of Seven-Source Matrix (Table 6.8) with a Smoothing Parameter of 0.1.[a]

Element/Source	Marine	Leaded Auto	Residual Oil	Ferro-manganese	Vegetative Burning	Urban Dust	Continental Dust
Nonvolatile C	-1.51	3.9	3.13	1.35	18.63	-2.53	4.33
Mg	5.15	-0.01	-0.01	0.0	-2.07	1.93	1.42
Al	-1.35	1.25	0.55	0.62	12.11	-0.16	21.25
Si	-0.75	0.73	0.9	0.61	3.03	23.78	28.47
Cl	40.73	3.03	0.02	0.35	14.09	-3.92	2.4
K	1.25	0.01	0.26	11.19	2.35	-0.64	2.41
Ca	1.43	1.15	1.54	1.15	1.37	7.17	-2.95
Ti	-0.02	-0.01	0.1	0.03	-0.06	0.74	0.81
V	-0.01	-0.01	3.48	0.02	0.03	-0.07	0.09
Cr	0.0	-0.01	0.04	0.04	-0.06	0.13	-0.04
Mn	-0.08	-0.07	0.02	18.51	0.57	-2.48	2.2
Fe	-0.18	2.16	2.98	2.11	0.44	6.2	7.75
Ni	-0.01	0.01	5.43	-0.01	0.03	-0.1	0.07
Zn	0.02	0.35	0.4	0.6	-0.22	0.34	-0.17
Br	0.17	5.24	0.01	0.14	0.26	-0.49	0.4
Pb	0.01	20.96	0.09	-0.08	-0.73	0.14	0.16

[a]Values are given as a percentage of total mass.

Equation (6.16) becomes

$$F = (A'G'GA)^{-1} A'G'GX \tag{6.18}$$

so by defining

$$B = GA$$

$$C = GX$$

the equation becomes

$$F = (B'B)^{-1} B'C \tag{6.19}$$

Using the singular value decomposition theorem, this weighted source profile matrix can be written as

$$B = UDV' \tag{6.20}$$

Table 6.10 V Matrix from Singular Value Decomposition of Source Composition Matrix in Table 6.8[a] (Values × 10,000)

Source/Eigenvector	1	2	3	4	5	6	7
Marine	105	8937	3634	1066	2210	719	613
Leaded auto	390	3057	-8709	3823	56	216	-104
Residual oil	9973	-394	478	384	145	37	-32
Ferromanganese	522	1664	-3254	-8861	2779	371	-102
Vegetative burn	44	952	162	-97	174	-8083	-5806
Urban dust	276	1763	-332	-1691	-6081	-4340	6168
Continental dust	210	1964	86	-1654	-7098	3891	-5278
Eigenvalue	7.795	2.977	2.175	1.525	0.941	0.0362	0.00935
Singular value	2.792	1.725	1.475	1.235	0.970	0.190	0.0967
Standard error in $\mu g/m^3$ (inverse of singular value)	0.358	0.580	0.678	0.810	1.031	5.26	10.34

[a]Columns are eigenvectors of the weighted matrix A^tWA, where the weights are the inverse squares of the uncertainties given in Table 6.8.

where B is the m element by n weighted sample composition matrix, U is an m by m orthogonal matrix, V is an n by n orthogonal matrix, and D is an m by n diagonal matrix. The columns of the U matrix are the eigenvectors of BB', the columns of V are the eigenvectors of $B'B$, and the diagonal elements of D are the square roots of the eigenvalues of $B'B$. The key to the least-squares fit is the inverse of the covariance matrix, $(B'B)^{-1}$. Then using (6.20), $(B'B)^{-1}$ can be written as

$$\begin{aligned}(B'B)^{-1} &= (VDU'UDV')^{-1} \\ &= V\ [d_i^{-2}]\ V' \end{aligned} \tag{6.21}$$

Thus small singular values, d_i, lead to large uncertainties in the regression results. Using the same sources listed in Table 6.8, the V matrix of the singular value decomposition can be calculated. It is presented in Table 6.10. This table also gives the singular values that are the diagonal elements of the D matrix in equation (6.20).

A question comes as to how small is small regarding the size of the singular values. An approach to identifying appropriately small values and which columns are interrelated has been outlined by Belsley, Kuh, and Welsch (1980). A condition index, $\eta(d)$, can be defined as the ratio of singular values

$$\eta(d) = \frac{d_{\max}}{d_k} \geqslant 1, \qquad k = 1, n \tag{6.22}$$

If columns are orthonormal, $\eta(d) = 1$. Weak dependencies are associated with condition indexes around 5 or 10 whereas moderate to strong dependencies are associated with condition indexes of 30 to 100.

The variance-covariance matrix of the least-squares estimator of **f** is $\sigma^2\ (B'B)$ where σ^2 is the common variance of the components of e in the model

$$\mathbf{x} = B\mathbf{f} + e \tag{6.23}$$

Thus, the variance-covariance matrix of **f**, $V(\mathbf{f})$, is

$$V(\mathbf{f}) = \sigma^2\ (B'B)^{-1} = \sigma^2\ VD^{-2}V' \tag{6.24}$$

where the V matrix has been obtained as described above. The variance in the kth component of **f** is then

$$\text{var}(f_k) = \sigma^2 \sum_{j=1}^{n} \frac{v_{kj}^2}{d_j^2} \tag{6.25}$$

where the d_j values are the singular values and v_{ij} are the elements of the V matrix.

It is then possible to examine the proportion of the variance of coefficients (two or more) concentrated in components with the same small singular value. The proportion of variance of the kth regression coefficient, $f(k)$, associated with the jth component of the decomposition will be called the variance-decomposition proportion. First, the fractions of variance are calculated

$$\phi_{kj} = \frac{v_{kj}^2}{d_j^2}, \qquad j, k = 1, \ldots, n \tag{6.26}$$

and for each row,

$$\phi_k = \sum_{j=1}^{n} \phi_{kj}, \qquad k = 1, n \tag{6.27}$$

The variance-decomposition proportions are then

$$\pi_{jk} = \frac{\phi_{kj}}{\phi_k}, \qquad j, k = 1, \ldots, n \tag{6.28}$$

Thus, a table of π values can be calculated. The collinearity problem can be identified as when (1) there is a singular value with a high condition index >5 and

Table 6.11. Condition Indexes and Variance-Decomposition Propositions for the Source Matrix in Table 6.8 Without Column Scaling

		Matrix						
Singular Values	Condition Index	var $f(1)$	var $f(2)$	var $f(3)$	var $f(4)$	var $f(5)$	var $f(6)$	var $f(7)$
279.2	1.00	0.000	0.000	0.968	0.005	0.000	0.000	0.000
172.1	1.62	0.287	0.063	0.004	0.013	0.000	0.000	0.000
147.5	1.89	0.065	0.697	0.008	0.069	0.000	0.000	0.000
123.5	2.26	0.080	0.192	0.007	0.117	0.000	0.000	0.001
97.5	2.88	0.056	0.000	0.002	0.117	0.000	0.008	0.016
19.0	14.7	0.153	0.026	0.003	0.054	0.334	0.112	0.121
9.67	28.9	0.431	0.023	0.008	0.016	0.666	0.878	0.862

(2) there are high variance-decomposition proportions (>0.50) for two or more regression coefficients. To illustrate this approach, Table 6.11 presents the singular values, condition indexes, and the variance-decomposition proportions for the source matrix of Table 6.8 divided by the uncertainties. The variance-decomposition proportions and condition indexes indicate potential problems with two of the independent variables. There are condition index values of 14.7 and 28.9. However for the 14.7 values, there are no proportions greater than 0.5. For the 28.9 value, three coefficients, agricultural burning, urban dust, and continental dust, have values greater than 0.5.

Belsley, Kuh, and Welsch (1980) clearly demonstrate the need for column scaling in order to fully separate the weak collinearities. If the source matrix of Table 6.8 is analyzed directly, without dividing by uncertainties but with scaling the columns vectors to unit length, the results presented in Table 6.12 are ob-

Table 6.12. Condition Indexes and Variance-Decomposition Proportions for the Source Matrix in Table 6.8 with Column Scaling

		Matrix						
Singular Values	Condition Index	var $f(1)$	var $f(2)$	var $f(3)$	var $f(4)$	var $f(5)$	var $f(6)$	var $f(7)$
1.56	1.00	0.008	0.016	0.030	0.011	0.017	0.001	0.001
1.29	1.20	0.056	0.017	0.002	0.000	0.030	0.001	0.001
0.367	4.24	0.776	0.013	0.171	0.015	0.776	0.000	0.001
0.060	26.1	0.137	0.034	0.033	0.000	0.171	0.998	0.997
0.997	1.56	0.015	0.083	0.154	0.538	0.002	0.000	0.000
0.974	1.60	0.008	0.517	0.068	0.333	0.002	0.000	0.000
0.905	1.72	0.000	0.321	0.542	0.103	0.003	0.000	0.000

tained. In this case there is one large condition index, 26.1, with high variance-decomposition proportions for urban and continental dust. There is also a value of 4.24 with high proportions for marine and agricultural burning. Since both of these sources are dominated by large values for chlorine this collinearity is quite reasonable and went undetected with the unscaled vectors. Using the vectors divided by the uncertainties and then scaling leads to much smaller condition indexes and variance-decomposition proportions, making the identification of collinearities more difficult.

An addition test can help to identify sources that cannot be estimated accurately. Each row of the *V* matrix given in Table 6.10 corresponds to a given source and can be interpreted as the coordinates of that source in the seven-dimensional space defined by the eigenvectors in *V*. Silvey (1969) has shown that accurate least-squares estimations can only be made in the directions of the eigenvectors corresponding to large eigenvalues. This subset of eigenvectors defines the "estimable space" and only those sources that lie virtually entirely within this space can be determined accurately. Those sources that do not lie within this space may be combined in linear combinations that are within the estimable space.

In Table 6.7, there are two very small eigenvalues so that it appears that the estimable space for this choice of source profiles is five or, in other words, there are two vectors with large condition indexes. By using the inverse of the uncertainties of the elemental concentrations in $\mu g/m^3$, the inverse of the singular values gives the uncertainty that can be associated with that eigenvector in $\mu g/m^3$. Silvey (1969) has shown that any source in the estimable space will have an uncertainty equal to or less than the uncertainty associated with the smallest eigenvalue. With the five-dimensional space, the maximum error in the example in Table 6.7 would be 1.03 $\mu g/m^3$. If, on the other hand, it was necessary to be certain that each source be estimated to be 0.75 $\mu g/m^3$, only three sources could be determined.

Using 1 $\mu g/m^3$ as the acceptable level of error, the length of a source in the estimable space is the square root of the sum of the squares of the first five eigenvector values for that source. For example, the marine source has a length of $[(0.0105)^2 + (0.8937)^2 + (0.3634)^2 + (0.1066)^2 + (0.2210)^2]^{1.2}$ or 0.9955 so it is well within the estimable space. Table 6.13 gives the lengths for each of the seven original sources. Clearly, the first four sources are within the estimable space and can be assessed to within 1 $\mu g/m^3$. However, it is not possible to assess the remaining three samples to the same level of precision. Silvey (1969) points out that it may be possible to develop linear combinations of the inestimable sources that would lie in the estimable space.

In this case, only one additional source can be included in the five-dimensional space. From the variance-proportion analysis of the scaled data, it would seem that the combination of urban and continental dusts provides the best possibility

Table 6.13. Length of Source Vectors in Five-Dimensional Estimable Source

Source	Length in Estimable Space
1. Marine	0.9955
2. Leaded auto	0.9997
3. Residual oil	0.99998
4. Ferromanganese	0.0002
5. Vegetative burn	0.0986
6. Urban dust	0.6567
7. Continental dust	0.7550

for an estimable source. This combination should be developed so it is estimable to 1 $\mu g/m^3$. It is desirable to calculate the combination that has the minimum variance. Henry (1983) has presented a method derived from Silvey (1969) which permits the calculation of such a combination.

If we suppose that there are p estimable sources, the linear combination can be written as

$$l = \sum_{i=1}^{p} x_i f_i \tag{6.29}$$

where f_i is the value of the ith source and x_i are the coefficients to be determined such that l has a minimum variance. Silvey (1969) has proven that if v_i is one of the n columns in the V matrix from equation (6.20) and

$$x = \sum_{i=1}^{n} \alpha_i v_i \tag{6.30}$$

then the variance in $(x'f)$ is given by

$$\text{var}(x'f) = \sum_{j=1}^{n} \left(\frac{\alpha_j}{d_j}\right)^2$$

where f is the weighted least-squares estimate of the source contributions. To illustrate, suppose we let

$$x = \begin{bmatrix} 0 \\ \cdot \\ \cdot \\ \cdot \\ 0 \\ 1 \\ 0 \\ \cdot \\ \cdot \\ \cdot \\ 0 \end{bmatrix} \tag{6.31}$$

where the 1 is in the ith position. Then the product $x'f$ is given by

$$x'f = f_i \tag{6.32}$$

and

$$\begin{aligned} \alpha_j &= x \cdot v_j \\ &= v_{ij} \end{aligned} \tag{6.33}$$

where v_{ij} is the ith component of v_j. The variance of f_i can then be expressed using the singular value decomposition as

$$\text{var}(f_i) = \sum_{i=1}^{n} \left(\frac{v_{ij}}{d_j} \right) \tag{6.34}$$

The variance in the linear combination [equation (6.29)] can also be computed in terms of the V matrix and the singular values. Using f_i as the $n \times 1$ vector of source contribution and x as the $n \times 1$ vector of coefficients to be determined, then equation (6.29) can be rewritten as

$$l = x' f_i \tag{6.35}$$

if the first p rows of f are the inestimable sources and all of the components of x beyond the pth are zero.

The coefficients can then be written as a linear combination of the columns of V

$$x = \sum_{k=1}^{n} \alpha_k v_k \tag{6.36}$$

Taking the minor product of (6.36) with v_j yields

$$\alpha_j = x \cdot v_j = \sum_{i=1}^{n} x_i v_{ij} \tag{6.37}$$

However, since x_i for i greater than p are zero,

$$\alpha_j = \sum_{i=1}^{p} x_i v_{ij} \tag{6.38}$$

and

$$\operatorname{var}(x'f) = \sum_{j=1}^{n} \left[\frac{\sum_{j=1}^{p} x_i v_{ij}}{d_j} \right]^2 \tag{6.39}$$

Define

$$Q = V_p \, [d_i^{-1}] \tag{6.40}$$

where V_p is a $p \times n$ matrix consisting of the first p rows of the V matrix corresponding to the p inestimable source and the d_i values are the corresponding singular values. Using this definition of Q, equation (6.39) can be expressed as

$$\operatorname{var}(x'f) = (Q'x)'(Q'x) = x'QQ'x \tag{6.41}$$

where the vector $\mathbf{x}$ has been truncated to its p nonzero values.

Equation (6.41) is a quadratic form and the value that minimizes this expression is the eigenvector of QQ' corresponding to its smallest eigenvalue. Courant and Hilbert (1953) show that the minimum value is equal to the smallest eigenvalue. Thus, an eigenvector analysis of the QQ' matrix and the use of the eigenvector with the smallest eigenvalue is the starting point for determining the minimum variance linear combination of inestimable sources.

In the specific example of the seven sources in Table 6.8, the least-variance combination of continental and urban dust is given by

$$0.758 \text{ (continental dust)} + 0.652 \text{ (urban dust)}$$

This combination can be estimated with a standard error 0.996 $\mu g/m^3$, whereas the individual sources could only be estimated to 5.9 and 6.8 $\mu g/m^3$, respectively.

After finding this minimum variance combination, it must be tested to be certain that it has sufficient length in the estimable space. It should be almost entirely in the space. If it is not, it is not estimable. The combination given in this example does lie entirely within the estimable space. The estimable combination is a combination of source contributions, f_i's. Ordinary least-square solutions are used to obtain these source contributions from which the estimable combinations are calculated. Thus, the separate estimable sources and the estimable combination can be determined to the desired level of precision by the following stepwise procedure taken from Henry (1983).

1. Calculate the singular values and their associated eigenvectors of the source concentration matrix weighted by realistic errors in the individual emission source concentration values.
2. Select the maximum error M_E that is desired in any source contribution.
3. Determine the dimension of the estimable space by determining the number of singular values d_i such that $1/d_i \leqslant M_E$. In addition, the number of large values (>5) of the condition index also suggest possible collinearities.
4. Calculate the variance decomposition proportions. The combination of a high condition index and two or more proportions >0.5 indicates a collinear condition.
5. Calculate the length of each source vector in the estimable space. For those sources for which their length is greater than 0.95, the source contributions can be calculated to an accuracy of at least M_E.
6. Choose the inestimable source for which linear combinations are required. Calculate the coefficients of the combinations with minimum variance $\leqslant M_E$ using the procedures outlined above.
7. Calculate the lengths of the linear combinations within the estimable space to ensure that they are $\geqslant 0.95$.

This step-by-step procedure allows the identification of the maximum number of sources or source types than can be accurately determined for a given source composition matrix, identification of those sources that can be separately quantified, and calculation of the linear combinations of inestimable sources that can be determined within the desired degree of accuracy.

OTHER APPLICATIONS OF THE MASS BALANCE APPROACH

The mass balance approach can be extended to other problems including reactive chemical species and light extinction. These extensions were suggested by

Friedlander (1981). In the paper by Miller, Friedlander, and Hidy (1972), the mass balance equation included a factor to account for changes in the composition of the emitted material during transport from the source to the receptor site. The mass balance equation is rewritten as

$$x_i = \sum_{j=1}^{p} \alpha_{ij} f_j a_{ji}, \qquad i = 1, m \tag{6.42}$$

where α_{ij} is the fractionation coefficient. The fractionation coefficient accounts for the transformation (loss or gain) of the particular species while in transit. If there is an unreactive species emitted with the reactive materials, the equation can be modified by multiplying each term in the summation in equation (6.42) by a_{jk}/a_{jk} where a_{jk} is the contribution of the kth unreactive species in material from the jth source:

$$x_i = \sum_{j=1}^{p} \alpha_{ij} f_j \frac{a_{ji}}{a_{ji}} a_{jk}$$

$$x_i = \sum_{j} \alpha_{ji} r_{jik} m_{jk} \tag{6.43}$$

where r_{ik} is the ratio of the concentration of the reactive species i to the concentration of unreactive species k in material from the jth source. It is important to recognize that there are three different periods for which reactive decay can occur; in the transport from source to receptor, in the sampling device on the filter or the impactor stage, and in the period between sample collection and laboratory analysis. The decay observed is the accumulation of the reactions that occur over these intervals.

Friedlander (1981) then provides a review of an approach that permits the atmospheric decay constants to laboratory reaction rate studies. Gordon and Bryan (1973) and Gordon (1976) have made detailed measurements of the atmospheric concentrations of 14 polycyclic aromatic hydrocarbons (PAH) at various locations in Los Angeles. Duval (1980) has used these data to resolve automotive and refinery emissions. He assumed that the PAH concentrations at a downtown Los Angeles site were dominated by automobiles and determined concentrations relative to lead. At a site near a refinery, he used lead as the tracer to remove the automotive component and determine the refinery component. He was able to estimate decay constants using emission data of Grimmer and Hildebrant (1975).

Another area of active effort has been in the use of mass balance methods to apportion sources of visibility degradation. The detailed development of the relationships between light extinction and the properties of the ambient aerosol are given by Friedlander (1977).

The light coming from an object to an observer is affected by absorption and scattering by the gas molecules and absorption and scattering by the particles or water droplets in the air. The total extinction coefficient is then the sum of these independent contributions

$$b_{\text{ext}} = b_{\text{ap}} + b_{\text{ag}} + b_{\text{sp}} + b_{\text{sw}} + b_{\text{sg}} \tag{6.44}$$

where b_{ext} is the total extinction coefficient, b_{ap} is the adsorption by particles, b_{ag} is the absorption by gases, b_{sp} is the scattering by dry particles, b_{sw} is the scattering by water associated with particles, and b_{sg} is the scattering by gas molecules. The Raleigh scattering by gases, b_{sg}, can be considered a constant. Thus, the remaining extinction coefficient, b_{ext}, consists of the effects of particles and their associated water on scattering and absorption. By using both heated and unheated nephelometers (Groblicki, Wolff, and Countess, 1981), the scattering from the associated water can be removed. Alternatively, if only unheated measurements and relative humidity are available, calculational procedures such as given by Cass (1979) or White and Roberts (1977) must be employed. Thus, the parameters to be apportioned are the scattering from particles after the effects of moisture have been eliminated, b_{sp}, and the absorption of light by particles, b_{ap}.

Beginning with Charlson and coworkers (1969) and Noll, Mueller, and Imada (1968), there have been a number of studies that have examined the relationship between visibility and airborne particle mass. When the relationships are examined with respect to fine particle mass, very precise proportionality is found between fine particle mass concentration and light scattering coefficient (Waggoner and Weiss, 1980; Heisler et al., 1980; Lewis, 1981; Groblicki, Wolff, and Countess, 1981). Since the fine particle mass is well correlated with light scattering, it seems reasonable to be able to apportion sources of light scattering material.

The absorption of light by particles is due almost entirely to elemental carbon in the fine particles (Heisler, Henry, and Watson, 1980). Therefore, apportioning the fine particle carbon to its sources will permit the apportionment of the absorption coefficient:

$$b_{\text{ap},i} = b_{\text{ap}} \left[\frac{EC_i}{EC_T} \right] \tag{6.45}$$

where $b_{\text{ap},i}$ is the absorption coefficient for particles from the ith source, b_{ap} is the total measured adsorption coefficient, EC_i is the concentration of elemental carbon attributed to the ith source, and EC_T is the total measured elemental carbon concentration.

The relationships between the amount of light scattering represented by b_{sp} and the aerosol composition and source contributions have been investigated using regression analysis for Denver, Colorado (Heisler, Henry, and Watson, 1980), Portland, Oregon (Shah et al., 1981a), and Houston, Texas (Dzubay et al., 1982). In these analyses, the relationship has been given as

$$b_{sp} = \sum_{i=1}^{m} \beta_i C_i \tag{6.46}$$

where β_i is the regression coefficient to be determined and C_i is the measured concentration of species i in the ambient aerosol. In order to remove some of the variance simply caused by the total mass of airborne particles, both sides of equation (6.45) can be divided by the measured total fine particle mass concentration, M_f,

$$\frac{b_{sp}}{M_f} = \sum_{i=1}^{m} \beta_i \frac{C_i}{M_f} \tag{6.47}$$

This analysis can help to identify which species are important to light scattering. However, it is desirable to determine the contribution of particle sources to the light scattering.

One approach to this problem is to assume that scattering per unit mass of suspended fine particles is the same for every kind of source of fine particles. The scattering from a particular source can then be obtained by

$$b_{sp,i} = b_{sp} \left[\frac{M_{f,i}}{M_f} \right] \tag{6.48}$$

where b_{sp} is the measured particle scattering coefficient, and $M_{f,i}$ is the fine mass concentration attributable to the ith sources obtained from the CMB analysis so that $b_{sp,i}$ is the scattering coefficient attributable to the ith source. Since the scattering depends on the size distribution and the refractive index, this assumption implies that all sources emit equally effective scattering particles. It is known that these properties do vary from source to source.

In order to try to examine the variation in scattering efficiency by materials from different sources, b_{sp} can be regressed against the mass contributions determined by the CMB analysis [equation (6.4)]. This approach has been used by Shah et al., (1981a). These approaches to apportioning light scattering have a major limitation in that much of the fine particle mass that is responsible for visibility reduction is due to secondary formation processes and is not emitted directly by sources. Thus, the light scattering can be attributed to sulfate, nitrate, and carbon species but not to specific sources. In Portland (Shah et al., 1981a), only agricultural burning was found to have a statistically significant effect on b_{sp}. In Denver (Heisler, Henry, and Watson, 1980), 45.6% of the measured extinction coefficient could not be assigned to known sources. Thus, more work is necessary to begin to further refine the relationships between visibility degrading species and source emissions.

PARTICLE CLASS BALANCE

As introduced in Chapter 4, the results of the scanning electron microscopic examination of a collection of particles can be expressed as the number of particles in each of the empirically defined categories. These particle/category values are then the new variables in the system to be apportioned. There are several variations in how these variables are used in a mass balance. In this chapter, it is assumed that authentic source materials can be obtained for the sources to be included in the mass balance.

Johnson and Twist (1983) describe an approach in which the source profiles are defined and the program obtained the best least-squares fit over all of the particles observed. Johnson and McIntyre (1983) report the application of a particle class balance to samples from Syracuse, New York. Twelve source types from 47 possible source signatures could be fit to the 23 ambient samples measured. The results obtained were considered to be reasonable based on the known source inventory in the Syracuse area although there is no direct way to provide verification.

A more detailed particle balance model has been suggested by Casuccio and Janocko (1981; Energy Technology Consultants, 1983). In this model, the source profiles and ambient samples are segregated into aerodynamic size categories and the least-squares fitting is made within size class. This model can be expressed as

$$M = \sum_{j=1}^{n} m_j \tag{6.49}$$

where M is the total mass of the sample, m_j is the mass in the jth aerodynamic size class, and n is the number of aerodynamic size classes.

$$m_j = \sum_{k=1}^{p} x_{kj} \tag{6.50}$$

where p is the number of sources and x_{kj} is the contribution of the kth source to the jth size class. The individual source contributions can then be expressed as

$$x_{kj} = \sum_{i=1}^{m} Q_i^{kj} = \sum_{i=1}^{m} \phi_t^j \frac{\beta_i^{kj}}{\xi_t^{ji}} \tag{6.51}$$

where m is the number of particle class types, Q_i^{kj} is the calculated contribution of particle type i from source k in size class j such that ϕ_t^j is the weight percent of a tracer particle type t in size class u, β_i^{kj} is the measured weight percent of particle type i in size range j from source k, and ξ_t^{ji} is the measured weight percent of tracer particle type t in size range j from source k. For each source for which authentic material is available, the value of $a_{ki}(t, j) = \beta_i^{kj} / \xi_t^{ji}$ is measured for each of the size categories. Thus for the mass in a given size range

$$m_j = \sum_{k=1}^{p} \sum_{i=1}^{m} \phi_t^j \frac{\beta_i^{kj}}{\xi_t^{ji}} = \sum_{k=1}^{p} \sum_{i=1}^{m} \phi_t^j a_{ki}(t, j) \tag{6.52}$$

and the total predicted mass is then

$$M = \sum_{j=1}^{n} \sum_{k=1}^{p} \sum_{i=1}^{m} \phi_t^j a_{ki}(t, j) \tag{6.53}$$

A chi-squared test is used to determine the best fit from the results for each tracer so that essentially the measured distribution is compared with the calculated distribution from the sum of source values. In this way, the choice of sources to be included is made and groupings of individual sources are obtained.

The calculation to determine the source contributions is then made by a least-squares solution to a set of simultaneous equations. The method attempts to find a combination of sources that minimizes the sum of squares of the differences between the calculated and measured particle mass concentrations. The source profiles are analyzed in this manner until a calculated distribution is obtained that approximates the measured distribution. The sources that contribute to the "best fit" solution are then presented with the contribution from each particle type reported for each source.

As an example, the use of particle class balance techniques to examine sources of airborne particles and lead in El Paso, Texas (Dattner et al., 1983) will be discussed. In El Paso, an air quality problem exists as shown by an annual geometric mean value of total suspended particle (TSP) concentration and quarterly lead concentrations significantly above the corresponding ambient air quality standards. In an effort to identify the sources of TSP and lead, the Texas Air Control

Board initiated a study whose primary approach utilized computer-controlled scanning electron microscopy. Both ambient filter samples and a large number of suspended source samples were taken in the El Paso area, particularly in the vicinity of a large smelter operation that was thought to be a major contributor to the observed TSP and lead levels. From the microscopic analysis and subsequent mass balance calculation, TSP and lead were apportioned as shown in Figure 6.10. These results are in quite reasonable agreement with two other apportionment studies (Radian, 1983; Trijonis, 1982) done for the same period primarily by examining the differences in lead levels with a period in 1980 when the smelter was closed because of a strike. They also agreed reasonably well with the tracer apportionment model developed and used by the Texas Air Control Board (Dattner et al., 1983) and presented in Chapter 8. Thus, the microscopic methods have compared well with some of the receptor models based on the concentrations measured in the collected particles.

Casuccio, Janocko, and Lucas (1982) have used this same approach in studies of the impact of a coking and iron production facility and to compare the receptor model results with a dispersion model. They examined a detailed emissions inventory and used the Industrial Source Complex (ISC) model (Environmental Protection Agency, 1979) to predict the impacts. This model uses a steady-rate Gaussian plume algorithm that takes into account particle settling and dry deposition. However, the user must input the particle size distributions and settling velocities for each major particle source. Thus, the use of CCSEM to assess the true particle size distribution and assist in better estimating the settling velocities should greatly improve the model prediction accuracy.

Since receptor models determine source impacts only for single ambient events, some attempt must be made to estimate the annual average contribution from the "short term" results, if they are to be compared with the annual results obtained with the ISC model. In previous studies (Casuccio and Janocko, 1981; Lucas, 1979), the annual average contribution to specific receptors was calculated by sampling when winds from the source(s) of interest to the downwind monitor were persistent. A series of such tests were conducted, from which an average contribution of each source was determined. This average contribution, obtained during persistent wind directions, was converted to an annual average by multiplying the average contributions by the percentage of time the wind was expected to be from the source(s) to the downwind monitor.

For example, in this study, an estimation of the total annual plant TSP contribution at a particular monitor was computed by relating receptor model results to a 5-year (1974-1978) wind rose obtained from meteorological data recorded at a nearby airport. For the plant to impact this monitor, the wind direction would have to be between 210° and 240°. Be relaing the receptor model results to the amount of time the wind is in this range (5-year wind rose to be approximately 26% during the year), it was determined that annual plant TSP contribu-

Table 6.14. Comparison of Dispersion Model (ISC) and CCSEM-Particle Balance Receptor Model for Annual Average TSP Contribution at Ambient Monitoring Site from a Coke and Iron Complex[a]

	Predicted Impact ($\mu g/m^3$)	
Source	Dispersion Model	Receptor Model
Blast furnace	<1	1
Plant roads	11	15
Other plant sources	7	2
Total	18	18

[a]From Casuccio, Lucas, and Janocko, 1982.

tion at the monitor was about 18 $\mu g/m^3$. A comparison of the ISC dispersion model and results to the receptor model estimated annual results at the monitor is presented in Table 6.14. It is likely that in the near term future microscopic studies will increase in their importance and be more common as a method to examine the impacts of particle sources.

SUMMARY

It can be seen from the examples presented that a mass balance approach can be a powerful method in resolving source contributions. It has been the most extensively used method. Mass balance methods have been developed to a point where a computer code developed under contract to the U.S. Environmental Protection Agency (USEPA) is available on the same basis as dispersion models (Williamson and DuBose, 1983). The program performs ordinary weighted and effective variance weighted least-squares regression as well as ridge regression. There are increasingly available libraries of source profiles. The appendix of this volume gives a partial compilation of the available data. A more complete compendium is currently being prepared under contract to USEPA. Thus, chemical mass balance can be used effectively in receptor studies. However, it is highly recommended that some of the regression diagnostic methods outlined in this chapter be employed to ensure that unreliable estimates are not obtained from sets of input profiles that have substantial collinearity problems.

CHAPTER

7

FACTOR ANALYSIS

Factor analysis is a collection of related mathematical techniques developed to examine the underlying structure that exists in large data sets. It is assumed that although many different parameters describing a system can be determined, such as elemental composition or particle size, there are really very much fewer causal forces that result in the state of the system as it is observed. It is this effort to identify these fewer causal forces and interpret their nature that factor analysis attempts to achieve.

For example, let us assume that we are collecting samples of airborne particles and that there are actually only two sources of particles; an iron works and automotive traffic. Each sample will be analyzed for iron, lead, and bromine. We will further assume that the iron works emits no lead or bromine and that automobiles are not responsible for any airborne iron. Each sample can be represented by a point in a three-dimensional space with axes representing the iron, bromine, and lead concentrations. In Figure 7.1 such data are presented assuming that the Br/Pb ratio from automobiles is 0.39 and using a random number generator to calculate a number of specific lead and iron values. As plotted, it is difficult to discern a pattern to the points. However, if the axes are rotated and replotted as in Figure 7.2, it can be seen more clearly that all of the points lie in a plane defined by the Fe axis and a line interrelating the lead and bromine. (Br = 0.39 Pb.) The objectives of factor analysis are then to identify the true dimensionality of the problem, the measures of interrelationship between the variables, and the importance of each of these actual causal factors to the observed parameter values for each sample.

Factor analysis was originally developed by psychologists to use the results of various tests to identify the nature of human intelligence. It has found wide use in the social sciences and education fields for interpreting large quantities of data regarding human behavior. It also was one of the first methods used to interpret the large quantities being obtained by the National Air Sampling Network (NASN) (Blifford and Meeker, 1967). In the last decade, there has been expanding interest in the development and use of statistical methods of interpreting environmental data since analytical methods have been developed that allow the cost-effective collection of very large and complex data sets. This chapter presents several forms

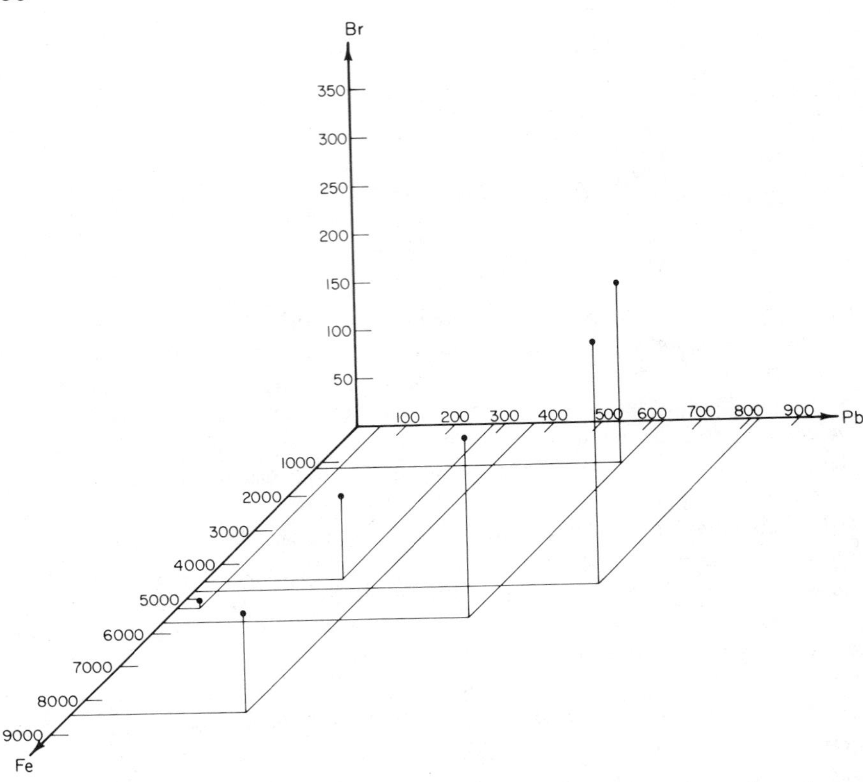

Figure 7.1. Artificial aerosol composition data plotted for automobile and steel plant emissions.

of factor analysis including principal components analysis, classical factor analysis, and target transformation factor analysis.

PRINCIPAL COMPONENTS AND CLASSICAL FACTOR ANALYSIS

Methodology

A common initial approach to interpreting data regarding an environmental system is to examine the correlation coefficients between the variables as defined in equation (5.55). This result will be useful when it is certain that the system under study is described by only two variables; one dependent, one independent. This coefficient has limited meaning when the system is truly multivariate. A high correlation between two variables may really be a manifestation of a strong de-

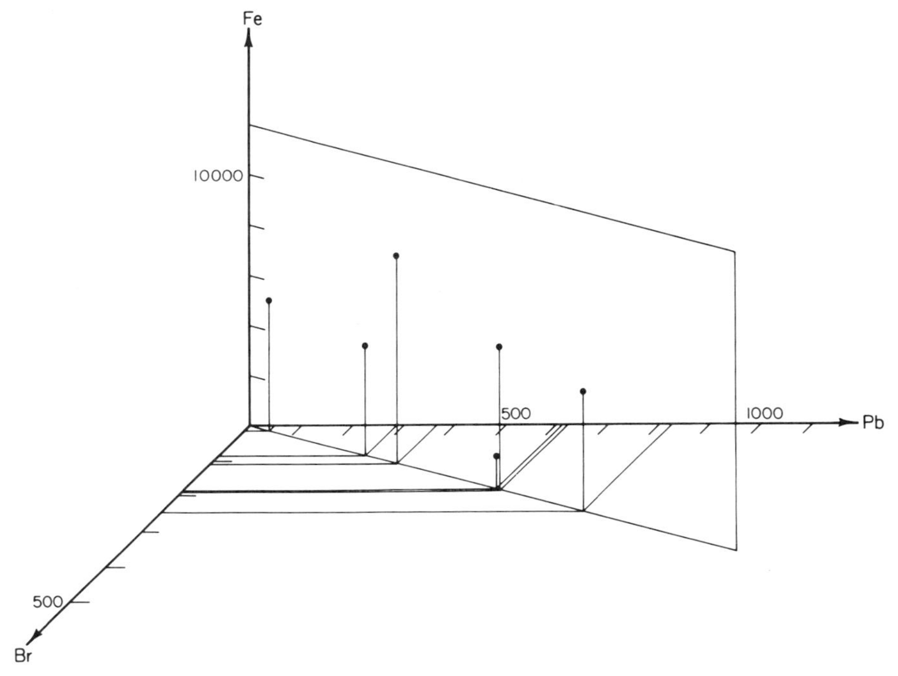

Figure 7.2. Artificial aerosol composition after axes rotation.

pendence of both variables on the same causal factor. In other cases, the two variables may be related but do not show a large pairwise linear correlation coefficient. It is important to remember that high correlation does not imply that a cause and effect relationship exists between these variables.

To examine the underlying relationships between the variables in the system, it is useful to assume that the measured parameters are related to a limited number of unerlying causal factors.

In the classical factor analysis model, the variables are assumed to be linearly related to some number of underlying factors

$$
\begin{aligned}
z_{1j} &= a_{11} f_{1j} + a_{12} f_{2j} + \cdots + a_{1m} f_{mj} + d_1 U_{1j} \\
z_{2j} &= a_{21} f_{1j} + a_{22} f_{2j} + \cdots + a_{2m} f_{mj} + d_2 U_{2j} \\
&\vdots \\
z_{nj} &= a_{n1} f_{1j} + a_{n2} f_{2j} + \cdots + a_{nm} f_{mj} + d_n U_{nj}
\end{aligned}
\tag{7.1}
$$

where the values of the n standarized variables in the system for the jth sample are z_{ij}, the values of the m factors common to all of the variables for the jth sample are f_{kj}, and the value of the factor unique to each variable is U_{ij}. The factors are derived from the matrix of linear correlation coefficients taken pairwise over all of the variables. If there are n variables measured for each sample, an n by n symmetric matrix of correlation coefficients can be calculated. The factor analysis tried to maximally reproduce this matrix. The reproduced correlation coefficient is calculated from the common factor coefficients, a_{ij}'s:

$$r_{ij} = a_{i1}a_{j1} + a_{i2}a_{j2} + \cdots + a_{im}a_{jm} \tag{7.2}$$

The factors are considered to be those fundamental parameters that govern the values of the observed data. They must be interpreted in terms that are meaningful to the data set under analysis. The interpretation of the factors is made by examining the dependence of the variables on the factors as indicated by the a_{ij} coefficients. These coefficients can be considered to be the correlation coefficient between the variable and the factor. Thus by examining which variables are influenced by a given factor, inferences can be made as to its nature.

In factor analysis the variance is apportioned between the common factors and the unique factor:

$$\text{system variance} = \overbrace{\text{common factor variance} + \text{specific variance}}^{\text{true variance}} + \text{error} \tag{7.3}$$

$$\underbrace{\text{specific variance} + \text{error}}_{\text{unique variance}}$$

The common and specific variance represent changes in the system because of variation in the causal factors. The error, however, changes the observed values of the system that are not related to changes in the system. It is clear that this form of data analysis will reflect sampling and measurement error as increased values of the unique factor. If the unique factor is found to be large, careful scrutiny must be made to determine whether this situation arises from a high specific variance or from error. The unique factor can indicate systematic error in the determination of particulate variables. The amount of common factor variance can be calculated from the factor coefficients

$$h_i^2 = a_{i1}^2 + a_{i2}^2 + \cdots + a_{im}^2 \tag{7.4}$$

where h_i^2 is called the communality of the ith variable. The unique variance of a variable is then calculated by subtracting the communality of that variable from unity:

$$d_i^2 = 1 - h_i^2 \tag{7.5}$$

We should distinguish between a classical factor analysis and principal component analysis. These terms are often interchangeable when they refer to somewhat different models. The goal of a components analysis is to maximally reproduce the total system variance or the sum of the squares of the standard deviations over all of the variables. It achieves this goal by setting up a linear model similar to equation (7.1) but *without the unique factor*. As discussed above, the factor analysis attempts to maximally reproduce the matrix of correlation coefficients. The components analysis is an orthogonal rotation of the variable axes so that the variance of the system is distributed among the components.

The major difference between the models is the presence or absence of the unique factor. However, in the principal component analysis, components can be found with a strong relationship to only a single variable. This single variable component could then be considered to be the unique factor. Many of the traditional factor analysis texts (Harman, 1976, for example) make much of the distinction. However, the two models really aim at the same result, and the differences are primarily differences in definition and semantics.

There are numerous ways in which the factor analysis can be done and the matrix of coefficients determined. Many of these different approaches are discussed in detail in Harman (1976). Each solution will produce somewhat different results, and it is often useful to obtain solutions by several different methods to observe if the results are invariant to the form of factor analysis employed. The invariance of the resulting coefficient matrix to the kind of program used adds credibility to the solution. The principal axis approach is most commonly employed. In this direct method, the matrix to be analyzed is diagonalized as described in Chapter 5 to give the matrices of eigenvalues and eigenvectors. In a classical factor analysis, the elements of the principal diagonal of the correlation matrix are replaced with estimates of the communality of that variable to yield a communality matrix. Various estimates are available for the communality (Harman, 1976). In a principal component analysis, the values of unity are retained as the elements of the principal diagonal.

One of the major problems in applying factor analysis is choosing the number of factors to be retained. There are the conflicting goals of trying to include as much variance as possible in the retained factors while achieving the maximum parsimony in the understanding of the system using the fewest number of factors possible. In theory, there should be a sharp distinction between the eigenvalues that contain the variance in the system owing to the actual governing processes and those that contain the variance owing to random error. In practice, the choice of the proper number of factors to retain in the analysis is often quite difficult. It can be useful to plot the eigenvalues as a function of factor number and look for sharp breaks in the slope of the line (Cattell, 1966). These breaks can provide an indication of the point of separation. A number of commonly employed methods have been reviewed (Duewer, Kowalski, and Fasching, 1976) and it appears that there are no uniformly applicable methods to determine the number of factors to be retained.

Algorithms can be found to perform the eigenvalue analysis in the statistical program packages available on most computer systems. Because of the many approaches of factor analysis to analyzing data, care must be taken before using these library routines until the user has ascertained from the documentation that these routines do what he wants. In most statistical packages, there are built-in assumptions that the user may or may not desire. The most common assumption has to do with determining the number of factors to be used. Many routines consider only the eigenvectors having eigenvalues greater than 1.0 as significant factors, unless the user explicitly sets other parameters, such as total included variance to override the default on the eigenvalues. Although the eigenvalues greater-than-one criterion sets a lower bound on the number of factors (Guttman, 1954), it does not provide a simultaneous upper limit (Kaiser and Hunka, 1973). In general, it appears that the best approach to principal component factor analysis is to set the parameters so that the computer codes will step through the analysis for various numbers of factors, and then decide what number of factors should be considered. A useful aid in making this decision is discussed later.

Results originally reported by Parekh and Husain (1981) illustrate the use of factor analysis. In this work, they had measured the concentrations of nine elements and soluble SO_4^{2-} in samples of airborne particulate matter collected at Whiteface Mountain in upstate New York during the summers of 1975–1977. They were particularly interested in periods of high sulfate concentrations and found 16 24-h sampling periods that they have designated as sulfate episodes. The correlation matrix obtained for those 10 variables and 16 samples is given in Table 7.1. The eigenvalues obtained from the diagonalization are given in Table

Table 7.1. Correlation Matrix for 1975–1977 Summer Sulfate Episode Samples from Whiteface Mountain, New York, from Parekh and Husain (1981)

	Na	K	Sc	Mn	Fe	Zn	As	Br	Sb	SO_4^{2-}
Na	1.0									
K	0.48	1.0								
Sc	0.03	0.46	1.0							
Mn	0.22	0.57	0.82	1.0						
Fe	0.21	0.61	0.72	0.88	1.0					
Zn	0.14	0.03	-0.26	-0.08	-0.03	1.0				
As	0.13	0.30	0.64	0.65	0.46	-0.12	1.0			
Br	0.04	0.28	-0.06	-0.03	0.08	0.04	-0.18	1.0		
Sb	0.08	0.15	0.07	0.13	0.09	0.62	0.07	0.27	1.0	
SO_4^{2-}	0.11	0.20	0.54	0.57	0.48	-0.07	0.68	-0.17	0.07	1.0

Table 7.2. Eigenvalues Analysis of Correlation Matrix Given in Table 7.1

Factor Number	Eigenvalue	Percent Variance	Cumulative Percent Variance
1	4.019	40.2	40.2
2	1.863	18.6	58.8
3	1.266	12.7	71.5
4	1.012	10.1	81.6
5	0.633	6.3	87.9
6	0.399	4.0	91.9
7	0.315	3.2	95.1
8	0.259	2.6	97.7
9	0.156	1.6	99.2
10	0.077	0.8	100.0

7.2. If only the eigenvectors are retained for eigenvalues greater than 1, then the set of four factors presented in Table 7.3 are obtained.

After the choice is made as to how many factors are to be retained, there still remains the problem of relating the resultant matrix of factor loadings, a_{ik}, to terms meaningful to the system under study. The results of the factor analysis itself are not unique. In addition, the initially calculated factors generally not interpretable in terms of physically real causalities and another transformation of these vectors is necessary to understand the nature of these factors. For ex-

Table 7.3. Unrotated Eigenvectors for the Parekh and Husain Data Set

	Factor Number				
Species	1	2	3	4	Communality
Na	-0.30	-0.39	-0.32	0.75	0.90
K	-0.66	-0.39	-0.46	0.12	0.81
Sc	-0.87	0.19	0.01	-0.23	0.84
Mn	-0.94	0.00	-0.00	-0.06	0.89
Fe	-0.87	-0.09	-0.14	-0.12	0.79
Zn	0.11	-0.75	0.50	0.14	0.85
As	-0.76	0.21	0.30	0.06	0.72
Br	0.01	-0.52	-0.50	-0.55	0.81
Sb	-0.14	-0.77	0.44	-0.23	0.86
SO	-0.70	0.20	0.38	0.06	0.68

Table 7.4. Varimax Rotated Factor Loadings for the Parekh and Husain Data Set

Species	Factor Number 1	2	3	4	Communality
Na	0.047	0.086	0.943	-0.071	0.90
K	0.478	0.020	0.628	0.436	0.81
Sc	0.901	-0.128	-0.015	0.104	0.84
Mn	0.913	0.006	0.210	0.103	0.89
Fe	0.815	-0.006	0.251	0.251	0.79
Zn	-0.152	0.897	0.144	-0.068	0.85
As	0.803	0.021	0.038	-0.274	0.72
Br	-0.079	0.126	0.141	0.890	0.81
Sb	0.134	0.887	-0.039	0.227	0.86
SO_4^{2-}	0.756	0.074	-0.006	-0.319	0.68

ample, the factor loadings given in Table 7.3 are difficult to physically interpret and so a further transformation of these vectors is often found to be useful. A rotation of the factor axes can be made to redistribute the variance in an effort to give a more interpretable structure (Table 7.4). There is substantial controversy over the validity of axis rotation. Blackith and Reyment (1971) strongly object to what they consider to be an arbitrary procedure and cite a number of references that agree with their position. Their purpose in utilizing multivariate analysis was to objectively classify objects into groups, and for that purpose rotation does not help. In fact, a rotation that does not retain the orthogonality of the components would hinder such use.

However, in a principal components or principal axis factor analysis, the program is designed to incorporate the maximum amount of variance into the first factor; the maximum amount of remaining variance into the second factor; and so forth. Thus, the mathematical method will arbitrarily produce a few factors containing most of the system variance and a repartitioning of the variance among the factors may lead to a more physically realistic set of factors. Thurston (1981) and Henry and Hidy (1981a) have taken the two sides of the rotation question. Thurston has suggested the principal components analysis given earlier by Henry and Hidy (1979) can be improved by including a rotation. Henry and Hidy (1981b) have responded that in all of the cases they have examined, they could unambiguously interpret the unrotated factors obtained. However, for the purpose of aiding the identification of the nature of the causal factors in a system, rotations have generally proven extremely useful. After the (A) matrix has been rotated the matrix of factor scores, (F), can be calculated from the standardized data.

In applications of factor analysis to environmental chemical studies, the matrix of factors is generally rotated in such a way as to maximize the number of values that are close to either zero or unity. This rotation criterion is called "simple structure" (see Hopke et al., 1976, appendix) and a varimax rotation (Kaiser, 1959) may be used to achieve it. Simple structure may not always be the most useful criterion for environmental application because an element may be present in a sample owing to several different causal forces; therefore, factor loadings should not necessarily have values of either 0 or 1, but some intermediate value. However, it has been the most widely used of the rotation schemes that are available. None has yet made a critical evaluation of the various orthogonal rotation criteria that exist to determine if any particular approach is more applicable to receptor model applications.

The varimax or any other orthogonal rotation of this type attempts to repartition the variance in the system over all of the factors retained. The examination of this variance *after* rotation can often be useful in determining the number of factors to retain. For this type of analysis, each of the original variables was standardized to have a variance of 1.0. Thus, a rotated factor that has a variance less than 1.0 explains less variance than one of the original variables and, therefore, seems unlikely to be important enough to keep. Thus, a criterion has been suggested (Hopke, 1983) for determining the factors to be retained as those having a variance $\geqslant 1.0$ after an orthogonal rotation. This approach has proven to be a useful rule of thumb for several data sets.

Hopke (1981) in his reanalysis of the results of Lewis and Macias (1980) found that it was valuable to separately analyze the compositions of fine and coarse particles as obtained by using a dichotomous sampler. It was also found that this criterion of unit variance after rotation yielded more easily interpreted factor loadings and a better understanding of the system than was obtained with the original solution.

Another example is the reanalysis of the Parekh and Husain results when applying this criterion. Table 7.5 taken from Hopke (1982) shows that the variance for the fifth factor remains above 1.0 even for substantial overspecification of of number of factors included in the rotation. The final five-factor solution for Parekh and Husain is given in Table 7.6. It can be seen that now the arsenic and sulfate are together and separated from the crustal elements. The other factors remain essentially the same.

One of the problems with a factor analysis is that normally a complete set of data is required (Harman, 1976), since missing data will provide unequal variances and degrees of freedom in the correlation matrix. An approach has been suggested by Heidam (1982) to treat missing data. There is some debate in the statistical literature over the analysis of incomplete data sets and it is suggested that one should proceed with caution to use approximation methods to solve missing data problems.

Table 7.5. Variance Contained in Factors Before and After Varimax Rotation[a]

Factor Number	Before	Number of Rotated Factors				
		4	5	6	7	8
1	4.02	3.81	2.87	2.98	2.89	2.79
2	1.86	1.64	1.90	1.65	1.69	1.09
3	1.27	1.41	1.63	1.63	1.62	1.08
4	1.01	1.30	1.27	1.28	1.17	1.06
5	0.63		1.12	1.12	1.12	1.00
6	0.40			0.53	0.55	0.99
7	0.32				0.46	0.98
8	0.26					0.79

[a]From Hopke (1982).

Table 7.6. Factor Loadings After Varimax Rotation for Parekh and Husain Data

Species	Factor Number					Communality
	1	2	3	4	5	
Na	0.09	0.08	0.07	0.96	-0.00	0.95
K	0.70	-0.00	0.04	0.51	-0.28	0.83
Sc	0.78	0.48	-0.11	-0.12	-0.00	0.86
Mn	0.85	0.43	0.03	0.09	-0.04	0.92
Fe	0.91	0.22	0.03	0.10	0.02	0.88
Zn	-0.06	-0.15	0.92	0.13	-0.12	0.89
As	0.37	0.78	-0.03	0.06	-0.10	0.76
Br	0.05	-0.14	0.09	0.04	0.97	0.96
Sb	0.07	0.14	0.87	-0.04	0.27	0.87
SO	0.24	0.85	0.02	0.06	-0.08	0.79

PREVIOUS APPLICATIONS

The first receptor modeling applications of classical factor analysis were by Prinz and Stratmann (1968) and Blifford and Meeker (1967). Prinz and Stratmann examined both the aromatic hydrocarbon content of the air in 12 West German cities and data from Colucci and Begeman on the air quality of Detroit. In both cases, they found three factor solutions and used a varimax rotation to give more readily interpretable results. Blifford and Meeker used a principal component analysis with both varimax and a nonorthogonal rotation to examine particle composition data collected by the National Air Sampling Network (NASN) during 1957-1961 in 30 U.S. cities. They were generally not able to extract much interpretable information from their data. Since there are a very wide variety of particle sources among these 30 cities and only 13 elements measured, it is not surprising that they were not able to provide much specificity to their factors. One interesting factor that they did identify was a unique copper factor for which they could not provide a convincing interpretation. It is likely that this factor represents the copper contamination from the brushes of the motors of the high-volume air samplers. This problem was identified to be a ubiquitous one by Hoffman and Duce (1971).

John et al. (1973) have also utilized an approach that can be considered as the initial steps of a factor analysis. They attempt to analyze their data by estimating the communalities in their system by the centroid method (Harman, 1976). This factor analysis procedure was popular when adequate computer facilities were unavailable. They did not complete the full factor analysis and were only able to draw limited conclusions from their data.

The factor analysis approach was then reintroduced by Hopke et al. (1976) and Gaarenstroom, Perone, and Moyers (1977) for their analysis of particle composition data from Boston, Massachusetts and Tucson, Arizona, respectively. In the Boston data, for 90 samples at a variety of sites, six common factors were identified that were interpreted as soil, sea salt, oil-fired power plants, motor vehicles, refuse incineration and an unknown manganese-selenium source. The six factors accounted for about 78% of the system variance. There was also a high unique factor for bromine that was interpreted to be fresh automobile exhaust. Large unique factors for antimony and selenium were found. These factors may possibly represent emission of volatile species whose concentrations do not covary with other elements emitted by the same source. Subsequent studies by Thurston and Spengler (1982) where other elements including sulfur were measured showed a similar result. They found that the selenium was strongly correlated with sulfur for the warm season (May 6 to November 5). This result is in agreement with the Whiteface Mountain results in Table 7.6 and suggests that selenium was primarily an indicator of longer range transport of coal-fired power plant effluents.

In the study of Tucson (Gaarenstroom, Perone, and Moyers, 1977), at each site whole filter data were analyzed separately. They found factors that were identified as soil, automotive, several secondary aerosols such as $(NH_4)_2SO_4$, and several unknown factors. They also discovered a factor that represented the variation of elemental composition in their aliquots of their neutron activation standard containing Na, Ca, K, Fe, Zn, and Mg. This finding illustrates one of the important uses of factor analysis; screening the data for noisy variables or analytical artifacts.

One of the valuable uses of this type of analysis is in screening large data sets to identify errors (Roscoe et al., 1982). With the use of atomic and nuclear methods to analyze environmental samples for a multitude of elements, very large data sets have been generated. Because of the ease in obtaining these results with computerized systems, the elemental data required are not always as thoroughly checked as they should be, leading to some, if not many, bad data points. It is advantageous to have an efficient and effective method to identify problems with a data set before it is used for further studies. Principal component factor analysis can provide useful insight into several possible problems that may exist in a data set including incorrect single values and some types of systematic errors. These uses are described by Dattner and Jenks (1981), Roscoe et al. (1982), and Hopke (1983).

Gatz (1978) applied a principal components analysis to aerosol composition data for St. Louis, Missouri taken as part of project METROMEX (Changnon et al. 1977; Ackerman et al. 1978). Nearly 400 filters collected at 12 sites were analyzed for up to 20 elements by ion-excited X-ray fluorescence. Gatz used additional parameters in his analysis including day of the week, mean wind speed, percent of time with the wind from NE, SE, SW, or NW quandrants or variable, ventilation rate, and rain amount and duration. At several sites the inclusion of wind data permitted the extraction of additional factors that allowed identification of specific point sources. An important advantage of this form of factor analysis is the ability to include parameters such as wind speed and direction or particle size in the analysis.

Another approach to including the meteorological data that are available has been developed by Thurston and Spengler (1981). They separately analyzed the elemental composition and the meteorological data. They then examined the correlations between the elemental factors and the meteorological factors. This approach assisted in the interpretation of their data from six different cities in the United States.

There have been a number of applications of principal components or factor analysis to identification of particle sources. Dattner (1978) has examined the sources of particles in a number of locations in Texas using X-ray fluorescence analysis on the Texas Air Control Board routine high-volume TSP samples. Sievering and coworkers (1980) have made extensive use of factor analysis in their interpretation of midlake aerosol composition and deposition data for Lake Michigan. As previously indicated, Lewis and Macias (1980) have analyzed particle composition data from Charleston, West Virginia. Heidam (1981) has examined the origins of particle samples collected in the arctic. Malissa, Puxbaum, and Wopenka (1981) have analyzed dichotomous sampler results for Vienna, Austria and Dutot, Elichegaray, and Vie le Sage (1983) have used this approach for analyzing a large particle composition data set for Paris, France.

In virtually all of the applications of factor analysis to airborne particle data, it has been used to directly analyze the volumetric, elemental concentrations

(ng/m^3). However, Henry (1978) has shown that the correlation between elemental concentrations from two uncorrelated sources

$$r(x_1, x_2) = e_{x1} e_{x2} / [(1 + e_{x1}^2 + e_m^2)(1 + e_{x2}^2 + e_m^2)]^{1/2}$$

where $e_x = \bar{x}_1 / \sigma_{x1}$, the ratio of the mean of x to the standard deviation of x and m is the total volumetric mass concentration (mg/m^3). Thus, for elements where there is a moderate to large ratio of mean value to standard deviations, a correlation of approximately 0.5 will be induced by the meteorology even if the sources of the elements are not directly correlated. Henry suggests that it would assist in removing the meteorological influence if the volumetric elemental concentrations were divided by the volumetric mass concentrations to yield gravimetric elemental values (ng/mg) as the input data. This normalization significantly changed the correlation matrix that was obtained for a small data set from Portland, Oregon (Henry, 1977). Although this suggestion has been made and appears to improve the resolution of the technique, it has not yet been utilized by other workers in the field.

Several reports have utilized components analysis to examine the relationship of sulfate with other particle components and air quality parameters. Henry and Hidy (1979 and 1981a) have used unrotated principal components analysis to examine data from Los Angeles, New York, Salt Lake City, and St. Louis. Tanner and Leaderer (1981) have examined seasonal variations in sulfur-containing aerosol in the New York area as have Lioy et al. (1981, 1982). Cox and Clark (1981) have examined ambient ozone patterns in the eastern United States. Factor analysis has also had a limited application to interpreting precipitation chemistry (Knudson et al., 1977; Baker, Camparoli and Harrison, 1981) and identifying sources of visibility degradation (Barone et al., 1978; Pitchford et al., 1981).

A problem with this form of analysis is that it deals with the standardized variables and apportions variance. This approach permits the inclusion of variables where the property is not linearly additive such as wind direction. However, it means that the analysis does not directly provide results analogous to the chemical mass balance method. Several investigators have used factor analysis to identify the key elements associated with a source and then perform a multiple regression analysis of the mass against those elemental concentrations. This approach will be discussed in the next chapter.

Thurston and Spengler (1982) have suggested another approach. Using a principal components model [equation (7.1) without the unique factor], the equation can be inverted to calculate the factor scores, f's from the standardized data, z's, and the factor loadings, a's. The factor scores will also be in standardized form. A value of zero means for that specific sample there is an average contribution of that source to that sample. Thus if all of the factor scores for a sample are zero, the elemental concentration will equal the value of the average concentration of that element for the data set. Thurston and Spengler have then realized that it is necessary to determine the values of the factor scores for the true zero concentrations. Thus, after performing the components analysis and obtaining the factor loadings and scores for all of the real samples, they calculate the factor scores for a dummy sample with zero concentration values for all elements. The real sample component scores could then be increased by adding the absolute value of the true zero score. The mass values of the samples can then be regressed against these rescaled factor scores to convert the score values into pollution source mass contributions ($\mu g/m^3$) for each sample.

In many applications it is desirable to be able to determine the composition of the source materials and mass contributions of each source to each sample in a more direct manner. To this end another factor analysis model has been developed and is presented in the next section.

Target Transformation Factor Analysis

The objective of this form of factor analysis is to obtain a determination of the sources of a sample that are directly comparable to the chemical mass balance:

$$x_{ij} = \sum_{k=1}^{p} a_{ik} f_{kj} \tag{7.6}$$

or rewriting in matrix form

$$X = AF \tag{7.7}$$

The problem with the conventional factor analysis is that is operates on the matrix of correlations about the mean [equation (5.55)]. In the calculation the mean value is subtracted from the raw data point. It is possible to obtain a result like equation (7.6) by utilizing the correlation about the origin [equation (5.59)] as the matrix to be diagonalized:

$$R_Q = (XV)'(XV) \tag{7.8}$$

where V is a diagonal matix whose elements are defined as

$$v_{ij} = \delta_{ij} \left(\sum x_{ij}^2 \right)^{-1/2} \tag{7.9}$$

where δ_{ij} is Kronecker's delta. With the order of multiplication given in equation (7.8), the correlations are between the samples. In the nomenclature defined by Rozette and Peterson (1975), the analysis of this matrix is termed a Q-mode analysis. If the order of multiplication is reversed,

$$R_R = (XV)(XV)' \tag{7.10}$$

then a matrix of correlations among the measured variables is obtained. The factor analysis of this matrix is termed an R-mode analysis.

In the factor analysis models previously discussed, the interrelations between measured species (R mode) have been utilized as a starting point for the analysis. However, following its introduction by Imbrie and Van Andel (1964), the alternative approach, Q-mode factor analysis, has been widely employed in geological studies (Meisch, 1976; Joreskog, Klovan, and Reyment, 1976). In a study of mass spectrometry data, Rozett and Peterson (1975) found that the Q-mode analysis gave a more sensitive response to the proper number of factors and was preferred over the R mode that examined the interrelationships between the variables. Thus, in the applications of TTFA, the Q-mode approach was employed. Commenting on Alpert and Hopke (1980), Redman and Zinsmeister (1982) point out that the two approaches should provide the same result but that the Q mode should be biased because of the inherent correlation between elements. In the response to these comments, Alpert and Hopke (1982) point out that the R mode is biased by the inherent serial correlations between samples in a set of airborne compositional data so that neither approach fulfills the assumption of independent samples drawn from a multivariate normal population. Therefore, one of the goals of this section will be to compare the results of R-mode and Q-mode analysis of several data sets. The analysis of the dispersion matrix proceeds as outlined in Figure 7.3. In the Q-mode analysis, the mass contribution matrix F is calculated directly from the matrix of eigenvectors while in the R-mode analysis the direct calculation is of the A matrix. The ultimate results should be the same and in the absence of error in the data, identical results are obtained. However, for real data, there are some differences obtained that will be presented later in this chapter.

After diagonalization the number of factors to be retained must again be determined. The same problems exist in this form of analysis as with the conventional factor analysis. It is generally useful to perform a classical factor analysis of the data to screen it for outliers and noisy variables using the approach of Roscoe et al. (1982) before beginning this form of analysis. The classical analysis will provide some information on the number of identifiable factors that can be obtained. In addition, a number of other tests are available to assist in making this decision, including the eigenvalues obtained from the diagonalization of the correlation matrix. To provide other tests, the data are reconstructed using one eigenvector, then two eigenvectors, and so forth with several measures of the quality of fit being calculated at each step. These measures include chi-squared

$$\chi^2 = \left[\sum_{i,j} \frac{(x_{ij} - \bar{x}_{ij})^2}{\sigma_{ij}^2} \right] \tag{7.11}$$

where x_{ij} is the reconstructed data point using p factors and σ_{ij} is the uncertainty in the value of x_{ij}. The Exner function (Exner, 1966) is a similar measure and is calculated by

$$E^p = \left[\frac{\sum_{i,j} (x_{ij} - \bar{x}_{ij})^2}{\sum_{i,j} (x^0 - \bar{x}_{ij})^2} \right]^{1/2} \tag{7.12}$$

where x^0 is a grand ensemble average value. The empirical indicator function suggested by Malinowski (1977) is also calculated:

For $n < m$,

$$\mathrm{RSD} = \left(\frac{\sum_{j=p+1} \lambda_j}{n(m - p)} \right)^{1/2} \tag{7.13}$$

$$\mathrm{IND} = \frac{\mathrm{RSD}}{(n - p)^2}$$

For $m < n$,

$$\mathrm{RSD} = \left(\frac{\sum_{j=p+1} \lambda_j}{m(n - p)} \right)^{1/2}$$

$$\mathrm{IND} = \frac{\mathrm{RSD}}{(m - p)^2} \tag{7.14}$$

where λ_j are the eigenvalues from the diagonalization. The functions given are for a Q-mode analysis. In an R-mode analysis, the positions of m and n in these equations would be reversed. This function has not proven as useful with particle composition data as Malinowski (1977) has found it to be with spectroscopy results because of miscalculations that have been made in prior reports. Further study of this function is needed. The root-mean-square error and the arithmetic average of the absolute values of the point-by-point errors are also calculated. All of these parameters can prove to be useful indicators of the dimensionality of the problem and permit the determination of the number of factors to be used in the subsequent portions of the analysis. These tests do not provide unequivocal

Q mode	R mode
Correlation Coefficient	
$R_Q = (XV^{-1})'(XV^{-1})$	$R_R = (XV^{-1})(XV^{-1})'$
where $v_{ii} = \left(\sum_{j=1}^{n} x_{2j}^2\right)^{1/2}$	
$v_{i \neq j} = 0$	
$i = 1, m$	
Diagonalization	
$Q_Q^{-1} R_Q Q_Q = \Lambda_Q$	$Q_R^{-1} R_R Q_R = \Lambda_R$
$Q'_Q = Q_Q^{-1}$	$Q'_R = Q_R^{-1}$
$Q'_Q (XV^{-1})'(XV^{-1}) Q_Q = \Lambda_Q$	$Q'_R (XV^{-1})(XV^{-1})' Q_R = \Lambda_R$
$[(XV^{-1})' Q_Q]' [(XV^{-1}) Q_Q] = \Lambda_Q$	$[(XV^{-1})' Q_R]' [(XV^{-1})' Q_R] = \Lambda_R$
If $A \equiv (XV^{-1}) Q_Q$, then $F = VQ'_Q$	If $F \equiv [(XV^{-1})' Q_R]' = Q'_R X V^{-1} = Q'_R V^{-1} X$, then $A = Q_R V$

Figure 7.3. Outline of the diagonalization process for the R- and Q-modes of factor analysis.

indicators of the number of factors that should be retained. Some judgment becomes necessary in evaluating all of the test results and deciding upon a value of p. In this manner, the dimension of the A and F matrices is reduced from n to p.

After the truncation of the A and F matrices to only p retained factors, it is necessary to interpret the nature of the factors for the system being studied. The A matrix produced by diagonalization generally cannot be directly associated with actual source profiles since it is one of an infinity of mathematically equivalent matrices that will diagonalize the correlation matrix. In order to relate this abstract matrix to one with physical significance, it is necessary to geometrically realign the factors axes of A with axes that represent real source emission concentration profiles. Such physically significant axes, called test vectors, are derived from existing knowledge of the relative elemental composition of actual source materials. The factor model is modified by the addition of the rotation matrix:

$$X = ARR^{-1} F \tag{7.15}$$

The realignment procedure, called target transformation (Malinowski and Howery, 1980), involves finding a rotation vector **r**, which aligns a column of the A matrix with the input test vector so as to maximize by least squares the overlap between a rotated axes of A and the test vector.

The difference between the rotated column of the A matrix and the input test vector **b** is given by epsilon,

$$\boldsymbol{\epsilon} = A\mathbf{r} - \mathbf{b} \tag{7.16}$$

In a least-squares fit, the quantity to be minimized is ϵ^2 given by

$$\epsilon^2 = (A\mathbf{r} - \mathbf{b})'(A\mathbf{r} - \mathbf{b}) \tag{7.17}$$

Rearranging yields

$$\epsilon^2 = (\mathbf{r}'A' - \mathbf{b}')(A\mathbf{r} - \mathbf{b}) \tag{7.18}$$

$$\epsilon^2 = \mathbf{r}'A'A\mathbf{r} - \mathbf{r}'A'\mathbf{b} - \mathbf{b}'A\mathbf{r} + \mathbf{b}'\mathbf{b} \tag{7.19}$$

The third term in this equation is a scalar quantity and thus equal to its transpose,

$$\epsilon^2 = \mathbf{r}'A'A\mathbf{r} - 2\mathbf{r}'A'\mathbf{b} + \mathbf{b}'\mathbf{b} \tag{7.20}$$

Taking the derivative and setting it equal to zero,

$$\frac{\partial \epsilon^2}{\partial \mathbf{r}} = 0 = 2A'A\mathbf{r} - 2A'\mathbf{b} \tag{7.21}$$

Rearranging,

$$\mathbf{r} = (A'A)^{-1} A'\mathbf{b} \tag{7.22}$$

Comparing this equation with equations (7.6) and (7.7) leads to

$$\mathbf{r} = (\Lambda)^{-1} A'\mathbf{b} \tag{7.23}$$

Each rotation vector aligns a column of A with an input test vector so as to maximize the least-squares overlap between the rotated axis of A and the test vector. The degree of alignment between test vector **b** and the rotated axes of A determines if **b** is a good representative of one of the sources in the data. The target transformation rotation given by equation (7.10) is dominated by the more abundant elements in the data. In order to resolve sources with similar elemental profiles, it is important that all elements have equal weight in the rotation since it is often the variation in the concentration of some trace elements which permits differentiation of sources with similar major-element compositions. A weighted target transformation scheme has been examined and found to enhance greatly the ability of the analysis to resolve sources with similar concentration profiles (Roscoe and Hopke, 1981a). The weighted target transformation rotation is given by

$$\mathbf{r} = (A'WA)^{-1} A'W\mathbf{b} \tag{7.24}$$

where W, the weight matrix, is a diagonal matrix that can have as its diagonal elements the inverse of the variance in the elemental concentrations or the inverse of the square of the average error of determination for the concentration values.

A suspected source of particles can then be tested by determining if a factor axes can be rotated as so to overlap the input vector. In addition to source profile test vectors, uniqueness test vectors that have all but one element set equal to zero and the remaining value set to unity are analyzed for all elements. An element is considered unique if its concentration in the data is decoupled from all other elements. The uniqueness test also determines if the concentration of an element is strongly related to any other element(s). The test vector predicted by the uniqueness test can then be used as a normalized source concentration profile for sources whose concentration profiles have not been measured. It has been found that it is possible to derive source concentration profiles from the unique vectors by an iterative process described below so that even less a priori knowledge of the source compositions is required (Roscoe and Hopke, 1981a). The initial vector has a value of 1.0 for one element and zero for the others. The resulting rotation vector is used to obtain the new values for $\mathbf{b}^*$.

$$\mathbf{b}^* = A\mathbf{r} \tag{7.25}$$

Then $\mathbf{b}^*$ is used as the input to the next iteration after setting all of the negative values to zero. This process is repeated until the average percent change in the values of $\mathbf{b}$ and $\mathbf{b}^*$ is less than 10^{-4}. At this point these iterated profiles are available to be grouped p at a time to form a matrix B.

When a number of possible sources have been identified, they are combined p at a time to form B, a matrix of test vectors, and R, a matrix of their corresponding rotation vectors. Substituting B for the rotated A matrix, AR, and calculating a new F matrix from

$$\bar{F} = R^{-1} F \tag{7.26}$$

leads to

$$\bar{X} = B\bar{F} \tag{7.27}$$

where $\bar{X}$ is the data reproduced with the chosen set of test vectors given by matrix B. We have found that the F values can be best calculated row by row using a least-squares fit to the concentration data (Severin, Roscoe, and Hopke, 1984)

$$\mathbf{f}_j = (B'WB)^{-1} B'W\mathbf{x}_j \tag{7.28}$$

The original data are compared point by point with $\bar{X}$ to determine how well the chosen set of test vectors represent the actual sources in the data. This comparison can also suggest further refinements to the test vectors which might improve the overall agreement. Repetition of the calculation of $\bar{X}$ with different combinations of test vectors allows the best possible set of source profiles to be determined. A computer code, FANTASIA (Hopke, Alpert, and Roscoe, 1983), has been written to perform all of these functions.

Since the analysis begins with these very arbitrary test vectors, the absolute concentrations of the iterated source profile vectors may not correspond exactly to those found in the system under study. Therefore, before the selected set of test vectors can be used to calculated the actual mass contribution of each source to each sample, they must be scaled to reflect the actual concentration of each element in the source material. The source vectors developed from the uniqueness test begin with a maximum concentration of unity and after iteration, therefore, require scaling to absolute concentration values. Scaling can be accomplished by again modifying the basic equation

$$x_{ij} = \sum_{k=1}^{p} \frac{b_{ik}}{s_k} s_k \bar{f}_{kj} \tag{7.29}$$

If we have included all of the sources that contribute to the total sample mass, then the sum of the scaled $\bar{f}_{kj}$ values should equal the total measured mass. Thus, the scaling coefficients for each source, s_k, can be calculated by means of a multiple linear regression of the form

$$M_j = \sum_{k=1}^{p} s_k f_{kj} \tag{7.30}$$

where M_j is the measured mass of sample j, f_{kj} is the mass matrix calculated in equation (7.28), and s_k is the scaling factor to be calculated by the regression analysis. Comparison of the predicted and measured sample masses gives indication of the quality of the fit with the selected set of test vectors. In addition, the regression coefficients should be positive and statistically significant. Finally, when the source profiles are scaled, a_{ik}/s_k, the sum for a given source should be less than 1,000,000 ppm. These values can assist in determining if too many or too few sources have been included in the analysis.

Although the resolution of data into separable components is useful for solving a variety of environmental problems, it was also necessary to have a procedure that would permit the estimation of the error in the elements of both the B and F matrices on the right side of equation (7.27). Two methods are available for making such a calculation; the "jackknife" method (Mosteller, 1971; Mosteller and Tuckey, 1977) and a method developed by Roscoe and Hopke (1981b) using the calculational framework described by Clifford (1973).

The jackknife method uses multiple determinations of B^* obtained by repeating the complete factor analysis several times with a reduced data set where one observation has been deleted each time. Therefore, to examine all of the possible combinations, the factor analysis must be done one time more than the number of observations in the study. The source profiles of the modified data, B_i^*, can be determined and used to calculate a better source profile in addition to an error in that profile. This determination is made by first forming a "pseudovalue" (Mosteller, 1971),

$$B_i^* = NB^* - (N - 1)B_i^* \tag{7.31}$$

where N is the number of observations in the total data set. The mean of the pseudovalues will give a better estimate of the source profiles,

$$\overline{B_i^*} = \frac{1}{N} \sum B_i^* \tag{7.32}$$

The variance, s^2, of B^* can then be calculated

$$s^2 = \frac{\sum (B_i^*)^2 - (\sum B_i^*)^2/N}{N(N - 1)} \tag{7.33}$$

A calculational method is described by Clifford (1973) and relys on the error obtained from comparing the reproduced data to the raw data. The derivation is summarized here; a more complete description can be found in Clifford.

Consider the formulation of the variance–covariance matrix of X in equation (7.28). If $\mathbf{dx}_i$ is the ith column of X and $\bar{\mathbf{x}}$ contains the means of each row, then the variance–covariance matrix of X can be formed by

$$M_x = \frac{1}{N} \sum \mathbf{dx}_i \, \mathbf{dx}_i' \tag{7.34}$$

where $\mathbf{dx}_i = \mathbf{x}_i - \bar{\mathbf{x}}$. For a least-squares solution to equation (7.28), a weight matrix, W, which is inversely proportional to M_x, is used:

$$W = \sigma^2 M_x^{-1} \tag{7.35}$$

After solving equation (7.28) using the weight matrix W, a variance–covariance matrix for F can be formulated,

$$M_f = \frac{1}{N} \sum \mathbf{df}_i\, \mathbf{df}_i' \tag{7.36}$$

where $\mathbf{df}_i = \mathbf{f}_i - \bar{\mathbf{f}}$ for $\bar{\mathbf{f}}$ containing the means of the rows in F. Since $\mathbf{f}$, and hence $\mathbf{df}$, were formed by the principle of least-squares,

$$\mathbf{df}_i = (B'WB)^{-1}\, B'W\, \mathbf{dx}_i \tag{7.37}$$

Substituting equation (7.37) into equation (7.36) and recognizing that $(B^t WB)$ is symmetric, yields

$$\begin{aligned} M_f &= \frac{1}{N} \sum (B'WB)^{-1}\, B'W\, \mathbf{dx}_i\, \mathbf{dx}_i' WA(B'WB)^{-1} \\ &= (B'WB)^{-1}\, B'WM_x\, WB(B'WB)^{-1} \end{aligned} \tag{7.38}$$

Substituting equation (7.35) into equation (7.38) produces a new representation for the variance–covariance matrix of F,

$$M_f = \sigma^2 (B'WB)^{-1} \tag{7.39}$$

Therefore, the variance in the f_i terms can be represented by the product of the total variance of the diagonal of $(B^t WB)$.

To determine the total variance of the analysis, assume that the original data matrix X has been reproduced to form a matrix X'. An error vector can then be formed from a column of the original and reproduced data, $\mathbf{x}_i$ and $\mathbf{x}_i'$, respectively:

$$\begin{aligned} \mathbf{e}_i &= \mathbf{x}_i - \mathbf{x}_i' \\ &= \mathbf{x}_i - B\mathbf{f}_i = \mathbf{dx}_i - B\, \mathbf{df}_i \end{aligned} \tag{7.40}$$

The weighted sum of the squares of the differences between the original and reproduced data can be formed by

$$s_i = \mathbf{e}_i' W \mathbf{e}_i \tag{7.41}$$

With equation (7.28), this yields

$$s_i = (\mathbf{dx}_i - B\ \mathbf{df}_i)' W\ \mathbf{dx}_i - (\mathbf{dx}_i - B\ \mathbf{df}_i)' W \mathbf{bf}_i \tag{7.42}$$

To determine the nature of the second term of equation (7.42), multiply both sides of equation (7.37) by $(B'WB)$ to get

$$(B'WB)\ \mathbf{df}_i = B'W\ \mathbf{dx}_i \tag{7.43}$$

Transposing equation (7.43) and noting that the weight matrix W is symmetric, yields

$$\mathbf{df}_i' B'WB = \mathbf{dx}_i' WB \tag{7.44}$$

Several manipulations of equation (7.44) produce

$$(\mathbf{dx}_i - B\ \mathbf{df}_i)' WB = 0 \tag{7.45}$$

and shows that the second term of equation (7.42) is zero. Thus equation (7.37) becomes

$$\begin{aligned} s_i &= (\mathbf{dx}_i - B\ \mathbf{df}_i)' W\ \mathbf{dx}_i \\ &= \mathbf{dx}_i' W\ \mathbf{dx}_i - \mathbf{df}_i' B' W\ \mathbf{dx}_i \end{aligned} \tag{7.46}$$

Substituting equation (7.33) into equation (7.46) yields

$$\begin{aligned} s_i &= \mathbf{dx}_i' W\ \mathbf{dx}_i - \mathbf{df}_i' (B'WB)\ df_i \\ &= \mathrm{Tr}(\mathbf{dx}_i\ \mathbf{dx}_i' W) - \mathrm{Tr}\ [\mathbf{df}_i\ \mathbf{df}_i' (B'WB)] \end{aligned} \tag{7.47}$$

For N observations, that is, columns of X, a mean values of s_i can be obtained

$$\bar{s} = \frac{1}{N} \sum [\mathrm{Tr}\ (\mathbf{dx}_i\ \mathbf{dx}_i' W) - \mathrm{Tr}\ [\mathbf{df}_i\ \mathbf{df}_i' (B'WB)] \tag{7.48}$$

Substituting equations (7.34) and (7.36) into equation (7.48) yields

$$\bar{s} = \mathrm{Tr}(M_x W) - \mathrm{Tr}(M_f A' W A) \tag{7.49}$$

Multiplying both sides of equation (7.35) by M_x and substituting that into equation (7.49) produces a simplified expression for $\bar{s}$,

$$\begin{aligned} \bar{s} &= \sigma^2 \, \mathrm{Tr}(I_N) - \sigma^2 \, \mathrm{Tr}(I_p) \\ &= \sigma^2 (N - p) \end{aligned} \tag{7.50}$$

where p is the number of rows in $\bar{F}$. Thus, we have established an expression for determining the total variance in the reproduction of the data of equation (7.28).

The derivation shows how errors in the $\bar{F}$ cofactor matrix may be estimated by comparing the original and reproduced data. This comparison produces the total variance of the data reproduced. The total variance can then be apportioned by multiplying it by the diagonal terms of the simple matrix, $(B^t WB)^{-1}$. The errors in the B cofactor matrix may also be determined by a similar procedure. This is done by initially taking the transpose of equation (7.28) and then following the same procedure.

This procedure has been tested by using the same simple geological data set. Jackknife and calculational methods of error analysis are compared in the report of Roscoe and Hopke (1981b). The calculational method is found to be quicker and easier to use than the jackknife method and therefore has an inherent advantage.

APPLICATIONS

Target transformation factor analysis has been applied to a number of subsets of samples from the Regional Air Pollution Study (RAPS) data base for the St. Louis, Missouri metropolitan area. Results have been reported for several locations (Alpert and Hopke, 1981; Liu et al., 1982; Severin, Roscoe, and Hopke, 1984). Alpert and Hopke also compared the utility of analyzing data from 10 sites over a shorter time frame against data from a single site and found the single site data much easier to analyze and interpret the results. It appears that the multitude of diverse sources in a complex airshed like St. Louis will not permit a resolution of sources over the whole region.

This method has not been widely applied by other groups. Stafford and Liljestrand (1983) have used TTFA to examine rainwater and particle composition data from eight sites in Texas. They found that the crustal components describing local conditions were useful in interpreting the precipitation chemistry data but the extent of their analysis was limited by the incomplete suite of elements determined in both types of samples. DeCesar and Cooper (1983) have also begun to explore the TTFA approach on the Portland aerosol data. However, extensive results have not been reported.

To illustrate the use of target transformation factor analysis in source resolution problems, two examples will be presented. These data sets include one of the data sets created by the National Bureau of Standards as described in Chapter 6 and by Gerlach, Currie, and Lewis (1983). The other is a set of elemental and mineral phase determinations performed on a set of coal samples (Cecil et al., 1981). In both cases it is possible to test the TTFA methodology against known or alternatively measured results.

One aspect of TTFA to be tested is the question of using an R- or a Q-mode analysis (see Figure 7.3). As described earlier, the difference between R and Q mode is in the order of matrix multiplication and, therefore, in whether the correlations between samples or between measured species is being examined. A correlation between elements normalizes the data for each element; destroying any information about the relative elemental concentrations in each sample. A Q analysis preserves this information by normalizing the data for each sample. Thus, an R analysis produces a simpler calculation, while a Q analysis facilitates the identification of sources by preserving the relative elemental concentrations. The disadvantage of Q analysis over R is that for airborne particulate data, it forces the diagonalization of large matrices. Limitations of many computers restricts the study to 175 samples or less at one time. If the matrix to be diagonalized were determined by the number of species determined rather than the number of samples, then a much smaller matrix would be analyzed. In order to test R- and Q-mode analysis, two data sets, an artificial data set (Gerlach, Currie, and Lewis, 1983) and a mineral matter in coal data set (Cecil et al., 1981), have been examined. These results have been reported by Hopke, Severin, and Chang (1983), Hwang, Severin, and Hopke (1984), and Roscoe, Chen, and Hopke (1984).

Artificial data were provided by the National Bureau of Standards (Gerlach, Currie, and Lewis, 1983). The set used in this study was the first of three and consists of 40 samples for which 19 elemental concentrations were provided. The data set was created using a contrived source geometry shown in Figure 6.7 and the RAM Gaussian-plume dispersion model. The details of the construction of these data are presented in the report of Gerlach, Currie, and Lewis (1983). The composition of the nine sources included in the data set are given in Table 7.7. In the data set, arsenic was found to be below detection limits for the vast majority of samples. In a classical factor analysis used to screen data (Roscoe et al., 1982), arsenic appears as a unique factor and is attributed to being a variable dominated by noise rather than containing unique source information. It

Table 7.7. Actual Compositions of Sources Used to Create N.B.S. Data Set No. 1

Element	Oil B	Incinerator	Soil	Road	Wood	Sandblast	Coal B	Steel	Basalt
C	200	40	66.4	429	600	5.0	8.0	150	0.1
Na	22	120	4.0	52.5	3.0	8.0	4.0	11	19
Al	5.4	4.2	222	62	7.0	7.5	140	13	88
Si	70	7.0	543	147	8.9	400	200	9	240
S	110	130	0.26	6.06	1.00	8.8	180	27	0.3
Cl	12	270	0.13	8.06	10.0	28	17	30	0.2
K	1.40	200	12.0	5.0	8.0	7.0	8.0	9.1	8.3
Ca	8.6	11.0	20.5	13.9	9.0	8.2	40	70	67
Ti	1.10	0.90	12.7	3.17	0.020	4.0	9.0	1.3	9
V	0.67	0.027	0.25	0.143	0	0.031	0.8	1.8	0.2
Cr	0.42	0.85	0.31	0.141	0.002	5.8	0.3	3.3	0.1
Mn	0.140	0.33	2.0	0.94	0.020	1.4	0.2	16	1.5
Fe	13.0	6.1	97	29.5	1.0	63	70	120	86
Ni	0.06	0.100	0.090	0.070	0.005	3.2	0.6	2.8	0.1
Cu	0.75	3.6	0.20	0.050	0	6.9	0.4	2.6	0.1
Zn	1.30	26	0.41	2.51	0.70	4.7	0.9	25	0.08
As	0.25	0.150	0.010	0.010	0	0.099	0.2	0.08	0
Br	0.39	0.82	0.010	31.3	0.080	0.047	0.01	0.1	0
Pb	3.8	17.0	0.26	90.1	0.100	5.8	1.2	7.6	0.01

was, therefore, eliminated from the data set. A classical factor analysis showed a very stable solution for seven factors using the criterion of variance greater than 1.0 after an orthogonal rotation (Hopke, 1981). This approach provides a useful rule of thumb for the maximum number of factors to retain.

The results of the *Q*-mode TTFA analysis are presented in Tables 7.8 and 7.9 and the *R*-mode values are given in Tables 7.10 and 7.11. In Tables 7.8 and 7.10 the results of the diagonalization and reproduction are presented. For the *Q*-mode analysis, examination of the chi-squared and the average percent error (Table 7.8) suggests the presence of six factors. After careful study of the six- and seven-factor solutions, the six-factor *Q*-mode solution seemed to be more satisfactory. Table 7.9 gives the refined source profiles developed from the set of unique vectors by the iterative process previously described.

Table 7.10 presents the *R*-mode diagonalization and reproduction results. There are indications of six to eight factors although the indicator function does properly show nine factors. The study of these possible solutions led to a seven-factor solution as the best fit to the data in agreement with the classical factor analysis. The refined source compositional profiles are given in Table 7.11.

This data set was created using nine source types. Neither *Q*- nor *R*-mode analysis correctly determined the presence of nine sources. The classical factor

Table 7.8. Diagonalization Results and Reproduction Summary for *Q*-Mode Analysis NBS Data Set No. 1

Factor	Eigenvalue	RMS Error	Chi-Square	Exner	Average Percent Indicator	Average Percent Error
1	3.63E + 01	4.48E + 01	3.20E + 02	0.3469	2.54E - 04	126.6
2	2.97E + 00	1.48E + 01	8.11E + 01	0.1525	1.30E - 04	70.9
3	3.26E - 01	1.09E + 01	6.30E + 01	0.1123	1.13E - 04	58.5
4	2.31E - 01	7.30E + 00	6.94E + 01	0.0710	8.45E - 05	51.5
5	8.13E - 02	4.65E + 00	8.18E + 01	0.0487	6.98E - 05	48.6
6	3.16E - 02	3.06E + 00	1.78E + 01	0.0366	6.41E - 05	37.2
7	2.13E - 02	2.14E + 00	1.55E + 01	0.0253	5.51E - 05	34.0
8	1.21E - 02	1.59E + 00	9.28E + 00	0.0156	4.33E - 05	25.2
9	5.62E - 03	7.19E - 01	1.04E + 01	0.0078	2.81E - 05	23.3
10	1.40E - 03	4.57E - 01	5.62E + 00	0.0038	1.88E - 05	14.3
11	2.63E - 04	3.38E - 01	2.80E + 00	0.0025	1.72E - 05	10.5
12	9.85E - 05	2.55E - 01	5.51E + 00	0.0018	1.81E - 05	10.2
13	6.97E - 05	1.17E - 01	1.70E + 00	0.0010	1.59E - 05	6.4
14	1.56E - 05	6.51E - 02	1.62E + 00	0.0007	1.98E - 05	5.1
15	9.50E - 06	4.14E - 02	1.43E + 00	0.0004	2.63E - 05	3.7
16	2.90E - 06	3.14E - 02	1.51E + 00	0.0003	5.46E - 05	2.9
17	2.21E - 06	1.94E - 02	1.68E + 00	0.0002	1.99E - 04	1.9

Table 7.9. Refined Emission Source Profiles (mg/g) *Q*-Mode Analysis of NBS Data Set No. 1

Element	Wood	Soil	Refuse	Oil	Sandblast	Motor Vehicle
C	464 ± 24	34 ± 10	33 ± 40	140 ± 40	3.6 ± 12.8	642 ± 30
Na	8.3 ± 9	4.3 ± 3.8	115 ± 16	12 ± 12	13 ± 5	84.7 ± 11.2
Al	0 ± 50	246 ± 21	76 ± 81	8.0 ± 63.1	83 ± 27	9.1 ± 59.6
Si	15 ± 80	518 ± 34	0.001 ± 0.646	37 ± 105	375 ± 44	11 ± 99
S	0.040 ± 0.123	12 ± 6	149 ± 23	87 ± 18	11 ± 8	47 ± 17
Cl	3.7 ± 5.1	4.2 ± 2.2	245 ± 9	6.2 ± 6.7	2.2 ± 2.8	16 ± 7
K	0.26 ± 7.96	9.6 ± 3.4	213 ± 14	7.9 ± 10.6	17 ± 5	0.51 ± 9.94
Ca	0 ± 18	7.3 ± 7.6	12 ± 31	41 ± 24	49 ± 10	54 ± 23
Ti	0.41 ± 2.43	11.3 ± 1.1	0.85 ± 4.11	4.6 ± 3.3	3.7 ± 1.4	8.5 ± 3.1
V	0.074 ± 0.216	0.30 ± 0.10	0.052 ± 0.364	1.94 ± 0.29	0.14 ± 0.12	0.007 ± 0.688
Cr	0.25 ± 0.49	0.78 ± 0.21	0.088 ± 0.821	0.11 ± 0.65	3.58 ± 0.27	0.78 ± 0.61
Mn	0.42 ± 0.33	2.20 ± 0.14	0.22 ± 0.55	0.36 ± 0.43	0.83 ± 0.18	1.1 ± 0.5
Fe	0.69 ± 3.38	85 ± 5.7	36 ± 23	4.4 ± 17.7	113 ± 8	6.9 ± 16.8
Ni	0.080 ± 0.330	0.29 ± 0.14	0.50 ± 0.56	0.93 ± 0.44	2.09 ± 0.18	0.71 ± 0.42
Cu	0.001 ± 0.398	0.18 ± 0.23	6.12 ± 0.92	0.63 ± 0.72	4.18 ± 0.30	0.98 ± 0.68
Zn	0.42 ± 1.66	1.2 ± 0.7	12 ± 3	0.52 ± 2.19	3.1 ± 1	5.4 ± 2.1
Br	0.014 ± 0.952	0.009 ± 0.400	0.021 ± 0.609	0.037 ± 0.257	0.010 ± 0.519	73.3 ± 1.2
Pb	0.27 ± 3.04	1.5 ± 1.3	13 ± 6	3.4 ± 4.1	0.35 ± 1.66	223 ± 4

Table 7.10. Diagonalization Results and Reproduction Summary for *R*-Mode Analysis NBS Data Set No. 1

Factor	Eigenvalue	RMS Error	Chi-Square	Exner	Average Percent Indicator	Average Percent Error
1	1.35E + 01	4.98E + 01	9.50E + 02	0.7514	2.81E - 04	180.7
2	1.92E + 00	3.46E + 01	4.05E + 02	0.5675	2.47E - 04	115.3
3	1.12E + 00	2.57E + 01	4.02E + 02	0.4244	2.17E - 04	97.6
4	1.01E + 00	1.89E + 01	1.34E + 01	0.2290	1.39E - 04	29.2
5	2.48E - 01	1.65E + 01	1.19E + 01	0.1453	1.06E - 04	23.1
6	4.46E - 02	1.32E + 01	1.33E + 01	0.1244	1.11E - 04	23.7
7	3.17E - 02	1.19E + 01	3.11E + 00	0.1071	1.19E - 04	14.5
8	2.46E - 02	8.33E + 00	2.20E + 00	0.0915	1.29E - 04	12.7
9	1.67E - 02	5.82E + 00	2.72E + 00	0.0791	1.45E - 04	11.6
10	1.08E - 02	3.50E + 00	3.35E + 00	0.0700	1.72E - 04	11.5
11	9.53E - 03	3.21E + 00	3.48E + 00	0.0608	2.09E - 04	10.5
12	9.24E - 03	2.54E + 00	4.30E + 00	0.0504	2.55E - 04	8.9
13	8.42E - 03	1.50E + 00	3.93E + 00	0.0384	3.07E - 04	6.6
14	5.15E - 03	1.26E + 00	3.47E + 00	0.0288	4.02E - 04	4.7
15	3.18E - 03	9.58E - 01	4.25E + 00	0.0207	5.93E - 04	3.5
16	1.44E - 03	7.40E - 01	7.70E + 00	0.0157	1.24E - 03	3.3
17	1.31E - 03	3.88E - 01	1.72E + 01	0.0091	4.07E - 03	3.0

analysis did not succeed in this either. There are several reasons for this result. The soil and basalt sources were colocated around the perimeter of the imaginary area around the receptor site (see Figure 6.7). The variation at the receptor site is then entirely due to meteorological conditions and they covary in the same manner. Thus, a factor analysis that operates on the variation will not be able to separate them. Since the variation in concentrations of soil and basalt is due entirely to the meteorology, the factor analysis would not be able to separate these sources even if there were substantial differences in their concentration profiles. The other source type missed in the analysis is the steel mill. It was included as a very weak source. It does not represent a unique source of any element and it does not contribute sufficient mass for its variation to be observed above the random variation built into the data set.

The *R*-mode analysis was able to obtain a seven-factor solution while it was not possible to do so using the *Q*-mode eigenvectors. The *Q* mode did not iteratively obtain as good source profiles as could be obtained with the *R*-mode approach. Both methods did give a reasonable description of the sandblast source composition. This source was not given to the analysts with the data as were the other profiles. The other unusual feature of the data set was the coupling of a

Table 7.11. Refined Emission Source Profiles (mg/g) *R*-Mode Analysis of NBS Data Set No. 1

	Wood	Oil-B	Incinerator	Coal	Soil	Sandblast	Motor Vehicles
C	591 ± 42	185 ± 26	35 ± 37	15 ± 51	68 ± 12	16 ± 25	573 ± 17
Na	4.0 ± 20.0	14 ± 13	64 ± 18	35 ± 25	0.006 ± 0.410	8.5 ± 11.8	54.8 ± 8.0
Al	10 ± 31	11 ± 20	0.002 ± 0.514	210 ± 40	166 ± 9	0.010 ± 0.552	44 ± 13
Si	12 ± 92	44 ± 57	0.53 ± 9.45	230 ± 120	466 ± 25	508 ± 54	73 ± 37
S	0.25 ± 6.51	120 ± 17	64 ± 24	180 ± 40	5.7 ± 7.2	0.025 ± 0.579	10 ± 11
Cl	16 ± 11	5.4 ± 6.7	196 ± 10	23 ± 14	0.35 ± 2.90	37 ± 7	16 ± 5
K	13 ± 5	2.9 ± 2.8	122 ± 4	17 ± 6	7.5 ± 1.3	8.5 ± 2.7	0.004 ± 0.650
Ca	14 ± 14	26 ± 9	1.6 ± 11.4	81 ± 16	41.4 ± 3.6	1.5 ± 7.7	13 ± 6
Ti	0 ± 3	2.7 ± 1.4	5.4 ± 2.0	16 ± 3	12.9 ± 0.7	2.7 ± 1.4	0.77 ± 0.90
V	0.00 ± 0.11	0.660 ± 0.068	0.25 ± 0.10	0.41 ± 0.14	0.315 ± 0.030	0.15 ± 0.07	0.003 ± 0.432
Cr	0.0 ± 0.4	0.026 ± 0.203	1.2 ± 0.3	0.35 ± 0.40	0.54 ± 0.09	5.49 ± 0.20	0.027 ± 0.130
Mn	0.0 ± 0.4	0.13 ± 0.20	1.3 ± 0.3	1.9 ± 0.4	1.97 ± 0.09	1.1 ± 0.2	0.61 ± 0.13
Fe	0.31 ± 9.32	35 ± 6	2.4 ± 8.2	143 ± 12	96.4 ± 2.6	45.6 ± 5.5	9.2 ± 3.8
Ni	0.00 ± 0.11	0.094 ± 0.066	0.12 ± 0.10	0.50 ± 0.13	0.196 ± 0.029	3.06 ± 0.07	0.004 ± 0.042
Cu	0.0 ± 0.7	0.40 ± 0.42	4.13 ± 0.59	0.014 ± 0.820	0.52 ± 0.19	6.63 ± 0.40	0.060 ± 0.269
Zn	0.45 ± 2.43	0.25 ± 1.52	10 ± 3	0.53 ± 2.97	0.28 ± 0.66	6.0 ± 1.5	4.2 ± 1.0
Br	0.12 ± 1.13	0.48 ± 0.71	0.37 ± 0.98	0.0 ± 1.4	0.89 ± 0.31	0.70 ± 0.67	52.4 ± 0.5
Pb	0.014 ± 0.614	3.4 ± 2.9	9.2 ± 4.1	1.6 ± 5.7	5.0 ± 1.3	5.9 ± 2.8	152 ± 2

motor vehicle composition with soil to give "road" material. Since the two sources were coupled, the factor analysis could not separate them and the analysis, therefore, deduced the composition of the road source rather than motor vehicles. For real data, a similar result might provide a better estimate of the true impact of traffic on airborne particulate loadings than trying to separate vehicle exhaust from reintrained soil.

Although the factor analysis did not fully resolve this data set, the *R*-mode analysis did provide reasonably good estimates of the impacts of the various sources. A comparison of the average results of the two forms of analysis with the actual values is made in Table 7.12. In the *R*-mode analysis, a coal source was resolved permitting much better predictions of the average mass values. In order to compare the ability to resolve individual samples, linear regressions between the mass values calculated by the factor analysis and the actual values were performed. Table 7.13 presents these results. For well-resolved sources like sandblasting and soil/basalt, excellent agreement was obtained as reflected by high correlation coefficients, a slope near 1.0, and a zero intercept. Coal burning, although resolved by the *R*-mode analysis, was not fit very well, presumably because of the substantial collinearity with the soil and basalt sources. These results indicate that the factor analysis obtained the major features of the data but did not always resolve the detailed structure. It must be remembered that part of the problem may be in the data set. The colocation and multiple location of sources may not be entirely realistic and may cause some of the difficulties encountered in the analysis. On the other hand, these data have been extremely valuable in testing our methodology and would seem to indicate that more intensive study of the *R*-mode approach is warranted.

A set of U.S.G.S. coal data (Cecil et al., 1981) from a study of the Upper

Table 7.12. Estimated Source Impacts ($\mu g/m^3$) Compared with True Values for NBS Data Set

Source	*Q* Mode	*R* Mode	Truth
Steel A	–	–	0.05
Oil B	2.3 ± 0.4	2.1 ± 1.0	2.0
Incinerator	1.8 ± 0.1	1.9 ± 0.1	1.3
Coal B	–	2.2 ± 0.7	2.4
Basalt	12.7 ± 0.5	12.5 ± 0.8	4.7
Soil	–	–	8.0
Road	3.0 ± 0.1	4.0 ± 0.1	7.1
Wood	7.4 ± 0.7	4.3 ± 0.5	3.3
Sandblast	5.2 ± 0.4	4.1 ± 0.2	4.2
Total	34.2	31.1	33.05

Table 7.13. Comparison of R-Mode Calculated Mass Contributions with the Actual Mass Values for Each Sample

Source	Correlation	Slope	Intercept
Wood	0.800	0.873 ± 0.133	642 ± 367
Oil B	0.867	1.208 ± 0.209	454 ± 584
Incinerator	0.968	1.502 ± 0.077	20 ± 84
Coal B	0.689	0.656 ± 0.191	383 ± 450
Soil/Basalt	0.949	0.957 ± 0.051	294 ± 572
Sandblast	0.990	1.029 ± 0.026	-195 ± 124
Road	0.992	0.588 ± 0.006	-67 ± 49

Freeport coal bed that extends in the vicinity of Indiana, Pennsylvania, has also been examined. The bed was sampled at two mines and 21 complete bench channel samples were obtained. The 21 samples were subsampled by facies to provide a total of 75 bench samples. Analysis of the samples included multielemental determinations of 78 elements in addition to other physical properties. X-Ray diffraction analysis (XRD) was also performed to ascertain the contribution of the mineral phases to the samples. This data set provides an excellent opportunity to test the ability of factor analysis to identify mineral phases and, since XRD results were available, they provided an opportunity to see how well the factor analysis results compare with those of standard minerological methods.

There were some difficulties in determining the number of factors. Table 7.14 gives the results obtained when the reproduction tests were used on the data. As can be noted, it is very difficult to determine the number of factors from this

Table 7.14. Reproduction Summary of USGS Coal Samples with All Elemental Data

Factor	Eigenvalue	RMS	Chi-Square[a]	Exner	Average Percent Error
1	72.8	248	225,000,000	0.173	146
2	2.10	49.1	9,110,000	0.0298	52.6
3	0.0467	23.4	2,140,000	0.0156	44.6
4	0.0143	10.5	447,000	0.00673	43.7
5	0.00204	6.80	195,000	0.00413	45.2
6	0.000641	4.37	84,000	0.00285	43.3
7	0.000260	3.34	51,100	0.00213	43.2
8	0.000163	2.32	25,700	0.00151	42.0
9	0.000085	1.62	13,000	0.00106	41.7

[a]Chi-square is not weighted by the error.

Table 7.15. Reproduction Summary of USGS Coal Samples with One Element Deleted

Factor	Eigenvalue	RMS	Chi-Square[a]	Exner	Average Percent Error
			Nitrogen Deleted		
1	72.8	254	230,000,000	0.173	149
2	2.10	50.0	9,280,000	0.0298	53.7
3	0.0468	23.6	2,150,000	0.0154	45.5
4	0.0143	10.3	423,000	0.00640	44.6
5	0.00197	6.31	165,000	0.00370	46.0
6	0.000638	3.44	51,100	0.00221	44.0
7	0.000175	2.47	27,300	0.00157	42.8
8	0.0000968	1.65	12,800	0.00106	42.4
9	0.0000411	1.25	7,670	0.00075	40.9
			Carbon Deleted		
1	63.6	167	100,000,000	0.412	70.5
2	8.80	53.2	10,500,000	0.196	46.1
3	2.10	26.9	2,790,000	0.0854	38.9
4	0.212	21.9	1,920,000	0.0644	39.4
5	0.176	10.1	420,000	0.0391	37.2
6	0.0675	6.27	169,000	0.0228	36.6
7	0.0310	2.81	35,500	0.00758	34.9
8	0.00247	1.69	13,400	0.00453	33.5
9	0.000712	1.33	8,590	0.00314	30.9
			Hydrogen Deleted		
1	72.8	254	230,000,000	0.173	149
2	2.10	50.0	9,280,000	0.0298	53.7
3	0.0468	23.6	2,150,000	0.0154	45.5
4	0.0143	10.3	423,000	0.00640	44.6
5	0.00197	6.31	165,000	0.00370	46.0
6	0.000638	3.44	51,100	0.00221	44.0
7	0.000175	2.47	27,300	0.00157	42.8
8	0.000096	1.65	12,800	0.00106	42.4
9	0.000041	1.25	7,670	0.00075	40.9

[a]Chi-square is not weighted by the error.

data. There is a slight indication of four factors, but it is not substantiated by all of the tests. Therefore, to learn more about the number of factors in the data, specific elements were deleted from the data set and the reproduction tests were repeated. Table 7.15 shows the results of the reproduction tests obtained when either carbon, nitrogen, or hydrogen were deleted. As can be noted from the table, as long as carbon was included in the data, the tests produced confusing results; however, when nitrogen and hydrogen were present, a uniform indication of seven factors was observed in all the tests. To carry the analysis further, the reproduction tests were again performed with carbon and nitrogen deleted from the data and then carbon and hydrogen deleted. The results of the tests, shown in Table 7.16, indicate the presence of only six factors. Therefore, it appears that hydrogen and nitrogen act as unique factors in the data by requiring one more factor to reproduce the data when they are present. This is not surprising

Table 7.16. Reproduction Summary of USGS Coal Samples with Two Elements Deleted

Factor	Eigenvalue	RMS	Chi-Square[a]	Exner	Average Percent Error
			Carbon and Nitrogen Deleted		
1	63.4	156	85,300,000	0.413	68.2
2	9.36	46.6	7,900,000	0.182	45.3
3	1.91	24.0	2,180,000	0.0724	39.2
4	0.215	10.5	432,000	0.0458	38.0
5	0.0965	6.59	177,000	0.0262	37.5
6	0.0413	2.96	37,000	0.00895	35.5
7	0.00349	1.90	15,900	0.00538	34.3
8	0.00115	1.41	9,090	0.00349	31.7
9	0.000494	0.984	4,640	0.00223	30.3
			Carbon and Hydrogen Deleted		
1	65.0	115	46,000,000	0.382	69.7
2	8.42	35.7	4,630,000	0.152	44.2
3	1.10	23.4	2,070,000	0.0841	39.6
4	0.259	11.2	491,000	0.0574	38.9
5	0.154	6.78	187,000	0.0322	37.0
6	0.0631	3.03	38,900	0.0108	35.2
7	0.00500	2.06	18,700	0.00658	34.3
8	0.00176	1.47	9,930	0.00419	31.2
9	0.00071	1.03	5,080	0.00268	30.2

[a]Chi-square is not weighted by the error.

Table 7.17. Reproduction Summary of USGS Coal Samples with Carbon, Hydrogen, and Nitrogen Deleted

Factor	Eigenvalue	RMS	Chi-Square[a]	Exner	Average Percent Error
1	63.3	72.8	18,100,000	0.413	70.2
2	10.7	17.5	1,080,000	0.122	44.5
3	0.558	12.8	603,000	0.0827	42.7
4	0.291	5.86	131,000	0.0509	41.5
5	0.162	2.42	23,400	0.0153	37.7
6	0.0130	1.12	5,170	0.0066	35.2
7	0.00127	0.891	3,440	0.0050	32.9
8	0.00091	0.681	2,100	0.0035	31.1
9	0.00039	0.341	547	0.0025	30.7

[a]Chi-square is not weighted by the error.

since coal, in various stages of development, will have different ratios between carbon, hydrogen, and nitrogen. The data used to obtain the results of Table 7.16 had only one element that could definitely be attributed to an organic if none of the major contributors to the organic phases were present. Carbon, hydrogen, and nitrogen were all deleted from the data, and the reproduction tests applied again. As can be seen in Table 7.17, there is a uniform indication for five or six factors as determined by all the tests. Because of the two successive large drops in the magnitudes of the eigenvalues between five, six, and seven factors, five factors could be assumed since the sixth factor was much less important (as indicated by the drop in the magnitude of the eigenvalue). If one considered that the organic elements were deleted, this sixth factor could be the organic phase due to contributions of minor elements.

To determine the nature of the mineral phases present in the coal, the data set with the carbon concentrations included was used. With carbon included in the analysis, six factors were assumed to be present. Unique vectors for each of the 45 elements were iterated until convergence, normalized, and clustered. We were ultimately able to identify an organic component and five mineral phases that could be identified as pyrite, calcite, kaolinite, illite, and quartz. Reasonable elemental profiles for each phase were obtained and are presented in Table 7.18. The amounts of each mineral phase determined by TTFA were then compared to the relative concentrations of the same phase as determined by XRD. The XRD results provided by U.S.G.S. are relative, not absolute, concentrations so we cannot directly compare the numbers on a 1 : 1 basis. These comparisons are shown in Figures 7.4–7.8 and the parameters for the linear regressions are given in Table 7.19. The correlation coefficients range from 0.63 to 0.85. Thus, quite

Table 7.18. Mineral Phases Determined by Factor Analysis in the USGS Coal Samples

Element	Organic	Pyrite	Calcite	Kaolinite	Illite	Quartz
C	863,000 ± 1000	0 ± 5000	140,000 ± 30,000	0 ± 6000	0 ± 4000	0 ± 4000
S	6870 ± 850	603,000 ± 17,000	0 ± 700,000	0 ± 30,000	0 ± 14,000	0 ± 12,000
Si	0 ± 300	0 ± 6000	0 ± 30,000	225,000 ± 8000	250,000 ± 5000	344,000 ± 4000
Al	0 ± 300	0 ± 5000	0 ± 30,000	217,000 ± 7000	188,000 ± 5000	0 ± 4000
Ca	0 ± 10	0 ± 200	480,000 ± 2000	0 ± 300	0 ± 170	0 ± 140
Mg	0 ± 120	0 ± 3000	0 ± 14,000	12,000 ± 4000	4800 ± 2000	6700 ± 1700
Na	0 ± 50	2900 ± 900	0 ± 5000	4000 ± 1300	1900 ± 800	2400 ± 700
K	0 ± 80	0 ± 1400	0 ± 8000	0 ± 2000	64,500 ± 1200	0 ± 1000
Fe	0 ± 9000	481,000 ± 16,000	0 ± 90,000	0 ± 30,000	0 ± 14,000	44,000 ± 12,000
Ti	0 ± 90	0 ± 1800	0 ± 10,000	9500 ± 2500	8900 ± 1500	9210 ± 1240
P	140 ± 70	0 ± 1400	0 ± 8000	0 ± 1900	0 ± 1200	0 ± 1000
As	0 ± 5	721 ± 88	0 ± 500	0 ± 130	0 ± 80	0 ± 70
B	0 ± 5	0 ± 90	0 ± 500	0 ± 130	94 ± 75	110 ± 70
Ba	0 ± 13	0 ± 300	0 ± 1400	930 ± 340	340 ± 210	410 ± 180
Be	0.0 ± 0.3	23 ± 5	0 ± 30	7.3 ± 6.8	6.0 ± 4.2	8.2 ± 3.5
Cd	0.00 ± 0.11	2.1 ± 2.0	0 ± 11	0 ± 3	0.0 ± 1.7	0.0 ± 1.4
Ce	0 ± 6	0 ± 110	0 ± 600	0 ± 150	190 ± 90	160 ± 80
Co	0.0 ± 1.2	63 ± 24	120 ± 140	16 ± 33	7.7 ± 19.8	1.9 ± 17
Cr	0 ± 3	0 ± 50	0 ± 300	140 ± 60	200 ± 40	96 ± 30
Cs	0.0 ± 0.4	0 ± 8	44 ± 41	12 ± 11	23 ± 7	7.0 ± 5.2
Cu	0 ± 5	0 ± 100	0 ± 600	150 ± 130	88 ± 79	94 ± 66
Eu	0.00 ± 0.12	0 ± 3	0 ± 13	0 ± 4	2.5 ± 1.9	2.9 ± 1.6
F	0 ± 30	940 ± 440	0 ± 3000	0 ± 700	980 ± 370	0 ± 400
Ga	0.0 ± 1.1	0 ± 20	0 ± 120	97 ± 28	31 ± 17	39 ± 15
Gd	0.0 ± 0.3	31.6 ± 4.7	0 ± 30	15 ± 7	8.3 ± 4.0	12 ± 4

Hf	0.00 ± 0.16	0 ± 4	0 ± 18	0 ± 5	11 ± 3	5.9 ± 2.3
Hg	0.00 ± 0.09	14.7 ± 1.8	0 ± 10	0 ± 3	0.0 ± 1.5	0.0 ± 1.3
La	0 ± 3	0 ± 60	0 ± 300	0 ± 80	100 ± 50	78 ± 38
Li	0 ± 8	0 ± 150	0 ± 900	380 ± 210	250 ± 130	190 ± 110
Lu	0.00 ± 0.04	1.5 ± 0.7	0 ± 4	0.0 ± 1.0	0.98 ± 0.56	0.85 ± 0.47
Mn	0 ± 10	0 ± 190	2500 ± 1100	0 ± 300	0 ± 160	210 ± 140
Nb	0.0 ± 0.5	0 ± 10	0 ± 60	23 ± 13	19 ± 8	28 ± 7
Ni	0 ± 3	0 ± 60	590 ± 320	140 ± 80	0 ± 50	81 ± 40
Pb	0 ± 3	42 ± 50	120 ± 280	3.2 ± 69.1	48 ± 43	28 ± 36
Sb	0.00 ± 0.17	18 ± 4	0 ± 19	0 ± 5	0 ± 3	0 ± 3
Sc	0 ± 0.6	0 ± 11	0 ± 60	24 ± 15	41 ± 9	24 ± 8
Se	0.0 ± 1.0	35 ± 19	200 ± 110	0 ± 30	0 ± 16	22 ± 14
Sm	0.0 ± 0.6	0 ± 11	0 ± 60	0 ± 15	12 ± 9	13 ± 8
Sr	0 ± 30	0 ± 500	4400 ± 2600	0 ± 700	0 ± 400	260 ± 330
U	0.0 ± 0.4	11 ± 7	0 ± 40	0 ± 10	8.6 ± 5.9	5.1 ± 5.0
V	0 ± 4	0 ± 80	0 ± 400	340 ± 100	120 ± 60	130 ± 60
Y	0.0 ± 1.5	0 ± 30	350 ± 160	0 ± 40	33 ± 24	51 ± 20
Yb	0.00 ± 0.20	0 ± 4	0 ± 30	0 ± 6	6.3 ± 3.3	5.3 ± 2.8
Zn	0 ± 7	390 ± 140	0 ± 800	250 ± 190	0 ± 120	230 ± 100
Zr	0 ± 4	0 ± 80	0 ± 500	140 ± 100	160 ± 70	220 ± 60

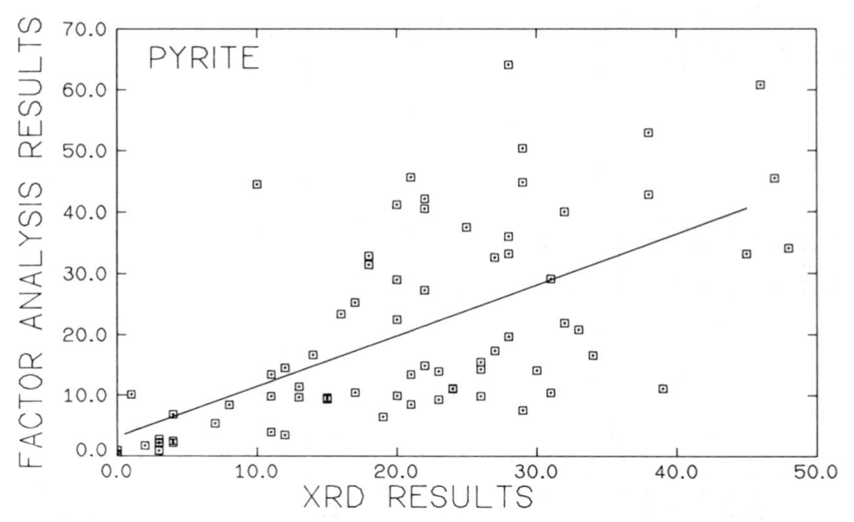

Figure 7.4. Scatter plot of the fraction of pyrite in the mineral matter compared to the relative XRD intensities for pyrite.

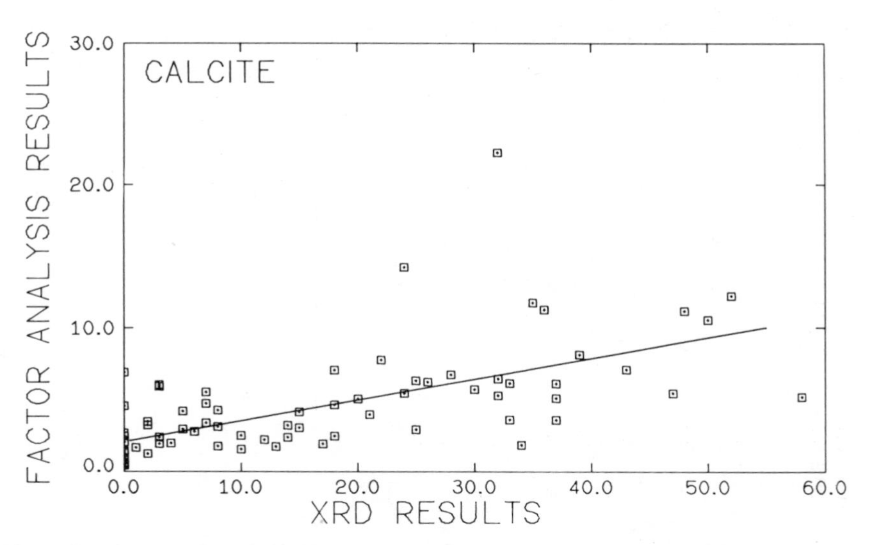

Figure 7.5. Scatter plot of the fraction of calcite in the mineral matter compared to the relative XRD intensities for calcite.

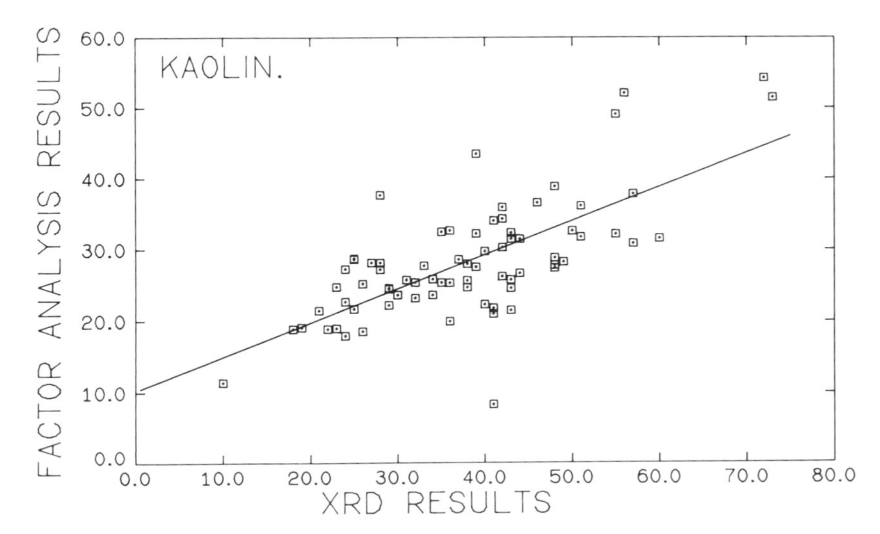

Figure 7.6. Scatter plot of the fraction of kaolinite in the mineral matter compared to the relative XRD intensities for kaolinite.

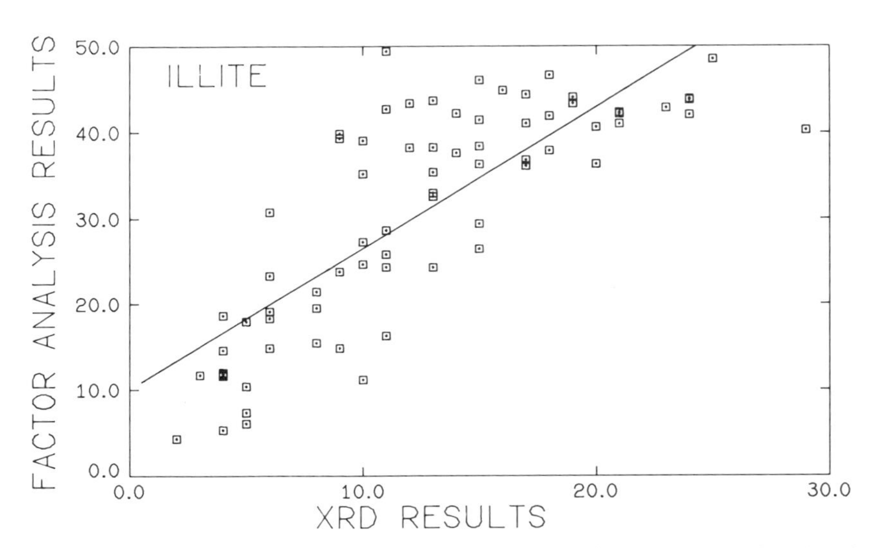

Figure 7.7. Scatter plots of the fraction of illite in the mineral matter compared to the relative XRD intensities for illite.

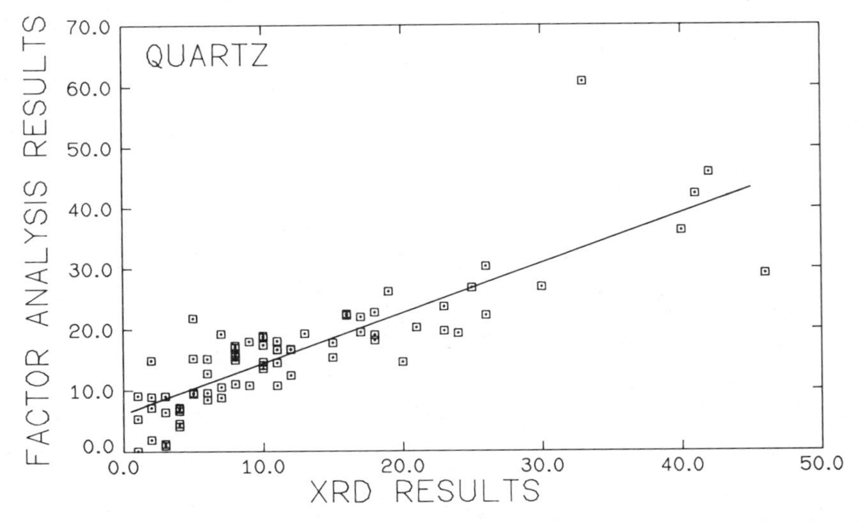

Figure 7.8. Scatter plot of the fraction of quartz in the mineral matter compared to the relative XRD intensities for quartz.

Table 7.19. Summary of Linear Correlation Coefficients, Slopes, and Intercepts for Mineral Phases Determined by XRD and Factor Analysis

	Pyrite	Calcite	Kaolinite	Illite	Quartz
Correlation	0.66	0.63	0.70	0.80	0.85
Slope	0.8309 ± 0.0088	0.1440 ± 0.0073	0.4766 ± 0.0097	1.644 ± 0.018	0.824 ± 0.011
Intercept	3.19 ± 0.22	2.12 ± 0.17	10.27 ± 0.38	10.08 ± 0.25	6.21 ± 0.18

good agreement between the TTF and XRD results were obtained although the precision in the XRD values preclude more detailed comparisons.

In developing the composition profiles for the mineral phases, we tried both R- and Q-mode analysis. The results presented are those from the Q-mode analysis, because it was impossible to develop iteratively reasonable concentration profiles using the R-mode approach. At this time it is not understood why the R-mode approach seems superior for the NBS data and the Q mode better for the coal data. Further study with similar known structure data sets is needed to identify the criteria to use in applying one form or the other. However, it is clear that both for created and real data sets a reasonable resolution of the important source contributions can be obtained with target transformation factor analysis. However, it is also clear that there are limits to the resolution obtained with this analysis and a number of questions regarding the application of TTFA methods need to be more fully answered.

SINGLE PARTICLE DATA

Another important application of factor analysis is in the interpretation of single particle data obtained by computer-controlled scanning electron microscopy. The problem with the particle class balance as described in Chapter 6 is that it requires that authentic source samples be available in order to define the profiles used in the fits. However, as with bulk composition data, it may be impossible to obtain the source samples or it may be of interest to obtain estimates of the profiles at the receptor site to compare with source samples to determine the degree of fractionation in transit. Factor analysis can prove useful in these applications.

Again as in the particle class balance, the input variables for each sample are the weight percent of particles in each of the empirically defined particle classes. Johnson and coworkers do not size segregate their samples or source profiles. They have used a *Q*-mode analysis of a combined data set of both the individual particle vectors and a set of source profiles. This approach has been applied to a set of samples from Houston, Texas, with the set of source profiles primarily obtained from Syracuse, New York. Johnson and Twist (1983) wanted to determine how well a source profile from a library could be applied to an unknown ambient particle sample. From a library of 52 sources and 3 Houston soil or road dust samples, 30 sources were selected as potential end members for the factor analysis by using a chi-squared minimum marginal homogeneity test. The test employs an $X \log X$ transformation of the number percent or weight percent by particle type categories as suggested by Kullback (1959). The 25 ambient samples were compared one at a time to each of the 55 library profiles and a homogeneity value calculated. The marginal or by-particle-class contributions to the total homogeneity were computed. Any library source profile for which the cumulative marginal χ^2 homogeneity for two-thirds of the 25 particle class types was less than the critical value (95% confidence level) was retained for inclusion in the apportionment calculations. Certain generic signatures (five in all) were also retained; these included secondary particles, sea salt, calcium sulfate, limestone, and organic debris.

After some adjustment in the particle classification following a discriminant analysis of a subset of ambient and source samples, a *Q*-mode factor analysis of the kind described by Full, Ehrlich, and Klovan (1981) was performed. This *Q*-model begins with a factor analysis of the samples and observation variables submitted to it. For a particular run or application there may be different dimensionalities to the mathematical solution of the vector analysis, depending upon variance of the input data. When utilizing the 58 variables for this work, the number of factors (end members) were assigned at the point where the eigenvalues fell below 1.00. This resulted in the specification of a 12-factor solution and explained 95% of the variability in the complete data set. The *Q*-model computations were carried out with summaries of the data (weight percent by particles

Table 7.20. *Q*-Mode Source Type Fractional Apportionments for Three Houston Aerosol Samples[a]

	Sample		
Source Type	1	2	3
Coal Flyash 1	—	—	—
Coal Flyash 2	—	—	—
Steel Mill	0.05	-0.01	0.0
Limestone	—	—	—
Seal salt	0.0	0.01	0.05
Cement plant	0.09	0.62	0.37
Organic debris	0.0	0.01	0.0
Auto (Pb)	0.01	-0.01	0.04
$CaSO_4$	0.05	0.03	0.03
Construction	0.49	0.15	0.35
Secondary particles	0.0	-0.01	0.0
Syracuse dust 1	-0.03	0.03	-0.01
Houston soil 2	0.37	0.29	0.21
Miscellaneous	-0.02	-0.09	-0.03

[a]Taken from Johnson and Twist (1983).

class histograms). When the individual object vectors were subjected to factor analysis (500 at a time), higher degrees of dimensionality were indicated. The number of factors seemed to be related to the number of elements monitored by the X-ray spectrometer. When unusual or trace elements like I, Sn, Cd, Mo, and Se were eliminated from the variables for factor analysis, 10–12 factors routinely had eigenvalues greater than 1.00. Thus, all *Q*-model experimental runs were made using 12 possible end members (factors identified with specific sources). Table 7.20 shows the source apportionment results for the Houston aerosol samples that were studied in detail.

It would appear from these results that a generic- or source-type apportionment of unknown ambient aerosol samples can be made using a source-type signature library and employing SEM methods. From an examination of the emission inventory for the Houston region, the types of apportionments shown in Table 7.20 would appear reasonable. Of course, the approach utilized here is dependent upon the textural quality of the source-type signature library—potential major source types must be on file. In the case of the Houston aerosol samples used in this example, the apportionment results indicate that much of the aerosol has calcium-rich sources (cement plant-like, construction/destruction, limestone, and $CaSO_4$). Based upon this type of SEM apportionment, it would seem that subsequent source samplings (as may be required in Houston) should concentrate on appropriate Ca-rich sources.

SUMMARY

It has been shown that factor analysis provides a set of powerful methods for examining data. First, it permits a screening of data sets to identify outlier samples or noisy variables. Once these samples and/or variables are eliminated from the data set, a reanalysis can be used to determine the number of estimable entities in the apportionment. Mass balance methods appear to give greater resolution in terms of the number of sources determined. However, there are questions of collinearity and stability of regression coefficients that have not been addressed in most published studies.

Factor analysis can be used to develop source profiles and mass concentrations. However, factor analysis depends on the independent variation of the source types over the set of samples available. In some cases where most of the variability comes from the meteorology and not from the source emission rates, colocated sources cannot be separated even with substantial differences in the source emission. In the factor analysis of the data of Kowalczyk, Gordon, and Rheingrover (1982), factor analysis encountered great difficulty in separating an oil-fired power plant in spite of it having a distinctive source composition (V and Ni). There are two major oil-fired plants in the Washington, D.C. area; one on the Delmarva peninsula and one adjacent to the Alexandria incinerator. In the first case, the easterly wind brings in the particles well mixed with sea salt. Since the samples are not size segregated, the two sources are inseparable. In the second case, the oil-fired plant and incinerator particles cannot be separated with a factor analysis. Thus, there are strengths and weaknesses for each of these two most widely employed mathematical methods.

CHAPTER

8

OTHER MATHEMATICAL MODELS

The preceding two chapters described in detail the chemical mass balance and factor analysis methods. These models have been the most widely used of the mathematical receptor models. A number of other mathematical approaches have been employed to obtain qualitative or quantitative information regarding sources of environmental samples.

FACTOR ANALYSIS/MULTIPLE REGRESSION MODEL

Several research groups have used a combination of methods to identify tracers for the sources of airborne particulate matter and then apportion the particulate mass. The major developers of this approach were Kleinman and coworkers at the New York University Medical Center (Kleinman, 1977; Kleinman, Kneip, and Eisenbud, 1976; Kleinman et al., 1980a, b). Their approach to particle sampling was to collect 7-day integrated samples. The particles were dissolved from the filters. Metals (Cd, Cr, Cu, Fe, Pb, Mn, Ni, V, and Zn) were determined by atomic absorption spectroscopy. Anionic species (NO_3^-, NO_2^-, and SO_4^{2-}) were determined using colorometric methods.

The first step in their mathematical treatment was to perform a classical principal components analysis to select the best predictor variables to be used in a multiple regression analysis to fit the total mass data. Their model can be expressed as

$$M_j = \sum b_k x_{kj} + b_0 \tag{8.1}$$

where M_j is the measured mass of the jth sample and x_{kj} is the value of the kth predictor variable measured in the jth sample. From the components analysis, they identified Cu, Pb, V, Mn, and SO_4^{2-} as the most likely predictor variables for about 100 samples taken in 1972–1973 (Kleinman et al., 1980b). The lead is attributed to motor vehicular emissions. Copper is considered indicative of refuse incineration particles. Sulfate is principally a secondary aerosol and is difficult to attribute it to particular sources. Vanadium is a well-known tracer for fuel oil combustion and it is suggested that manganese arises from wind-blown soil. They obtained the following multiple regression result

$$M_j = 12.0[\mathrm{Pb}] + 54.0[\mathrm{Cu}] + 103[\mathrm{V}] + 1.66[\mathrm{SO_4^{2-}}] + 420[\mathrm{Mn}] + 26.8 \quad (8.2)$$

where all of the concentrations are in $\mu g/m^3$. The multiple correlation coefficient for this data set was 0.76 ($p < 0.001$) and the equation leads to a standard deviation of the residuals of ± 15 $\mu g/m^3$. An average of about 27 $\mu g/m^3$ is not apportioned and is in part related to sea salt and other emissions for which none of the elements are good predictors.

There have been a number of reports on the development and other applications of this approach to the New York City aerosol (Kleinman, Bernstein, and Kneip, 1977; Kleinman, Kneip, and Eisenbud, 1976; Kleinman et al., 1980a; Kneip, Mallon, and Kleinman, 1983) including the specific investigation of sources of carbonaceous aerosol (Shah et al., 1981b) and polycyclic aromatic hydrocarbons (Daisey, 1983). Recent application was made to Newark, New Jersey (Morandi, Daisey, and Lioy, 1983).

Dattner (see Gerlach, Currie, and Lewis, 1983) has used a similar approach in the analysis of the first NBS simulated data set as well as the intercomparison data set for Houston, Texas (see Dzubay and Stevens, 1983). This approach is being used in the analysis of the X-ray fluorescence elemental data taken in the Texas Air Control Board's ambient TSP monitoring network (Dattner et al., 1983).

Multiple linear regression has also been used for many other air quality data analyses without preselecting the variables with a factor analysis. For example, the polycyclic aromatic hydrocarbon concentrations in urban aerosols have been predicted using multiple linear regression (Butler, Crossley, and Colwill, 1982). Visibility degradation as a function of particle size and composition has also been examined. These methods can be useful for interrelating measured parameters. However, they depend too much on the availability of unique tracers to be applicable to areas with a variety of sources of the measured elements.

Thus, a problem does arise in the use of the elemental concentrations as the predictor variables for mass in a multiple linear regression analysis. Since more than one source type can emit a particular element, there will be some collinearity built into such an analysis. An alternative approach would be to use the principal components analysis to generate a set of linearly independent variables and perform the linear regression against the totally noncorrelated component scores. Henry and Hidy (1979; 1981a) perform just such an analysis by regressing measured ambient sulfate levels against the scores obtained from a principal component analysis of gaseous precursor and meteorological data. This approach then uses the patterns of measured species rather than a specific one to represent a single source. However, it really becomes identical to the approach described by Thurston and Spengler (1982) that was discussed in the previous chapter.

One of the other important aspects of the NYU work was the consideration of the problem of dispersion. Kneip et al. (1971) demonstrated a relationship between measured suspended particulate matter levels and atmospheric stability.

As discussed earlier, the volumetric concentration of material can vary depending on the wind speed and the mixing height. To normalize this effect, Kleinman (1977) suggests the use of a "ventilation factor" that is the product of the 7 A.M. mixing height times the wind speed. The average value of these ventilation factors represents the average dispersion volume and is termed the dispersion factor. For the 1972-1974 period in New York, a value of 4200 m^2/sec was obtained. Individual species and mass concentrations can be dispersion normalized by multiplying their value times the monthly average dispersion factor and dividing by 4200. This normalization should permit the comparison of season to season and year to year values with the effects of varying dispersion removed. This approach has not been widely used because in most cases, data from a significantly limited time period has been used so that there is little systematic variation in the dispersion factor. However, it appears to be an approach worth considering if data spanning more than a few months is to be used in an analysis.

TIME-SERIES ANALYSIS

The assumption underlying the use of time-series analysis for source identification is that different source types and even different specific sources of a given type may have regular periodicity to their emissions. Then, this periodicity could lead to the separation of the influence of the particular source's emissions from the remainder of the specific species being studied. In addition, the concentrations of airborne species as in many of environmental studies will be serially correlated because of meteorological effects. Again, through the examination of these interrelationships, it may be possible to remove some of the meteorological effects and make the effects of specific source variations more easily discerned.

There are a number of methods available to examine the variations in a sequential series of measurements of the properties of the system at a fixed location. These methods include spectral analysis, autoregression analysis, transfer function models, and intervention analysis. These methods have been used extensively in environmental studies including meteorology (Pasquill, 1962), atmospheric turbulence and electric and magnetic fields (Israelson and Oluwafemi, 1975), and temperature (Barn, 1976) to cite a few. Murray and Farber (1982) have recently used a Box–Jenkins autocorrelation approach to examine the relationship between sulfate and visibility and Bilonick and Nichols (1983) have looked at variations in acid precipitation levels. However, there are very few applications to the source identification and apportionment of measured environmental parameters that is the province of receptor models. The two principal published results in the field are Zinmeister and Redman (1978, 1980) and Hwang and Hopke (1983). In both cases, a univariate time-series analysis was made. In other words, only one measured variable was examined as a function of time with no examination of the covariation in time of multiple variables.

The first step in the analysis is to remove any constant direction trends in the data. A time series to be statistically analyzed is assumed to vary around a given mean value with no discernible trend. Thus, the series has to be examined to determine if there is a statistically significant trend in the sequence. A straight line can be fit to the data and its slope examined to determine if it is different from zero. If it is, then the line should be subtracted from the data. It is also common practice to remove the mean value from the series so it is centered on zero. There are other ways that the data may be treated so as to prepare them for proper time-series analysis but they are beyond the scope of this discussion. The interested reader can find full discussions in Box and Jenkins (1976) or Chatfield (1980).

The autocorrelation of the value of the variable at time t with the value of the same variable at time $t + k$ is estimated by

$$r_k = \frac{1}{N} \sum_{t=1}^{N-k} (Z_{t+k} - \bar{Z})(Z_t - \bar{Z})/s_x^2 \tag{8.3}$$

where $\bar{Z} = \Sigma_{t=1}^{N} Z_t/N$ and k one called the lag. If there were no periodic processes, r_k would be zero for all $k > 0$. The partial autocorrelation function, ϕ_{kk}, may be obtained from the following system of equations.

$$\begin{aligned} r_j &= \phi_{k1} r_{j-1} + \phi_{k2} r_{j-2} + \cdots + \phi_{kk} r_{j-k} \\ j &= 1, \cdots, k \end{aligned} \tag{8.4}$$

where $r_i = r_{-i}$.

The value of a variable at time t may depend on the values during the p prior time periods. The autoregressive model of order p fits the data with a series of the form

$$\begin{aligned} x_t &= b_1 x_{t-1} + b_2 x_{t-2} + \cdots + b_p x_{t-p} + e_t \\ t &= 1, 2, \cdots, N \end{aligned} \tag{8.5}$$

where b_i are the constant parameters to be fit and e_t is a sequence of uncorrelated and identically distributed random variables with mean zero and constant variance. This equation can be rewritten as

$$b(B)\, x_t = e_t \tag{8.6}$$

where $b(B)$ is a polynomial in the backshift operator, B

$$b(B) = 1 - b_1 B - b_2 B^2 - \cdots - b_p B^p \tag{8.7}$$

This model is termed an autoregressive model of order p, AR(p). For AR model of order p, the autocovariances, $\phi_{kk} = 0$, for all $k > p$.

In order to better fit the data, Box and Jenkins (1976) suggest the possible addition of a moving average model of order q, MA(q). This model can be represented by

$$\begin{aligned} x_t &= e_t - \theta_1 e_{t-1} - \cdots - \theta_q e_{t-q} \\ &= (1 - \theta_1 B - \cdots - \theta_q B^q)\, e_t \end{aligned} \tag{8.8}$$

where $|\theta_i| < 1$ and for an MA series of order q, $\phi_{kk} = 0$ for all $k > q$.

Another way of analyzing the periodic components in a time sequence is called spectral analysis. In effect, the series is fitted to a series of sine and cosine functions:

$$x_t = \sum_{i=1}^{m} (C_i \sin w_i t + D_i \cos w_i t) + e_t \tag{8.9}$$

where C_i and D_i are the amplitude of the sine and cosine, respectively, with frequency, w_i. The possible frequencies can be found by performing a Fourier transformation of the data from time to frequency space and looking for important frequencies as indicated by their spectral intensity, $I(w)$:

$$\begin{aligned} I(w) &= \frac{1}{N\pi}\left[\left(\sum_{t=1}^{N} x_t \sin w_i t\right)^2 + \left(\sum_{t=1}^{N} x_t \cos w_i t\right)^2\right] \\ i &= 1, M \end{aligned} \tag{8.10}$$

By examining $I(w)$ as a function of w, the important frequencies appear as peaks. The most significant frequencies can then be included in the summation in equation (8.9). The residuals in this equation ought to be such that their autocorrelation function should be less than or equal to $2/N$.

Spectral analysis has been applied by Hayas et al. (1982) to analysis of airborne particulate levels over a 1 year period. Seasonal, meteorological, and emission factors can be identified. These results indicate that some of the nonsource factors may be identifiable and therefore their influences can be removed from the data. Further study of this approach is needed to test its utility.

These various models can be combined with the desired final result being a sequence of residuals that cannot be distinguished from white noise, that is, a series of independent and identically distributed random variables with mean zero and variance, σ_e^2.

The autocorrelation function of the residuals can be calculated using equation (8.4) and a statistic, Q_k, can be formed by

$$Q_k = N \sum_{i=1}^{k} r_i^2(e)$$

Q_k should be approximately distributed as chi-squared and can be tested to determine if it is. Another approach is to use a Kolmogorov–Smirnov test of the goodness of fit to determine the randomness of e_t. The details of this test are given by Miller, Jr. (1981), Miller (1956), and Wine (1976).

Zinmeister and Redman (1978, 1980) used an autoregressive moving average, ARMA, model to fit sequences of single-element composition data taken in 2-h time intervals using a Florida State streaker sampler and proton-induced X-ray emission analysis. Their objective was to fit the general trends of the data and examine the residuals for the random source impacts on the particle levels. Thus, little information can be derived regarding the periodic sources of particles in the system. They are able to extract a considerable amount of information from the residuals in terms of source effects that seem to be reasonable for the airshed.

Hwang and Hopke (1983) used a combination of autoregression and spectral analysis to examine time sequences at two sites in St. Louis. The 2-hr time sequences for Pb, Br, Cu, Zn, and S were examined at each site. The sequences for RAPS site 103 are shown in Figures 8.1 to 8.5. The *S* series at both sites were very similar and had relatively long lag times that were significant in predicting the current value. Thus, sulfur appears to be from a source outside the region and its variation primarily dependent on the meteorology. The copper and zinc sequences were characterized by relatively low values with occasional and nonperiodic spikes of very high concentration values. Time-series analysis is not suited to this kind of pattern and little information was derived from these data. Site 103 near East St. Louis, Illinois, is near a major highway. The Pb and Br values at this location showed clear diurnal variations that could be readily interpreted as rush hour traffic. Spikes in the lead values corresponding to high emissions from the Cu/Zn source were found and they were carried through into the residual series after removal of the periodic components. Even with these high values included, the residuals from the Pb and Br correlated well at this site. At a site in south St. Louis, Missouri (site 105), the lead values were apparently also affected by a smelter to the south of the city. The effect of traffic was more diffuse and readily identifiable periodic components are not present in these data. Thus, for some cases, time series is useful in confirming the postulated sources but in other more complex cases, it may not provide many additional insights.

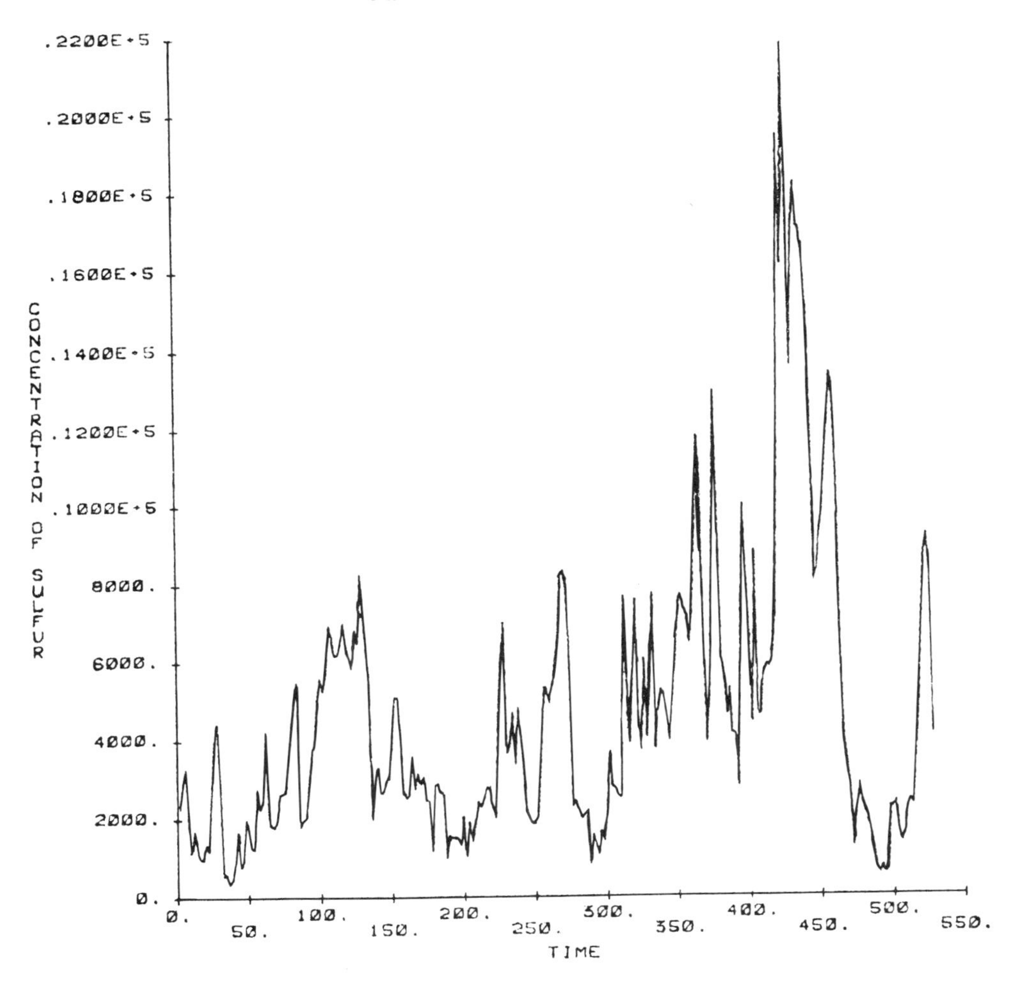

Figure 8.1. Concentration of sulfur (ng/m^3) at RAPS site 103 as a function of time (hours) after 0:00 hr on July 12, 1975. Each point represents a 2-hr interval.

LINEAR PROGRAMMING

The methods described previously have often fit models to data by the minimization of the sum of squares of the residual differences between the fitted values and the original data point. This use of least squares is particularly appropriate for data demonstrating a symmetric distribution of errors and the distribution is typically assumed to be normal. This approach then may overestimate the significance of points with large error. An alternative approach is to minimize the sum

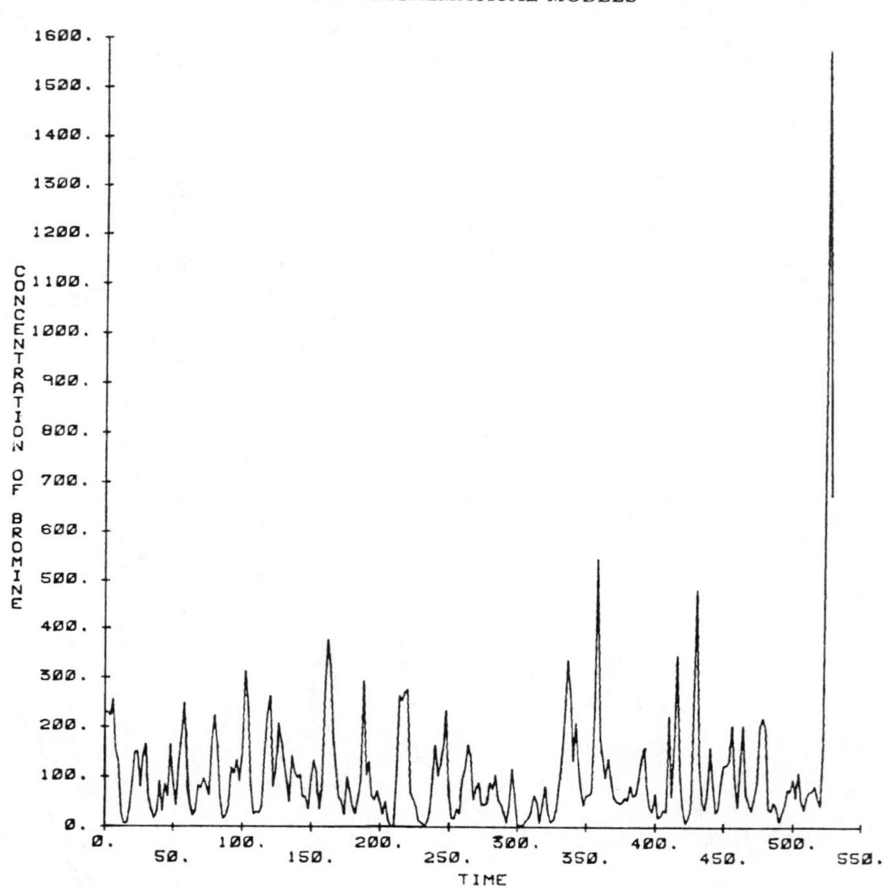

Figure 8.2. Concentration of bromine (ng/m^3) at RAPS site 103 as a function of time (hours) after 0:00 hr on July 12, 1975. Each point represents a 2-hr interval.

of the absolute differences between the observed and predicted values. The details of linear programming are discussed in a number of texts (e.g., Hadley, 1962). The use of linear programming was first suggested by Henry (1978) and the only application of it to receptor modeling was by Hougland (1983).

The approach suggested by Henry is to maximize the sum of mass fractional contributions from the various sources. The mass fractional contribution, s_{kj}, is the mass contribution, f_{kj}, in mg/m^3 divided by the total sample mass, m_j:

$$s_{kj} = f_{kj}/m_j \tag{8.11}$$

The sum of these fractional contributions should be 1 for any jth sample if all of

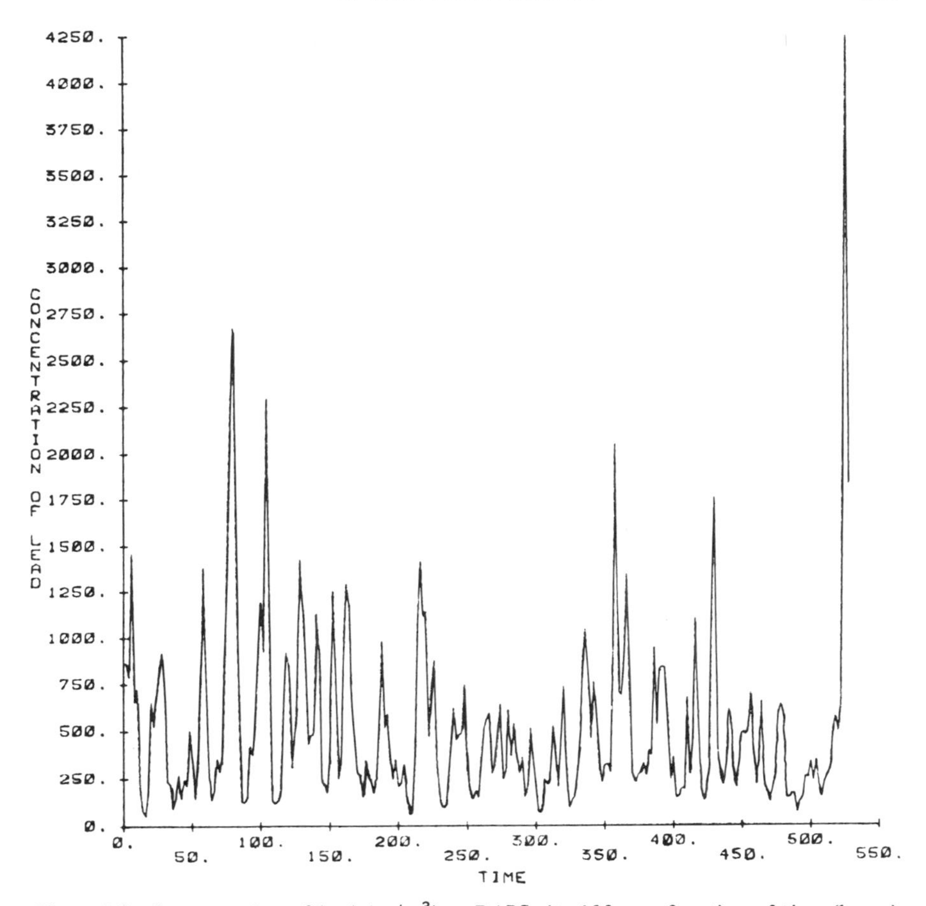

Figure 8.3. Concentration of lead (ng/m^3) at RAPS site 103 as a function of time (hours) after 0:00 hr on July 12, 1975. Each point represents a 2-hr interval.

the sources have been identified and included in the analysis. Thus, there is a constraint on the value of the s values so that

$$\sum_{k=1}^{p} s_{kj} \leqslant 1 \qquad \text{for all } j \tag{8.12}$$

Furthermore, each individual source contribution cannot be greater than the total sample mass loading so that

$$0 \leqslant s_{kj} \leqslant 1 \qquad \text{for all } j \text{ and } k \tag{8.13}$$

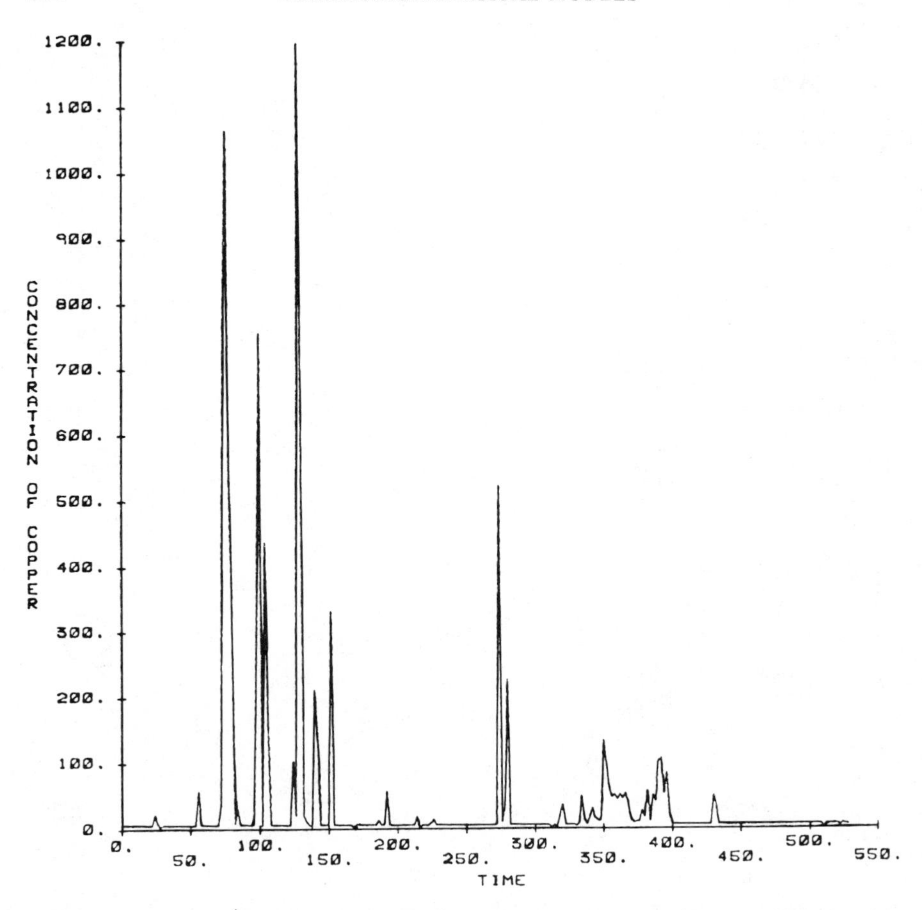

Figure 8.4. Concentration of copper (ng/m^3) at RAPS site 103 as a function of time (hours) after 0:00 hr on July 12, 1975. Each point represents a 2-hr interval.

In addition, it is also necessary that the values of the mass fractions yield the measured elemental concentrations when combined with source profile concentration values [equation (5.4)].

$$c_{ij} = x_{ij}/m_j = \sum_{k=1}^{p} a_{ik} f_{kj}/m_j = \sum_{k=1}^{p} a_{ik} s_{kj} \tag{8.14}$$

Thus, the mass balance assumption imposes the constraint that

$$c_{ij} + e_{ij} \geqslant \sum_{k=1}^{p} a_{ij} s_{kj} \tag{8.15}$$

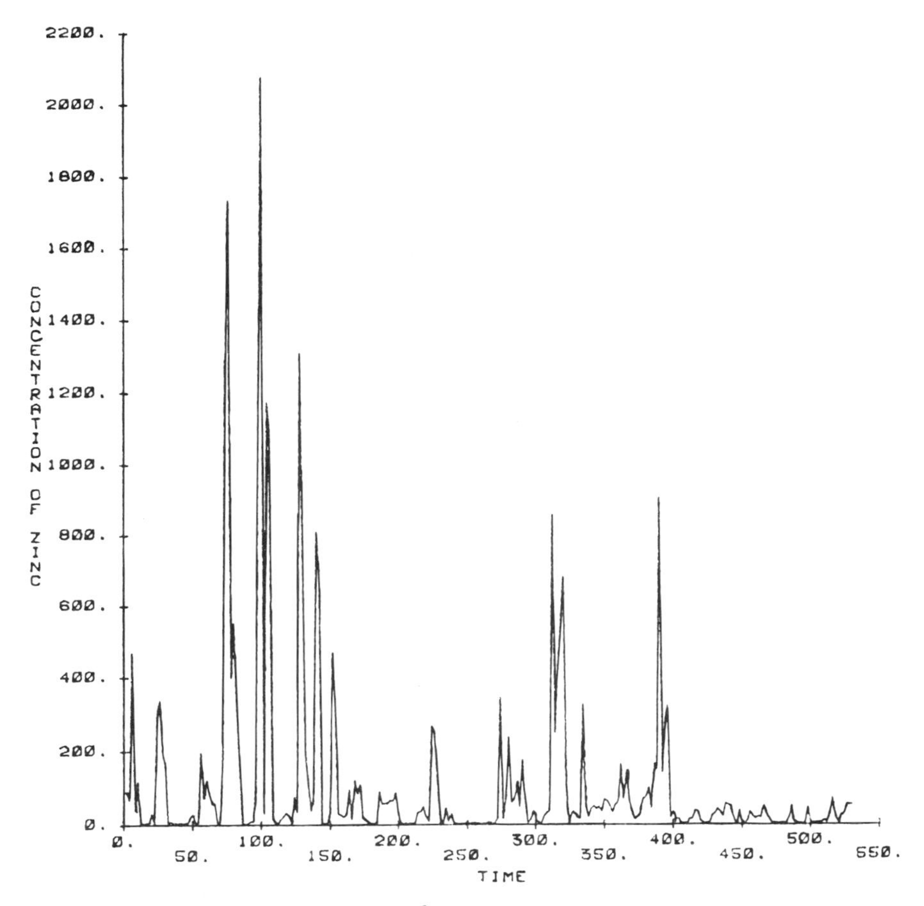

Figure 8.5. Concentration of zinc (ng/m^3) at RAPS site 103 as a function of time (hours) after 0 : 00 hr on July 12, 1975. Each point represents a 2-hr interval.

where e_{ij} is the uncertainty in the value of c_{ij}. There are no reports applying this approach to any receptor modeling problem.

Hougland (1983) has taken a somewhat different approach to describing the linear programming objective. He has reanalyzed the data from Washington, D.C. originally reported by Kowalcyzk, Choquette, and Gordon (1978). Unfortunately, he used only the six sources that the original authors suggested rather than the seven sources subsequently used by Kowalczyk, Gordon, and Rheingrover (1982) on a more extensive set of Washington, D.C. data. Hougland creates two error variables for each elemental concentration. One variable, error$i$+, represents the error between the observed value and mass balance prediction when it is positive and a second variable, error$i$−, has a value when the error is negative (i.e., when predicted exceeds the observed value). These two variables are exclusive so that if one has a nonzero value, the other must be zero. The objective is then to minimize the sum of errors

$$\sum(\text{error}i+ + \text{error}i-)$$

subject to the constraint imposed by the mass balance

$$x_{ij} = \sum_{k=1}^{p} a_{ik} f_{kj} + \text{error}i+ + \text{error}i- \tag{8.16}$$

for each element i. The results of this approach are given in Table 8.1 and show that the linear programming model gives quite similar results to the least-squares regression. Further studies of this method are needed to fully test its capabilities and determine its merit relative to the standard regression approach.

CLUSTER ANALYSIS

In many areas of study there is a need to be able to group objects that have characteristics similar to one another objectively. Cluster analysis is a technique that makes such groupings and displays the pattern of groups so that the investigator may see both the relationship between objects as well as between the groups of objects. The first large-scale efforts in this area came in sorting species into consistent sets of taxa, and a large literature has been developed in numerical taxonomy (Sneath and Sokal,1973).

The basic concepts of clustering can be best considered in geometric terms. In the space whose dimensionality is defined by the number of variables, each sample is represented by a point. It is then of interest to see which points are physically close enough to one another to form a group or cluster, just as the stars in the night sky are grouped into constellations. However, we would like a quantitative measure of cluster membership as well as a definition of the relationships between groups.

Several quantitative measures of distances between points or measures of sample dissimilarity can be defined. One measure of distance is merely the extension of the simple Euclidean distance to an m-dimensional space. The Euclidian distance (ED) is given by

$$\text{ED}_{jk} = \left[\sum_{i=1}^{m} (x_{ji} - x_{ki})^2 \right]^{1/2} \tag{8.17}$$

Table 8.1. Chemical Element Balances of Washington, D.C. Aerosols by Least-Squares and Linear Programming

Element ID	Observed[a] Value	Least-Squares Value	Least-Squares L/S[b]	Linear Programming Value	Linear Programming L/S
Elements used in fitting					
Al	1680 ± 1100	1940	1.2	1798	1.1
Na	470 ± 470	460	1.0	467	1.0
Fe	1260 ± 940	1340	1.1	1260	1.0
V	54 ± 57	55	1.0	54	1.0
Zn	150 ± 90	150	1.0	150	1.0
Pb	1329 ± 1706	1380	1.0	1329	1.0
Mn	27 ± 16	22	1.2	18	1.5
As	6 ± 5	5	1.2	6	1.0
Remaining Elements					
K	510 ± 350	340	1.5	466	1.1
Mg	440 ± 380	210	2.1	236	1.9
Ca	770 ± 500	280	2.8	302	2.5
Ba	27 ± 24	39	1.4	41	1.5
Cl	140 ± 75	870	6.2[c]	843	6.0
Br	190 ± 220	500	2.6[c]	480	2.5
I	9.3 ± 7.2	4.7	2.0[c]	5.2	1.8
Sc	0.6 ± 0.63	0.47	1.3	0.57	1.1
Ti	120 ± 95	120	1.0	147	1.2
Cr	11 ± 9	3.0	3.7	3.4	3.2
Co	1.1 ± 1.0	1.0	1.1	1.1	1.1
Ni	27 ± 28	11	2.5	11.1	2.4
Cu	13 ± 12	6.7	1.9	7.2	1.8
Se	3.5 ± 2.7	1.2	2.9[c]	1.4	2.4
Cd	3.5 ± 2.2	1.7	2.1	1.7	2.1
Sb	9.7 ± 8.5	2.3	4.2	2.3	4.2
La	1.9 ± 1.7	1.2	1.2	1.9	1.0
Ce	3.4 ± 3.1	2.3	1.5	2.8	1.2
Th	0.32 ± 0.24	0.33	1.0	0.38	1.2
Average L/S			2.0		1.8

[a] Concentrations in ng/m^3. Uncertainty = ±1 standard deviation.
[b] Ratio of larger to smaller of observed values and prediction from least squares or linear programming.
[c] Not included in L/S averaging.

Other functions have also been used including squared Euclidian distance (SED),

$$\text{SED}_{jk} = \sum_{i=1}^{m} (x_{ji} - s_{ki})^2 \tag{8.18}$$

mean character difference (MCD),

$$\text{MCD}_{jk} = \frac{1}{m} \sum_{i=1}^{m} (x_{ji} - x_{ki}) \tag{8.19}$$

mean Euclidian distance (MED),

$$\text{MED}_{jk} = \frac{1}{m} \left[\sum_{i=1}^{m} (x_{ji} - x_{ki})^2 \right]^{1/2} \tag{8.20}$$

and the mean squared Euclidian distance (MSED),

$$\text{MSED}_{jk} = \frac{1}{m} \sum_{i=1}^{m} (x_{ji} - x_{ki})^2 \tag{8.21}$$

An alternative approach is to consider a vector drawn from the origin to each of the n points. A measure of the similarity between two samples could be the cosine of the angle between their respective vectors (Imbrie, 1963). The cosine is given by

$$\cos \theta_{jk} = \frac{\sum_{i=1}^{m} (x_{ji})(x_{ki})}{\left(\sum_{i=1}^{m} x_{ji}^2 \sum_{i=1}^{m} x_{ki}^2 \right)^{1/2}} \tag{8.22}$$

With only positive valued data, this measure has range of 0–1. A value of 0.0 signifies there is nothing to common between the two samples, 1.0 shows identical samples, and 0.7071 (cos 45°) shows that the two vectors are about as similar as columns of random digits. The $\cos \theta$ is clearly just the correlation about the origin between the jth and kth samples that can be given this geometrical interpretation. Alternatively, the standard Pearson product-moment correlation coefficient can also be calculated between any pair of samples summed over the value of the various variables that have been determined for both samples. All of these parameters have been used for clustering in one of a wide variety of fields.

It is essential that the variables be standardized before calculating the dissimilarity measures. If the original variables are used, then trace species will span a much smaller size space than minor or major constituents. Thus, the major species will dominate the distance calculation. It is the variation within the available range that is important, not the magnitude of the range.

A question may be raised as to the number of parameters that should be included in the calculation of the dissimilarities between samples. Sneath and Sokal (1973) indicate that the greater the number of parameters included in the calculation, the more reliable the classification will be. To try to pick a subset of the data may bias the result. Certain elements may provide better discrimination between certain groups at the expense of poorer separation of others. The advantage of multielemental techniques like instrumental neutron activation or X-ray fluorescence is that they can provide the analysis for a large number of elements and it is best to use as many methods as are available with reasonable accuracy.

The distance measures, $\cos\theta$, or the correlation coefficient can be calculated. It must be remembered that the distance parameters are dissimilarity measures while $\cos\theta$ and the correlation about the mean both indicate increasing similarity with increasing value. They must be transformed before being used in a clustering routine based on dissimilarities. Several approaches have been used including taking the negative value of the $\cos\theta$ or correlation coefficient, $1 - r_{jk}$, $1 - \cos\theta$, $1 - r_{jk}^2$, or $1 - \cos^2\theta$. The matrix of dissimilarity measures is then passed to a program that will systematically group the individual points into clusters.

The clustering process can be hierarchical or nonhierarchical. In a hierarchical clustering, one starts with n clusters, each containing one sample, and sequentially merges clusters together based on some clustering criterion. Once a pair of samples is grouped in a cluster, they can never be assigned to different clusters. In nonhierarchical clustering, a number of clusters is assumed a priori and the samples are grouped into that number as best as can be according to the chosen criterion. These techniques are discussed in detail by Massart and Kaufman (1983). There are a number of possible clustering criteria. These criteria can be divided into those that cluster points or groups by the distance between the groups and those that define the size of the cluster which results from the clustering process.

The "nearest neighbor" criterion is that the clusters having the minimum number of dissimilarities between points a in cluster A and points b in cluster B are combined.

$$C_1(A, B) = \min d(a, b) \tag{8.23}$$

where the minimum is over all points a in A and b in B. This minimum between cluster distance is supported by Jardine and Sibson (1971) and is criticized by Lance and Williams (1967).

Alternatively, the measure of the distance between clusters A and B can be the largest of the dissimilarities between the points of A and the points of B.

$$C_2(A, B) = \max d(a, b) \tag{8.24}$$

where the maximum is over all points a in A and b in B. The clusters with the smallest values of C_2 are then combined. The same result is obtained by defining the size of cluster A as the maximum of the dissimilarities between pairs of points in A:

$$C_2(A) = \max d(a, a') \tag{8.25}$$

where the maximum is over all of the points a and a' in the combined cluster A. This criterion is quite popular in the behavioral sciences. It is referred to by Johnson (1967) as the cluster diameter. It can also be called the maximum between cluster distance or maximum within cluster distance.

Instead of using either the minimum of the distances between clusters, the average distance can be employed as the criterion.

$$C_3(A, B) = \frac{1}{n_A n_B} \sum d(a, b) \tag{8.26}$$

where n_A is the number of points in A, n_B is the number of points in B, and the summation is over all points a in A and b in B. This measure would not be as sensitive to a badly determined single parameter that might skew the results.

The other criteria depend on the measures of the size of the resulting cluster following combination. The object is to combine the clusters that result in the smallest value of the clustering criterion. The most straightforward of these size parameters is the mean of the distances between all pairs of distinct points within the cluster

$$C_4(A) = \frac{1}{n_A(n_A - 1)} \sum_{a \neq a} d(a, a') \tag{8.27}$$

where n_A is the number of points in A and the summation is over all pairs of points in A such that $a \neq a'$. If, for example, the input matrix were correlation coefficients, then $C_4(A)$ would be the mean interitem correlation.

$$C_5(A) = \frac{1}{n_A^2} \sum d(a, a') \tag{8.28}$$

where now the summation is over all points in A. The distance measures between a point and itself will be zero while the other measures, with the exception of the MCD, will all be greater than zero. Thus, in most cases, the effect of using this criterion will be to decrease the apparent cluster size, and this reduction will be relatively larger for smaller clusters. This criterion biases the results toward smaller clusters. However, this criterion has an interesting feature when used in conjunction with the SED. For this case, the cluster size is the mean squared distance of the points from the centroid of the cluster and the hierarchical clustering method tries to minimize this cluster spread. This cluster size could also be considered as the within-cluster variance, which is being minimized as clusters are combined.

If $C_5(A)$ is multiplied by n_A, the next criterion is obtained,

$$C_6(A) = n_A C_5(A) = \frac{1}{n_A} \sum d(a, a') \tag{8.29}$$

where the summation is the same as for C_5. This criterion will bias the clustering even more heavily toward clusters with small numbers of points. The interesting use of C_6 is again with the SED dissimilarity measure. The cluster size becomes the within-cluster sum of the squares.

Finally, a criterion can be established that calculates a distance between clusters A and B based on the increase in the total size of the combined cluster C over the sizes of A and B.

$$C_7(A, B) = C_6(C) - C_6(A) - C_6(B) \tag{8.30}$$

This criterion does not have the bias toward small numbers of points that C_6 does. Again using the SED measure of dissimilarity, the criterion is then the increase in within-cluster sum of the squares that comes from merging clusters A and B. This increase would be minimized at each stage in the sequential clustering process.

There are then several different dissimilarity or similarity measures and several ways of combining the points into clusters based on criteria regarding the measures. How does one choose the "best" measure and clustering criterion? It is impossible to provide a general answer applicable to all problems. In some cases it may be possible to decide a priori, but, in most cases, it is necessary to expeximent with the possibilities until some experience is gained for various kinds of problems.

Applications

Cluster analysis to identify and group similar entities has not been widely used in receptor modeling. Hopke et al. (1976) used a hierarchical aggregative cluster analysis to group samples of unsized airborne particles taken in the Boston area and analyzed for 18 elements by instrumental neutron activation analysis. Sites in Boston were chosen at a number of locations across the area and they are shown

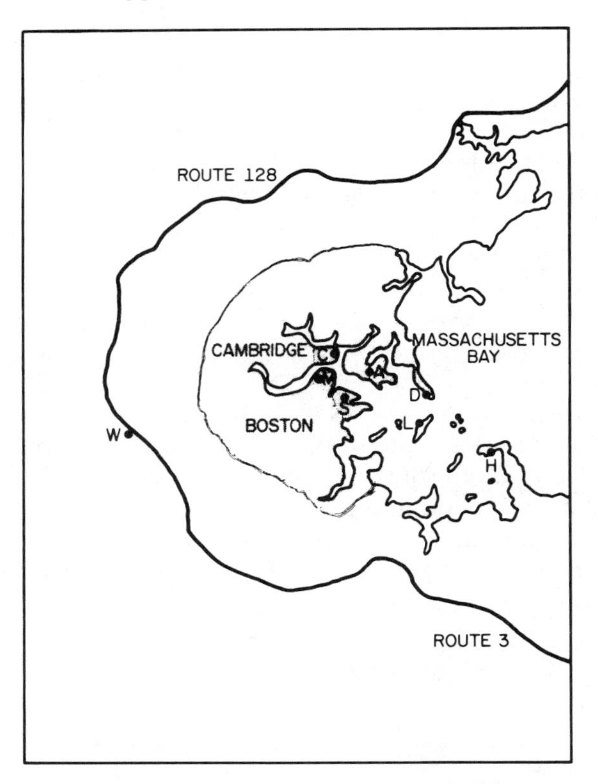

Figure 8.6. Map of Metropolitan Boston showing the location of the sampling sites. W = Wellesley, M = Massachusetts General Hospital (Downtown Boston), C = Charlestown (Boston Naval Shipyard), A = Logan International Airport, D = Deer Island, S = South Boston (Annex of Boston Naval Shipyard), L = Long Island, H = Hull. Taken from Hopke et al. (1976) and used with permission. Copyright 1976 Pergamon Press Ltd.

in Figure 8.6. Two series of sample collection were made. For the first series, samples were taken at Hull, Long Island, Massachusetts General Hospital, and the South Boston Annex of the Boston Naval Shipyard, from March 25, 1970 to April 27, 1970. The second series included the sites at Deer Island, Logan International Airport, the Boston Naval Shipyard, Massachusetts General Hospital, and Wellesley for the period of May 12, 1970 to June 29, 1970.

The cluster pattern or dendogram is shown in Figure 8.7. The dendogram is interpreted by examining the way in which the samples group together to form the clusters and the order in which the clusters combine to form larger groups. To assist in interpreting the clusters, average values for each of the 18 elements that were measured in each sample were calculated for each of the clusters labeled with a letter in the figure. These average values are given in Table 8.2.

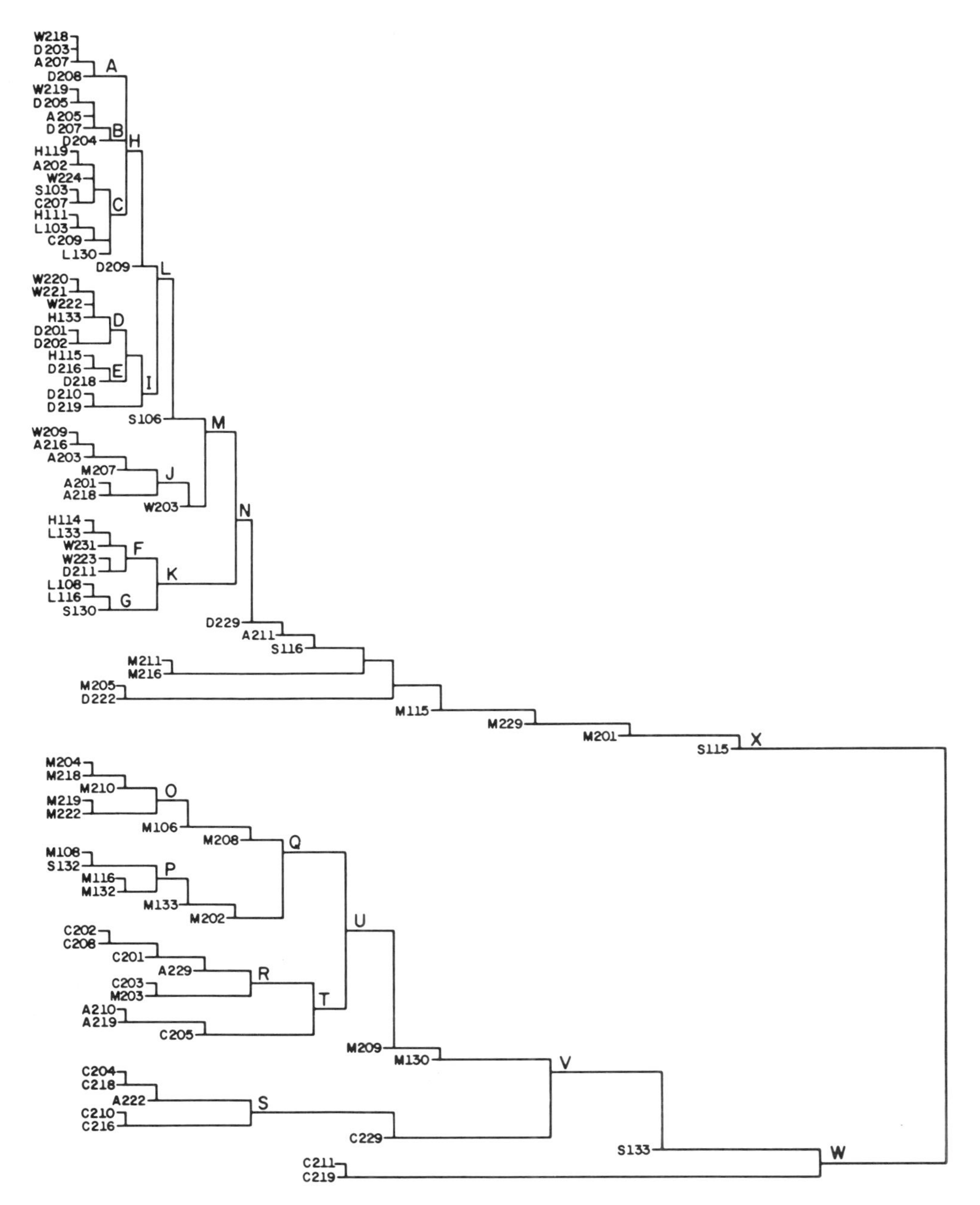

Figure 8.7. Dendogram for the cluster analysis of the Boston aerosol sample analyses. The horizontal distance is proportional to the degree of dissimilarity between the clusters. The first letter of the sample designation indicates the sample site using the same lettering code as Figure 8.6. Taken from Hopke et al. (1976) and used with permission. Copyright 1976 Pergamon Press Ltd.

Table 8.2. Average Factor Scores for Clusters

	Factor Number					
Cluster	1	2	3	4	5	6
H	-0.6261	-0.3434	-0.3073	-0.7111	-0.5461	-0.0785
I	-0.2575	0.0960	-0.8568	-0.6118	-0.1140	-0.3000
J	-0.8346	-0.4880	0.0786	0.4731	0.1816	0.5126
K	-0.3439	-0.1258	-0.1664	-0.3745	1.0174	-0.6591
L	-0.4932	-0.1207	-0.5204	-0.6802	-0.3924	-0.1599
M	-0.5881	-0.2088	-0.3241	-0.4095	-0.3524	-0.0697
N	-0.5456	-0.1944	-0.2967	-0.4034	-0.1142	-0.1722
Q	-0.1528	-0.2592	-0.1633	0.9630	-0.4730	-0.0865
R	-0.4208	-0.3177	1.8922	0.2552	0.4505	-0.2367
S	1.8084	-0.1795	0.5181	-0.7904	-0.4073	-0.5662
T	-0.0644	0.0094	1.4573	-0.1542	0.2171	0.1580
U	0.0640	-0.1493	0.4997	0.5059	-0.1907	0.0135
V	0.4693	-0.1407	0.3174	0.5876	-0.2432	-0.0756
W	0.7398	-0.1592	0.3352	0.4187	-0.3000	0.0400
X	-0.4282	0.0921	-0.1939	-0.2424	0.1737	-0.0234

The cluster pattern is divided into two major groups, *W* and *X*, with the samples in cluster *X* generally having lower average concentrations than those in cluster *W*. Cluster *H* has the lowest average concentrations of any of the clusters and appears to include those samples when the particulate loading was lowest. The levels of concentration appear to increase as one moves from top to bottom in the dendogram.

In trying to interpret these data, Hopke et al. have also performed a classical factor analysis and calculated the factor scores. The factor loading pattern is given in Table 8.3. The six factors were identified as soil, sea salt, oil-fired power plants, motor vehicles, an unknown Se/Mn source, and refuse incineration. The Se/Mn source is likely to reflect the secondary sulfate aerosol being transported into the Boston area (Thurston and Spengler, 1982). Because sized samples were not taken and analyzed, these six sources were all that could be resolved. For each of the samples, the factor scores could then be calculated so that the important of each source to each sample could be assessed. Hopke (1976) has shown that it is useful to calculate the average factor scores for each cluster to help interpret the cluster pattern. These average factor scores for the Boston dendogram are given in Table 8.4.

Factor 2, attributed to sea salt aerosol, is found to be generally uniform across the region. The sea salt aerosol is thus fairly uniformly dispersed throughout the region. Group I has the largest positive score value for this factor although the

Table 8.3. Varimax Rotation of Principal Factor Pattern for Boston Airborne Particles

	Factor Loadings					
Element	a_1	a_2	a_3	a_4	a_5	a_6
Na	0.169	0.816	0.084	0.074	0.063	0.016
Cl	-0.007	0.835	-0.040	0.077	-0.027	0.020
Br	0.132	0.121	0.130	0.592	0.163	0.037
Al	0.890	0.105	0.235	0.104	0.236	0.190
Sc	0.900	0.093	0.125	0.090	0.252	0.258
V	0.307	0.042	0.722	0.254	0.064	0.203
Cr	0.764	-0.117	0.444	-0.113	0.086	-0.079
Mn	0.607	0.126	0.161	0.168	0.558	-0.013
Fe	0.902	0.078	0.246	0.013	0.224	0.171
Co	0.657	0.068	0.587	0.212	0.228	0.101
Zn	0.382	0.011	0.313	0.062	0.236	0.614
Se	0.173	-0.009	0.057	0.122	0.683	0.161
Sb	0.450	-0.051	0.016	0.169	0.191	0.334
La	0.803	0.061	0.203	0.510	-0.017	0.097
Ce	0.704	0.094	0.156	0.467	0.106	0.196
Sm	0.846	0.104	0.194	0.424	0.010	0.150
Fu	0.827	0.206	0.139	0.135	0.153	0.258
Th	0.824	0.057	0.159	0.227	0.272	0.092

sodium and chlorine levels are found to be as high in a number of other clusters. This cluster consists primarily of samples collected at Deer Island and Hull. These sites are at the outer edge of the harbor and are thus expected to have major sea salt contributions. The Wellesley samples that were included in this group may represent sampling periods with strong easterly winds that result in an increased sea salt component.

The group labeled *J* has the largest positive value for the factor identified as incineration (factor 6) and also a relatively large value for factor 4 that was attributed to automotive exhaust. Samples in this cluster included four from Logan International Airport, and one each from Wellesley and downtown Boston. The Boston area's second largest incineration facility is in Sommerville, Massachusetts (Morgenstern, Goldfish, and Davis, 1970) and had a burning rate in 1965 of 62,400 tons per year of the 357,860 tons per year incinerated in the area (Massachusetts, 1968). Previous reports have indicated little trace element injection by aircraft (Conry, 1974), but large amounts of automotive emissions near airports. Thus, the incinerator represents a major contribution to the elemental concentrations at this site.

Table 8.4. Average Values for Clusters

Element	*H*	*I*	*J*	*K*	*L*	*M*	*N*	*Q*	*R*	*S*	*T*	*U*	*V*	*W*	*X*
Vanadium ($\mu g/m^3$)	0.391	0.209	0.647	0.582	0.316	0.412	0.442	0.877	1.30	1.05	0.21	1.02	1.02	1.03	0.547
Aluminum ($\mu g/m^3$)	0.325	0.715	0.700	0.918	0.467	0.498	0.571	1.37	1.52	2.72	1.70	1.51	1.77	2.01	0.839
Manganese (ng/m^3)	10.6	16.5	14.1	36.6	12.8	13.1	17.2	28.5	31.0	38.0	31.3	29.6	31.8	33.4	23.2
Chlorine ($\mu g/m^3$)	0.375	0.395	0.225	0.270	0.429	0.393	0.372	0.362	0.348	0.300	0.326	0.347	0.381	0.381	0.580
Sodium ($\mu g/m^3$)	0.802	1.66	0.888	1.46	1.16	1.07	1.14	1.32	1.32	1.68	1.89	1.55	1.60	1.59	1.46
Bromine (ng/m^3)	105	116	268	115	110	145	140	234	288	154	232	233	239	232	160
Samarium (ng/m^3)	0.043	0.095	0.095	0.131	0.062	0.068	0.079	0.261	0.196	0.382	0.214	0.242	0.296	0.333	0.122
Lanthanum (ng/m^3)	0.302	0.695	0.701	0.930	0.442	0.489	0.566	2.08	1.37	2.60	1.46	1.82	2.18	2.43	0.876
Selenium (ng/m^3)	0.488	1.01	0.791	1.95	0.670	0.690	0.910	0.854	1.52	1.18	1.33	1.05	1.13	1.17	1.26
Cerium (ng/m^3)	0.572	1.24	0.945	1.74	0.806	0.836	0.992	2.93	2.41	4.40	2.47	2.74	3.38	3.52	1.46
Thorium (pg/m^3)	44.4	94.9	117	133	62.0	69.7	80.8	210	239	491	264	232	307	121	345
Chromium (ng/m^3)	0.5	0.5	0.2	1.7	0.5	0.5	0.7	2.0	8.6	12.7	7.6	4.3	5.6	6.8	1.5
Europium (pg/m^3)	0.21	18.5	18.8	24.8	12.8	13.5	15.5	39.6	30.1	75.2	38.2	39.0	48.3	54.9	22.7
Antimony (ng/m^3)	2.49	2.43	7.39	3.25	2.41	3.17	3.19	12.5	11.6	12.9	12.2	12.4	13.4	14.1	6.15
Scandium (ng/m^3)	0.060	0.135	0.103	0.183	0.087	0.090	0.106	0.291	0.291	0.586	0.329	0.307	0.376	0.433	0.162
Zinc ($\mu g/m^3$)	0.072	0.075	0.208	0.115	0.073	0.096	0.098	0.110	0.376	0.251	0.319	0.196	0.203	0.249	0.148
Iron ($\mu g/m^3$)	0.302	0.587	0.518	0.757	0.409	0.427	0.484	0.977	1.33	2.68	1.43	1.16	1.47	1.73	0.714
Cobalt (ng/m^3)	0.384	0.452	0.685	0.789	0.404	0.481	0.534	1.03	1.64	1.81	1.55	1.25	1.36	1.48	0.721

Cluster *K* is strongly associated with the unidentified fifth factor, with an average value for this factor score of greater than one. Subgroup *F* has an average factor score for factor 5 to 1.42. Sites represented in this cluster are Hull (1), Long Island (3), Wellesley (2), Deer Island (1), and South Boston (1). This disparate collection of sites does not provide additional insights into the nature of this factor other than its association with sites out of the downtown area. These areas have few known sources of particles other than automobiles and the burning of fuel oil for home heating. There are several small incinerators at Wellesley, Dedham, Weymouth, Waltham, Brookline, Newton, and Watertown. The facilities would not be expected to affect sites such as Long Island and Hull. Home heating fuel has not been found to contain large quantities of either Mn or Se that are strongly related to the fifth factor (Cahill, 1974), but, in the absence of other sources of these elements, there may be sufficient heating oil burned to produce the observed Mn and Se concentrations. Five of the eight samples in cluster *K* were taken in March and early April while heating was still necessary. All three samples in subgroup *G*, which also has a positive value for factor 5, were taken in the first series.

In the lower major cluster, *O*, the samples have a large positive value for Factor 4. This factor was attributed to aged automobile exhaust particles. This cluster consists almost entirely of downtown Boston samples, including only one South Boston sample among the 13 samples. The downtown Boston site is adjacent to a major highway and several additional major commuter traffic routes converge nearby. Thus, motor vehicle emissions are expected to be high in this area. Downtown Boston has few industrial particle sources. Although there is some residual fuel burning for heating of large buildings it is not a major influence on the particulate composition. From these results it can be concluded that the primary anthropogenic sources of airborne particulate matter in downtown Boston is motor vehicles.

The score of factor 3, residual fuel burning, is large for cluster *T*. These samples were taken primarily in Charlestown (5), with some from nearby areas such as the airport (3) and downtown Boston (1). There are several major oil-fueled, electric generating plants in the area of East Cambridge and Charlestown. Subgroup *S* also has a large score for factor 3 and again consists mainly of Charlestown samples.

The combination of factor and cluster analysis is of considerable value in helping to interpret the sources of airborne particulate matter. Probably more could be said about the interpretation, especially if one conducted a detailed survey of the types of sources present in the local areas near the sampling stations. However, our purpose here has been to introduce the application of these techniques and show their capabilities.

Cluster analysis has been used to trace the sources of the chemical composition of precipitation (Slanina et al., 1983). The analysis sorted the 26 sampling periods into four major groups. Two major clusters each contained two subclusters. The major clusters represented days that could be clearly associated with clean maritime air or more polluted continental air. The subgroups could not be fully interpreted in terms of meteorological regimes or other sources because of length of the sampling periods (2 to 3 days). However, the clustering did separate the data into interpretable groups and suggested additional experiments that could provide further understanding.

Another common use of clustering is in archaeological applications. A problem in the interpretation and understanding of artifacts found at a site is in the identification of the number of sources of the material found at a location. For example, obsidian pieces are found at a site where a workshop was thought to exist. There are several sources of obsidian that might have been traded to this site in exchange for finished goods. Are the shards found at the workshop similar or dissimilar to the material found at the sources? This problem is made much easier if authentic source materials can be found.

Charlton, Grove, and Hopke (1978) used the compositions of both obsidian artifacts and sources to show that a recently rediscovered source at Paredon in the central Valley of Mexico was important in the supply of materials to workshops that had been thought to be using material from a single source at Otumba. Of the 90 samples analyzed from Chalcatzingo, 10% were from the early Formative Period and this analysis showed that 8 of the 9 were from previously unreported Paredon source rather than the Otumba source. These results required that the obsidian trade networks of this location and time be reanalyzed in light of a second obsidian source demonstrated by the cluster analysis.

Often the ancient source has been worked out or its location lost so that it is not possible to obtain such source material. The ability to determine the number of distinctly different types of material and their average properties can often still provide useful insights into problems being studied. This approach has been used with many archaeological problems in categorizing sources of ceramics, metal, or stone objects.

Cluster analysis can also be used to portray graphically the interrelationships among variables as well as between samples. Gaarenstroom, Perone, and Moyers (1977) looked at both nearest neighbor clustering and nonlinear mapping. In the dendograms shown in Figures 8.8 and 8.9, the concentrations of elements in airborne particles at two different sites in the Tucson, Arizona area are examined. Site 2 is an urban location and site 11 is a rural location. The matrix of pairwise correlations about the mean were used as the dissimilarity measure and $1 - r_{ij}$ was used as the similarity measure. It can be seen [Figure 8.3(a)] that at the urban location (site 2), there is a tight clustering of Ca, K, Fe, Mg, Na, Cr, Rb, Ti, Al, Si, Sr, Li, Mn and total mass and Ni, Co, and Cs are also associated in this group. NH_4^+ and SO_4^{2-} show a strong relationship at both the rural and urban sites and both are apparently related to nonferrous metals. At the urban site, the lead

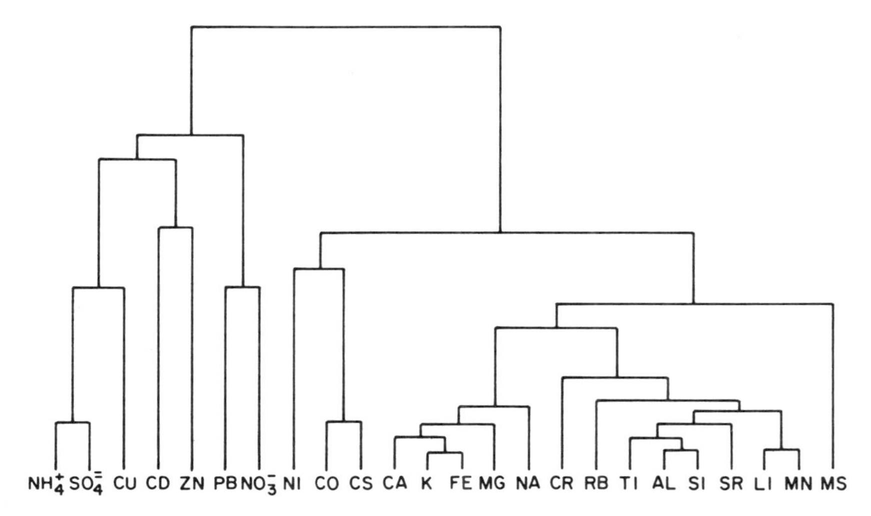

Figure 8.8. Dendogram for the cluster analysis for the elements measured in aerosol samples from an urban location in Tucson, Arizona. Taken from Gaarenstroom, Perone, and Moyers (1977) and used with permission. Copyright American Chemical Society.

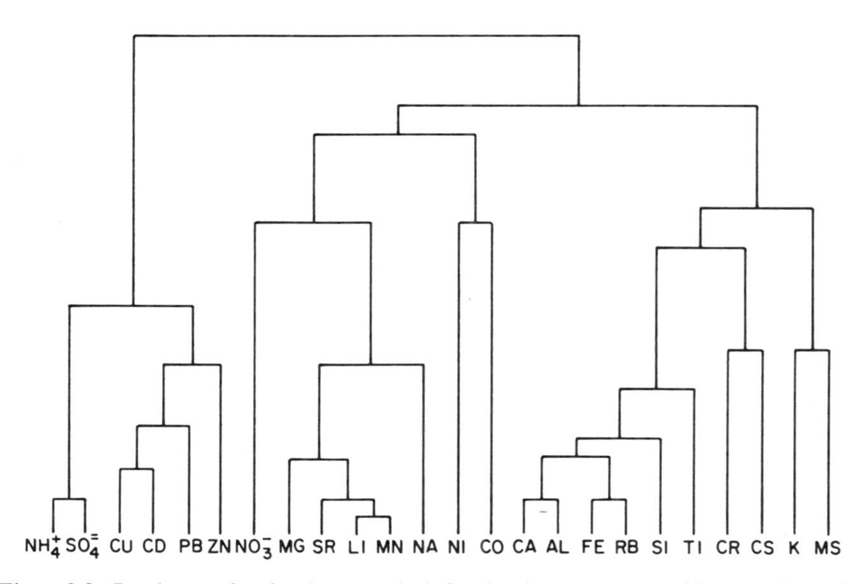

Figure 8.9. Dendogram for the cluster analysis for the elements measured in aerosol samples from a rural location near Tucson, Arizona. Taken from Gaarenstroom, Perone, and Moyers (1977) and used with permission. Copyright 1977 American Chemical Society.

separates from the Cd, Cu, and Zn, and clusters with NO_3^-, probably reflecting the strong effect of motor vehicle traffic in urban Tucson. The rural cluster tends to group Pb with the other nonferrous metals. In both cases and nonferrous metals probably originate from copper smelters in the region that are far enough away to permit development of partially neutralized secondary sulfate. In the absence of a strong motor vehicular source, the lead behaves as the other elements in the copper emissions. Thus, the pattern of elemental associations does help in interpreting the origins of the particle samples.

An alternative approach to aggregative clustering is to start with a single cluster containing all of the samples and sequentially separating into smaller groups. Gether and Seip (1979) have used a one-way splitting program with interactive graphics methods to analyze a number of air pollution data sets. They have been able to identify unusual cases or exceptional values. Their approach provides some useful exploration of the data as well as an interesting method of displaying the results.

Another way to portray the multidimensional picture is through nonlinear mapping. In this approach, the n-dimensional space is projected onto a plane in a way so as to minimize the mapping error. The mapping error is defined as

$$E = \left(\sum_{i<j}^{n} d_{ij} \right)^{-1} \sum_{i<j}^{n} (d_{ij} - d_{ij}^*)^2 / d_{ij} \tag{8.31}$$

where d_{ij} is the distance between points i and j in the n space and d_{ij}^* is the corresponding distance in 2 space. Nonlinear mapping has also been applied to a number of different problems (Kowalski and Bender, 1972). Gaarenstroom, Perone, and Moyers (1977) also display their results in this way. The patterns are not very different from the cluster patterns previously discussed. Fordyce (1975) and Neustadter, Fordyce, and King (1976) applied nonlinear mapping to interpretation of airborne particulate data in Cleveland, Ohio. From using a combination of techniques, they identify soil, automotive exhaust, metallurgy, and specific point sources as the major contributors of airborne particles in this region. The use of clustering methods can be helpful in selecting groups of similar samples or measured variables and in examining some of the structure in the data set. Clustering is a valuable method, particularly in conjunction with other techniques like factor analysis.

WIND TRAJECTORY ANALYSIS

One of the key problems in the application of receptor models to the source apportionment of aerosol mass is the determination of the composition of particles from a single source at the receptor site. There can be changes in the composition of particles from the point where they are emitted to the point where they are collected. New methods for determining these point source profiles are needed.

Rheingrover and Gordon have developed a method to identify samples strongly affected by single point sources (1984). This approach, called wind trajectory analysis, can select samples from a large data base such as the one obtained in the Regional Air Pollution Study (RAPS) of St. Louis, Missouri. Samples that were heavily influenced by major sources of each element are identified first according to the following criteria:

1. Concentration of the element in question $X > X + Z_{cr}\sigma$, where $\bar{X}$ is the average concentration of that particular element for each station and size fraction (coarse or fine particle size fraction), Z_{cr} is typically set at about three for most elements, and σ is the standard deviation of the concentration of that element.
2. The standard deviation of the 6 or 12 hourly average wind directions for most samples, or minute averages for 2-hr samples, taken during intensive periods is less than 20°.

Samples strongly affected by emissions from a source were identified through observations of clustering of mean wind directions for the sampling periods selected with angles pointing toward the source. The RAPS data have been screened according to these criteria. With wind trajectory analysis, specific emissions sources could be identified even in cases where the sources were located very close together (Rheingrover and Gordon, 1984). A compilation was made of the selected impacted samples for further analysis.

Rheingrover and Gordon used a simple linear regression of the measured elements against the particular element used to select the samples. Thus, those elements strongly correlated to the element observed to be in above average concentration could be identified. The linear relationships between the pairs of elements and between the criterion element and total mass permitted an estimation of the elemental composition of the point source. Thus, the composition of the source at the receptor site could potentially be identified. However, not all of the observed elements were well correlated and it is sometimes difficult to sort the specific source profile from other source profiles.

Target transformation factor analysis, TTFA, for aerosol source apportionment offers an alternative method of analysis for these data sets. Chang et al. (1982) have reported the results of such analyses on these data sets. To illustrate this method, the data set for fine-particle zinc samples whose wind trajectories point toward a copper products plant in the East St. Louis, Illinois area is analyzed. Some of the trajectories are shown in Figure 8.10. The details of the analysis have been given by Chang et al. (1982). The results for this analysis indicated motor vehicles, secondary sulfate, soil or fly ash, and the copper prod-

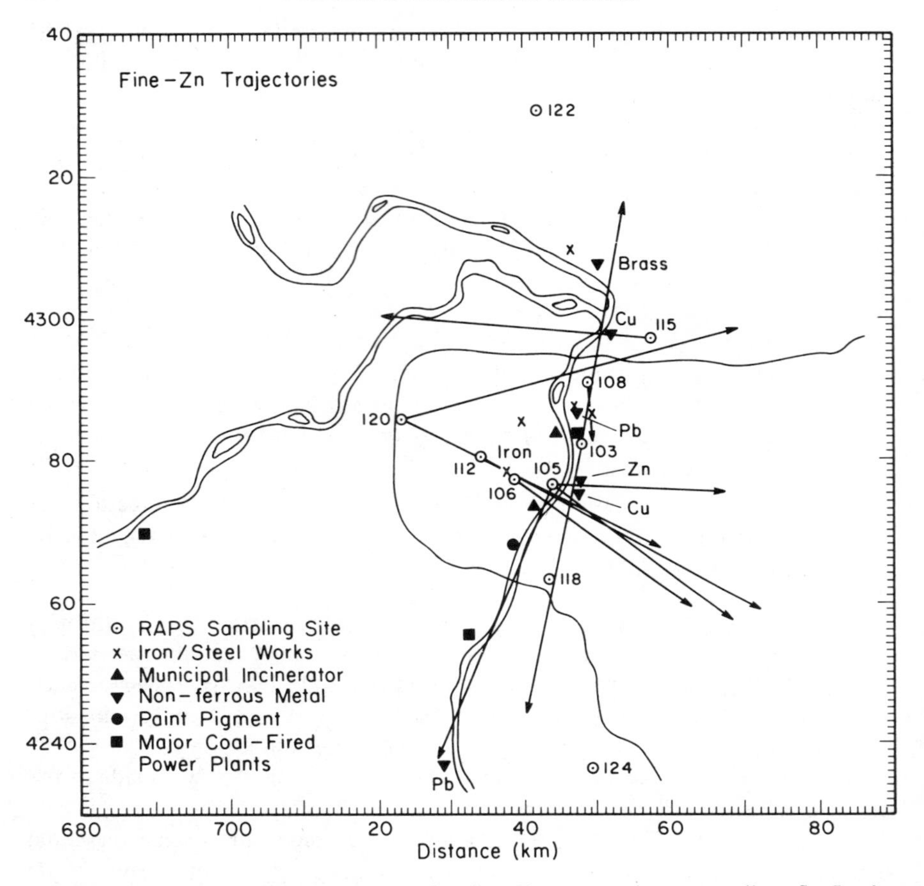

Figure 8.10. Major trajectories observed for fine Zn sources in metropolitan St. Louis, Missouri. Taken from Rheingrover and Gordon (1984).

ucts plant as the major mass contributors. The calculated source profiles and the associated uncertainties that best reproduced this data set are listed in Table 8.5. The motor vehicle factor is the one with a strong dependence on lead and bromine. The ratio of bromine to lead of 0.24 is typical of values reported in the literature (Dzubay, Stevens, and Richards, 1979; Chu and Macias, 1981). The concentration of lead in motor vehicle exhaust will vary from city to city and is dependent on the ratio of leaded gasoline to unleaded and diesel-powered vehicles. The values of 11.5% lead is quite similar to that obtained in other studies of the RAPS data at other sites (Alpert and Hopke, 1981; Liu et al., 1982; Severin, Roscoe, and Hopke, 1984). The concentration of sulfur was not related to any other elements and presumably represents a secondary sulfate aerosol resulting from primary emissions of sulfur dioxide. The sulfur concentration of 19.1% is

Table 8.5. Refined Source Concentration Profiles (mg/g) for Fine Particle Fraction Zn Trajectory

	Sulfate	Cu Plant	Motor Vehicle	Flyash/ Soil
Al	0.0 ± 1.3	0.0 ± 5	0.0 ± 9	61 ± 15
Si	0.0 ± 1.2	0.0 ± 5	0.0 ± 8	195 ± 13
S	191. ± 1	47.5 ± 1.1	0.0 ± 1.9	71 ± 4
K	2.1 ± 1.5	8.5 ± 5.1	3.6 ± 9.0	26 ± 16
Ca	0.0 ± 0.8	6.5 ± 2.6	0.0 ± 5	37 ± 8
Ti	0.56 ± 0.25	0.0 ± 0.9	0.0 ± 1.6	4.3 ± 2.7
Mn	10.0 ± 0.6	41.0 ± 2.2	0.0 ± 4	9 ± 6.5
Fe	0.87 ± 1.17	2.5 ± 4.2	0.0 ± 8	68 ± 13
Ni	0.19 ± 0.08	2.35 ± 0.29	0.0 ± 0.5	0.7 ± 0.86
Cu	0.0 ± 0.7	108 ± 3	0.0 ± 30	0.0 ± 8
An	0.0 ± 4	175 ± 14	0.0 ± 30	0.0 ± 50
Se	0.08 ± 0.063	0.82 ± 0.23	0.45 ± 0.40	0.0 ± 0.7
Br	0.24 ± 0.34	1.2 ± 1.3	28 ± 2.2	4.3 ± 3.8
Cd	0.0 ± 0.18	5.03 ± 0.64	0.0 ± 1.2	0.0 ± 2
Sn	0.0 ± 0.7	21.2 ± 2.5	0.0 ± 5	0.0 ± 8
Pb	5.2 ± 1.3	101 ± 5	115 ± 9	8.5 ± 14.2

a bit lower than what would be expected when the material consists of only ammonium sulfate. The soil/flyash factor could potentially represent both soil particles as well as materials orginating from the combustion of coal. Differentiating soil and coal fly ash, due to the similarity of their elemental profiles, is a problem often encountered in aerosol source resolution work. Reliable data for other elements, such as arsenic, might permit the resolution of the soil and coal fly ash contributions.

The copper products plant factor is associated with elements Cu, Zn, Pb, and S. Gatz (1978) noted the presence of a zinc/lead factor and attributed it to smelters located primarily in the East St. Louis and Granite City areas. Rheingrover and Gordon (1984) indicate that the zinc smelter contributes primarily to the coarse samples while the copper products plant emits fine particles that are actually higher in zinc than copper. Therefore, this factor is attributed to the copper products plant. In this data set, secondary sulfate aerosols account for 68.3% of the mass of fine Zn trajectory samples. The copper products plant contributes 14.0%. Motor vehicle emissions account for 10.5% and the fly ash/soil represents the remaining 7.2%. The wind trajectory/TTFA method thus represents a useful approach when source information for the area is lacking or suspect, and if there is uncertainty as to the identification of all of the sources contributing to the measured concentration at the receptor site.

DISPERSION/RECEPTOR HYBRID MODELS

In many cases, in addition to the ambient air quality parameters at a site, meterological data are also known. It is then possible to utilize these data to run a dispersion model in reverse to locate pollution sources and estimate their emission strengths. This approach was first suggested by Yarmartino and Lamich (1979). They employed a network of 18 carbon monoxide (CO) monitors for Williams Air Force Base, Arizona, for collecting 37 hr of data during a two-week period, as well as a five-station network that operated over a 13-month period. A conventional Guassian-plume dispersion model is used to couple the source and receptor locations. The model solves for a set of source strengths, Q_j, that yield a minimum residual between the observed and model predicted concentrations. This condition leads to a set of equations

$$B_i = A_{ij} Q_j$$

where

$$A_{ij} = \frac{R_{ikt} R_{jkt}}{(\Delta x_{kt})^2}$$

$$B_i = \frac{x_{kt} R_{ikt}}{(\Delta x_{kt})^2}$$

where x_{kt} is the concentration at location k at time t, Δx_{kt} is the corresponding measurement error, and R_{jkt} is the value of the dispersion coefficient coupling source j with receptor k during time period t. The equations are solved with a least-squares algorithm requiring only positive Q_j values (Lawson and Hanson, 1974).

The results with the small data set (37 times periods) did not work well and failed to indicate the presence of any on-base sources. This failure was apparently due in part to high background concentrations and large measurement errors. Using the larger data sets with fewer sites, the model indicated a number of possible area sources, each of which was found to have real emissions. The solution did not provide an emission density map since the significance of a source is defined by its receptor proximity and meteorological parameters as well as its emission magnitude.

The extension of this approach to point sources with bouyant plumes from elevated stacks was presented by Yarmartino (1983). He examined the data from a 31-station network with hourly SO_2 values. He found good resolution of ground-level area sources but rather poor determinations of point source strengths. It seems that with a more sophisticated dispersion model and a more extensive pre-investigation of the network's sensitivity and resolution functions, a significant improvement in predictive power could be achieved.

Cooper (1983) also explored the use of the dispersion model estimates of the transport coefficients to infer source strengths from the information about concentrations of species in particulate matter measured at a site in the suburbs of Boston. Using a similar approach to that discussed above, the results were comparable to the small data set results of Yarmartino and Lamich. There has not been a really systematic study of these methods with large, well-validated data sets. There are a variety of options as to the choice of dispersion models and these possibilities need to be explored. There exists the potential for using hybrid source–receptor models but much more extensive development, testing, and sensitivity analysis will be needed before this approach becomes an accepted tool.

Another approach to a hybrid model is the microinventory–regression method developed by Pace (1979). The basic concept is that particulate matter levels at a particular site are primarily caused by sources within a very small area around the monitoring site. Emissions from point sources within 5 miles and area source emissions within 1 mile are inventoried. The area around the site is divided into a number of sectors, typically 9, and the inventory for each sector is defined. The measured total mass is then regressed against those inventories. This approach does not provide any characterization of particles but it can provide a predictive equation that can be used to estimate the effects of air quality management strategies or the addition of new sources. If multiple samplers are located in an area, it is possible to estimate the relative effects of point and area sources within the region. The method has been able to predict TSP with a standard error of 13.6 $\mu g/m^3$ or about 17% of the average measured value. It can thus provide a useful tool in assessing the particle sources at a site where information about the emission sources in the area is available.

SUMMARY

There have been a number of other mathematical approaches to solving receptor modeling problems. However, in most of the cases reported in this chapter, only limited investigations have been made into the potential applications of these methods and further work will be needed to determine fully their utility. The attempts to couple dispersion and receptor models appears to be an area that does need additional study. The results of the preliminary studies do suggest that more extensive efforts may lead to useful results. There may be additional areas where cluster analysis may prove useful, for example, in the problem of grouping particles on the basis of scanning electron microscopy results. Currently, the approach is to assign the particles to empirically defined particle type categories. It appears that an objective grouping can be achieved using clustering methods. Particles that are not assignable to any category can be identified and it will not be necessary to make a priori judgments on the number and nature of particle type classes to be used. Thus, further efforts in the use of these methods may improve and expand the tools available to the receptor modeler.

CHAPTER

9

APPLYING RECEPTOR MODELS TO ENVIRONMENTAL PROBLEMS

In the previous chapters, analytical and mathematical methods for approaching the resolution of environmental problems through receptor modeling were presented. There is a question of how to develop an effective strategy to put these methods together to solve particular problems within the constraints of resources that can be committed to a given program.

Watson (1983) points out in his analysis of receptor models that the real world is a very complex entity with many more variables and larger scales than is reasonably possible to measure or model. Table 9.1 provides order of magnitude estimations for the number of significant physical variables and the temporal and spatial scales that exist for reality, measurement, and models (dispersion and receptor). It is clear from these comparisons that only a very small portion of the problem is amenable to measurement and analysis. However, it is certainly possible to choose that portion so that subsequent actions can have a significant effect on the overall system.

The concepts of how to approach the development of such strategies have been outline by Watson (1981, 1983) and Gordon et al. (1984). Watson (1983) steps through a sequence of assumptions that are made to establish a conceptual model, the measurements needed to use the model, and the constraints that arise because of problems that occur with either the model or the measurements. To resolve the problems, the conceptual model is modified and/or new or additional measurements are made. This analysis and framework for reviewing these models proves useful in more fully recognizing the underlying concepts of receptor modeling.

Gordon et al. (1984) try to establish a practical framework for approaching a problem to be solved through receptor models. They suggest that there are three possible levels of detail in the receptor modeling that can be employed. These approaches are choosen depending on the difficulty of the problem to be solved and the level of resources that are available to gather and analyze the data.

The first level (level I) of study primarily consists of utilizing the data that may already exist relating to the problem. For example, in the case of nonattainment of a TSP standard, there may be data collected by state or federal agencies as part of their routine monitoring efforts. Although in many cases only the TSP

Table 9.1. Comparisons of Measurement, Source, and Receptor Model Components to Reality on the Urban Scale[a]

	Reality	Measurement	Dispersion Model	Receptor Model
Air volume (m^3)	10^{11}	10^3	10^7	10^3
Number of variables	10^3	50	50	50
Range of pollutant concentrations (g)	10^{-15}–10^{-3}	10^{-9}–10^{-3}	10^{-6}–10^{-3}	10^{-9}–10^{3}
Number of variable interactions	10^4	0	10	0
Number of pollutant sources	10^3	10	10^2	10
Range of pollutant emissions (g/sec)	10^{-2}–10^3	1–10^3	1–10^3	1–10^3
Emissions time scale (sec)	10^2	10^7	10^7	10^7
Emissions spatial scale (m)	Vert. 10	10	10	10
	Horz. 100	100	100	100
Dispersion time scale (sec)	10^2	10^2	10^3	10^4
Dispersion spatial scale (m)	Vertical 0–1000	10, 100	10	10
	Horizontal 100, micro	1000	1000	1000
	10^4, meso	Few	None	None
	10^6, macro	Few	None	None

[a]Watson (1983).

masses may be recorded, in many instances some elemental data may be available from U.S. EPA programs like the National Air Sampling Network (NASN) or the Inhalable Particulate Network (IPN). A number of states perform routine multielemental analyses of their TSP sampling such as the program in Texas (Price et al., 1982). Over the past decade, a number of short-term studies have focused on certain areas of the country and have collected extensive data sets. It is often possible to find data for many airsheds.

There is also a growing library of source composition profiles. The profiles presented in the appendix of this volume represent much of what is currently available; it is expected that many more will become available in the near future. The profiles developed using dilution sampling and size selective samplers so that the source and receptor samples are taken with similar physical characteristics will be particularly useful. Since there will be differences in transport characteristics between sub- and supermicron particles, the ability to measure the composition of individual modes of the size distribution should improve the receptor model's accuracy and resolution.

The ambient data can be screened using a classical factor analysis and the general quality of the data can be assessed as described by Roscoe et al. (1982). The use of factor analysis requires data for a reasonable number of samples to be available. It is best to have 40 or more but successful analyses have been run with less than 20 samples. It must be remembered that fewer samples mean the sensitivity of the analysis for minor sources will be reduced as will the ability to identify noisy variables or unusual samples. The factor analysis can also provide a strong indication of the number resolvable sources and their general characteristics. At this point, a target transformation factor analysis (TTFA) can be performed. This analysis can provide a much better identification of the source types and quantitative apportionments can be obtained.

For areas where the major sources are known and an elemental profile of those source types can be found in the literature, a chemical mass balance (CMB) can be performed. An emission inventory for the area can be helpful in identifying the possible major point sources. Data on emissions are also available in the Fine Particle Emissions Information System (FPEIS) and the National Emission Data System (NEDS) of the U.S. EPA. These data are available in computer-compatible format and can then be automatically searched for location, emission rate, and in some cases, particle composition data. Likely sources can be identified by plotting sources on a map of the area and considering prevailing wind directions. Area sources are usually near the site location and the microinventory techniques of Pace (1979) can be useful in assessing these impacts.

The CMB analysis is performed on one sample at a time so only one ambient sample is the minimum number required. However, to provide more confidence in the conclusions regarding significant sources, a more representative set of samples is needed. The CMB analysis requires the assumption that literature source profiles are sufficiently good to permit reasonably accurate source apportionment. Problems with this assumption were discussed in Chapter 6.

Such a level I effort will provide good indications of the most probable major sources as well as semiquantitative estimates of their impacts. If abundant, high-quality ambient data are available, it may be possible to obtain quite good values for the major source impacts using a TTFA approach. Possible problems with resolution of the TTFA method were discussed in Chapter 7. The level I effort can also serve as the necessary first step in planning a more detailed study involving the actual procurement of data as part of the study. The level I results can be used to focus the study on the most probable sources so that the maximum return is obtained for the higher resource costs of a higher level study.

The level II study is distinquished from the first level by the inclusion of field studies. In these studies, sampling and analysis of ambient air and/or sources is performed. The field work should be carefully planned and targeted at taking the specific types of samples and using the specific analytical and mathematical methods to provide source identification and quantification. The choice of sampler and collection media should fit the requirements of the analytical method. The output of the analyses should give the data needed by the mathematical methods. At this level of project, only a few grab samples of source materials can be obtained. A systematic source sampling program is reserved for the highest level of available resources. Therefore, it is critical that the source and ambient sampling programs be well designed. This approach will tend to make the most cost-effective use of the limited additional resources required.

The sampling systems employed can be optimized for the particular problem as discussed primarily in Chapter 2. Different analytical methods have different sampling requirements and often these requirements are mutually exclusive. For example, to perform elemental analysis by X-ray fluorescence or neutron activation analysis, it is desirable to have as much sample as possible at least to a uniform, complete layering across the filter. For optical or electron microscopy, it is desirable to have a lightly loaded filter so that each particle is separate and distinct from one another. All of these methods work best with pore-type filters that are made of a carbon-based material. Thus, for carbon or organic compound analysis, it is preferable to have samples taken on glass or quartz fiber filters. In all cases, the mass of sample should be determined whether it is a source or an ambient sample. In addition, the other features can then be determined using the analytical methods chosen. In this level project, compromises must be made determining those features of a sample that will give the most definitive information on sources. Flexibility to alter the plan after some samples have been taken and analyzed may be very valuable in arriving at the most credible results within the resource limits set.

Another point to be made on sampling is to take potentially more samples than can or will be analyzed and then selecting the ones to be scrutinized based on other results. These other results could be sample mass or meteorology. If the impacts of sources that cause exceedance of a standard are of interest, it may be expedient to examine only samples from days when the standard level is exceeded. Since TSP compliance is measured with a standard high-volume sampler and other sampling devices are best for receptor model sampling, it may be useful to have side-by-side samplers. It is then critical to ensure that one sampler does not impact the other such as with fine copper particles from motor brush wear. Simple gravimetric analysis of the TSP samples for mass will then allow the selection of the samples from the days with high mass loadings for further study.

Another useful method of selecting samples for analysis is by the meteorological conditions so that the set of samples ultimately analyzed will be representative of the average meteorological conditions. The Portland study illustrates this concept well (Watson, 1979). A limited number of samples can then give results that are as representative of a whole year's samples as possible with a limited size data set. In both cases discussed here, it is necessary that sampling be much more extensive than the analytical program. Although field costs can be substantial, the sample collection is usually only a small fraction of the analytical costs per sample. Archived samples may also serve as useful baseline samples for future studies if and when additional resources become available.

It is probably impractical to use highly sophisticated source sampling methods within the constraints of this level of study. For particle source sample, a reasonable approach is to utilize a standard EPA Method 5 (Rom, 1972) modified to have a particle sizing device such as a cascade impactor or virtual impactor in place of the filter. It is important to measure the stack gas velocity profile so that isokinetic sampling is assured. Although the usual Method 5 approach is to obtain multiple samples across the stack so the integrated mass emissions can be estimated, a single sampling location away from the wall may suffice for the purposes of these kinds of receptor modeling studies (Gordon et al., 1984).

The final and highest level of study requires a more sophisticated and expensive source sampling approach. It involves the sampling of area as well as point stack sources in an effort to identify nonducted emissions. In addition to more sources sampled with more complex systems, enough source samples should be obtained to permit the determination of the variability in the composition of the emitted material. It is important to be able to know the change in composition with time over the course of the study so that the results obtained will have the maximum reliability.

Stack samples should be obtained in a way such that the emissions samples are comparable to the ambient samples. As has been discussed, there are important compositional variation with size and thus, sized source samples will probably be helpful in providing better inputs to the receptor models. Cooling is also important so that condensable species will have been collected at the source in the same manner as they will appear to be at the ambient site.

Similarly, area sources such as fugitive emissions from large-materials handling operations or sources such as agricultural burning should also be sampled in a manner similar to the ambient sampling. It may be possible to obtain samples of the material from a facility such as a coal-fired power plant, smelter, or cement plant. It can be suspended in the laboratory and collected by a sampler to yield samples with characteristics analogous to the ambient samples. It may also be possible to use wind trajectory methods to isolate fugitive emission impacted samples and derive from them the source profiles. The careful sampling of potential fugitive emission sources can often become a key component in the kind of complete source resolution that is desired at this level of program.

Motor vehicle emissions can be sampled in areas where other sources have limited impacts such as immediately adjacent to highly traveled roadways or in tunnels. There is a major difficulty in accurately assessing motor vehicle impacts as fewer and fewer vehicles burn leaded fuel. Although there is a major benefit to reducing airborne lead exposure particularly in urban areas, the continuing transformation of the automobile fleet to nonleaded fuel means that the unique tracers of motor vehicles are being reduced in concentration per unit emitted mass averaged over the entire fleet. Thus, careful attention to the value used in a model and the uncertainty in the resultant mass should be a key part of evaluating motor vehicular source impacts.

An important point to be reemphasized is that there are real limitations to the resolvability of sources. In Chapter 6 the concept of collinearity and the inability to resolve similar composition sources was presented. Therefore, before proceeding very far into a major study, it will be prudent to begin to examine the types of sources that are likely to impact a site and determine if, in fact, they are resolvable with the kind of data being collected. If they can be separately quantified, it is reasonable to proceed to do so accurately. If they cannot be resolved with the kinds of data being obtained, it becomes important to assess that constraint before proceeding further. It may be sufficient for the purposes of the particular study to resolve some linear combination of sources only. On the other hand, it may also be important to look for other kinds of analyses that can provide the data needed to give the level of resolution desired.

Receptor modeling has developed rapidly over the past decade and will continue to develop as both analytical and mathematical methods are improved. These techniques have proven useful in solving problems of air quality as well as potentially being useful in a variety of other research areas. This volume has reviewed these methods to data and the current review will become outdated. However, it is hoped that it has proven useful in introducing the conceptual framework and presenting most of the currently used methods of receptor modeling.

REFERENCES

Ackerman, B., S. A. Changnon, G. Dzurisin, D. L. Gatz, R. C. Grosh, S. D. Hilberg, F. A. Huff, J. W. Mansell, H. T. Ochs, M. E. Peden, P. T. Schickedanz, R. G. Semonin, and J. L. Vogel. 1978. Summary of METROMEX, Volume 2: Causes of Precipitation Anomalies. Illinois State Water Survey Bulletin 63, Urbana, IL. 395 pp.

Adams, D. F., S. O. Farwell, M. R. Pack, E. Robinson. 1981a. Biogenic Sulfur Gas Emissions from Soils in Eastern and Southeastern United States. *J. Air Pollut. Control Assoc.* **31**:1083–1089.

Adams, D. F., S. O. Farwell, E. Robinson, M. R. Pack, and W. L. Bamesberger. 1981b. Biogenic Sulfur Source Strengths. *Environ. Sci. Technol.* **15**: 1493–1498.

Alary, J., P. Bourbon, J. Esclassan, J. C. Lepert, and J. Vandaele. 1983. Zinc, Lead, Molybdenum Contamination in the Vicinity of an Electric Steelworks and Environmental Response to Pollution Abatement by Bag Filter. *Water, Air and Soil Pollution* **20**:137–145.

Alpert, D. J. and P. K. Hopke, 1980. A Quantitative Determination of Sources in the Boston Urban Aerosol. *Atmospheric Environ.* **14**:1137–1146.

Alpert, D. J. and P. K. Hopke, 1981. A Determination of the Sources of Airborne Particles Collected During the Regional Air Pollution Study. *Atmospheric Environ.* **15**:675–687.

Alpert, D. J. and P. K. Hopke. 1982. Response to Comments on a Quantitative Determination of Sources in the Boston Urban Aerosol. *Atmospheric Environ.* **16**:1568.

Appel, B. R., P. Colodny, and J. J. Wesolowski. 1976. Analysis of Carbonaceous Materials in Southern California Atmospheric Aerosols. *Enivron. Sci. Technol.* **10**:359–363.

Appel, B. R., E. M. Hoffer, E. L. Kothny, S. M. Wall, and M. Haik. 1979. Analysis of Carbonaceous Material in Southern California Atmospheric Aerosols. 2. *Environ. Sci. Technol.* **13**:98–103.

Appel, B. R., Y. Tokiwa, and E. L. Kothny. 1983. Sampling of Carbonaceous Particles in the Atmosphere. *Atmospheric Environ.* **17**:1787–1796.

Aras, N. K., W. H. Zoller, G. E. Gordon, and G. L. Lutz. 1973. Instrumental Photon Activation Analysis of Atmospheric Particulate Matter. *Anal. Chem.* **45**:1481–1490.

Armstrong, J. A., P. A. Russel, L. E. Sparks, and D. C. Drehmel. 1981. Tethered Balloon Sampling Systems for Monitoring Air Pollution. *J. Air Pollut. Control Assoc.* **31**:735–743.

Arnold, E. and R. G. Draftz. 1979. Identification of Sources Causing TSP Non-Attainment: Analysis of Historical High-Volume Filters, Decatur and Quad Cities Areas. Report No. 79/21, Illinois Institute of Natural Resources, Chicago, IL.

Baker, M. B., D. Camparoli, and H. Harrison. 1981. An Analysis of the First Year of MAP35 Rain Chemistry Measurements. *Atmospheric Environ.* **15**:43–55.

Barn, W. C. 1976. The Power Spectrum of Temperatures in Central England. *Q. J. Royal Met. Soc.* **432**:464–466.

Barone, J. B., T. A. Cahill, R. A. Eldred, R. G. Flocchini, D. J. Shadoan, and T. M. Dietz. 1978. A Multivariate Statistical Analysis of Visibility Degradation at Four California Cities. *Atmospheric Environ.* **12**:2213–2221.

Belsley, D. A., E. Kuh, and R. E. Welsch. 1980. *Regression Diagnostics, Identifying Influential Data and Sources of Collinearity.* Wily, New York.

Bennett, R. L. and K. T. Knapp. 1982. Characterization of Particulate Emissions from Municipal Wastewater Sludge Incinerators. *Environ. Sci. Technol.* **16**:831–836.

Bertin, E. P. 1975. *Principles and Practice of X-Ray Spectrometric Analysis*, 2nd Edition. Plenum, New York. 1079 pp.

Biermann, A. H. and J. M. Ondov. 1980. Application of Surface-Deposition Models to Size-Fractionated Coal Fly Ash. *Atmospheric Environ.* **14**: 289–295.

Biggins, P. D. E. and R. M. Harrison 1979. The Identification of Specific Chemical Compounds in Size-Fractionated Atmospheric Particulates Collected at Roadside Sites. *Atmospheric Environ.* **13**:1213–1216.

Bilonick, R. A. and D. G. Nichols. 1983. Temporal Variations in Acid Precipitation over New York State—What the 1965–1979 USGS Data Reveal. *Atmospheric Environ.* **17**:1063–1072.

Blackith, R. E. and R. A. Reyment. 1971. *Multivariate Morphometrics.* Academic Press, London. 412 pp.

Blifford, I. H. and G. O. Meeker. 1967. A Factor Analysis Model of Large Scale Pollution. *Atmospheric Environ.* **1**:147–157.

Block, C. and R. Dams. 1976. Study of Fly Ash Emission During Combustion of Coal. *Environ. Sci. Tech.* **10**:1011–1017.

Bogen, J.. 1973. Trace Elements in Atmospheric Aerosol in the Heidelberg Area, Measured by Instrumental Neutron Activation Analysis. *Atmospheric Environ.* **7**:1117–1125.

Bowen, H. J. M. 1966. *Trace Elements in Biochemistry.* Academic Press, New York.

Box, G. E. P. and G. M. Jenkins. 1976. *Time Series Analysis: Forecasting and Control*, Revised Edition. Holden-Day, San Francisco. 575 pp.

Braman, R. S. 1983. Chemical Speciation, in *Analytical Aspects of Environmen-*

tal Chemistry. D. F. S. Natusch and P. K. Hopke, eds. Wiley, New York. pp. 1–59.

Britt, H. I. and R. H. Luecke, 1973. The Estimation of Parameters in Nonlinear, Implicit Models. *Technometrics* **15**:233–247.

Broekaert, J. A. C., B. Wopenka, and H. Puxbaum. 1982. Inductively Coupled Plasma Optical Emission Spectrometry for the Analysis of Aerosol Samples Collected by Cascade Impactor. *Anal. Chem.* **54**:2174–2179.

Brosset, C., K. Andreasson, and M. Ferm. 1975. The Nature and Possible Origin of Acid Particles Observed at the Swedish West Coast. *Atmospheric Environ.* **9**:631–642.

Budiansky, S. 1980. Dispersion Modeling. *Environ. Sci. Technol.* **14**:370–374.

Butler, J. D., P. Crossley, and D. M. Colwill. 1981. Scanning Electron Microscopy and X-Ray Fluorescence Analysis of a Fractionated Urban Aerosol. *Science of the Total Environment* **19**:179–194.

Butler, J. D., P. Crossley, and D. M. Colwill. 1982. Predicting Polycyclic Aromatic Hydrocarbon Concentrations in Urban Aerosols by Linear Multiple Regression Analysis. *Environ. Pollut.* (Ser. B) **3**:109–123.

Cadle, S. H., P. J. Groblicki, and P. A. Mulawa. 1983. Problems in the Sampling Analysis of Carbon Particulate. *Atmospheric Environ.* **17**:593–600.

Cadle, S. H., P. J. Groblicki, and D. P. Stroup. 1980. Automated Carbon Analyzer for Particulate Samples. *Anal. Chem.* **52**:2201–2206.

Cahill, R. A. 1974. A Study of the Trace Element Distributions in Petroleum. M.S. Thesis. University of Maryland, College Park.

Cahill, T. A. 1975. Ion-Excited X-Ray Analysis of Environmental Samples, in *New Uses of Ion Accelerators*. J. F. Ziegler, ed. Plenum, New York, pp. 1–71.

Cahill, T. A. and P. J. Feeney. 1973. Contribution of Freeway Traffic to Airborne Particulate Matter. Crocker Nuclear Lab., Universtiy of California, Davis. Report No. UCD-CNL169 (June).

Cass, G. R. 1979. On the Relationship Between Sulfate Air Quality and Visibility with Examples in Los Angeles. *Atmospheric Environ.* **1**:1069–1084.

Cass, G. R. and G. J. McRae. 1981. Source-Receptor Reconciliation of South Coast Air Basin Particulate Air Quality Data. Report to the California Air Resources Board Under Contract A9-014-31.

Casuccio, G. S. and P. B. Jancoko. 1981. Quantifying Blast Furnace Cast House Contributions to TSP Using a Unique Receptor Model. Paper No. 81.64.2. Air Pollution Control Association, Pittsburgh, PA. 16 pp.

Casuccio, G. and P. Janocko. 1983. Analysis of Ambient Fine and Coarse Dichot Filters by Computer Controlled Scanning Electron Microscopy (CCSEM). Report to Dr. Robert K. Stevens, USEPA, by Energy Technology Consultants, Murraysville, PA. 53 pp.

Casuccio, G. S., P. B. Janocko, R. J. Lee, and J. F. Kelly. 1982. The Role of Computer Controlled Scanning Electron Microscopy in Receptor Modeling. Paper No. 82-21.4. Air Pollution Control Association, Pittsburgh, PA. 19 pp.

Casuccio, G. S., P. B. Janocko, R. J. Lee, J. F. Kelly, S. L. Dattner, and J. S. Mgebroff. 1983. The use of Computer Controlled Scanning Electron Microscopy in Environmental Studies. *J. Air Pollu. Control Assoc.* **33**:937–943.

Casuccio, G. S., P. B. Janocko, and J. H. Lucas. 1982. A Comparison of the Particulate Matter Emission Inventory to Dispersion Modeling and Receptor Modeling Results at the Coke and Iron Division of Shenango Inc. Energy Technology Consultants, Murraysville, PA. 16 pp.

Cattell, R. B. 1966. *Handbook of Multivariate Experimental Psychology.* Rand McNally, Chicago, pp. 174–243.

Cecil, C. B., R. W. Stanton, and F. T. Dulong. 1981. Geology of Contaminants in Coal: Phase I Report of Investigation USGS Open-File Report No. 81-953-A.

Chang, S. N., P. K. Hopke, S. W. Rheingrover, and G. E. Gordon. 1982. Target Transformation Factor Analysis of Wind-Trajectory Selected Samples. Paper No. 81-21.1. Air Pollution Control Association, Pittsburgh, PA. 14 pp.

Changnon, S. A., R. A. Huff, P. T. Schickendenz, and J. L. Vogel. 1977. Summary of METROMEX, Volume 1: Weather Anomalies and Impacts. Illinois State Water Survey Bulletin 62, Urbana, IL. 260 pp.

Charlson, R. J., N. C. Ahlquist, N. Selridge, and P. B. MacCready, Jr. 1969. Monitoring of Atmospheric Aerosol Parameters with the Integrating Nephelometer. *JAPCA* **19**:937–942.

Charlton, T. H., D. C. Grove, and P. K. Hopke, 1978. The Paredon, Mexico, Obsidian Source and Early Formative Exchange. *Science* **201**:807–809.

Chatfield, C. 1980. *The Analysis of Time Series: An Introduction.* Chapman and Hall, London. 268 pp.

Chow, J. C., V. Shortell, J. Collins, J. G. Watson, T. G. Pace, and R. Burton. 1981. A Neighborhood Scale Study of Inhalable and Fine Suspended Particulate Matter Source Contributions to an Industrial Area in Philadelphia. Paper No. 81-14.1. Air Pollution Control Association, Pittsburgh, PA. 16 pp.

Chow, J. C., J. G. Watson, and J. J. Shah. 1982. Source Contributions to Inhalable Particulate Matter in Major U.S. Cities. Air Pollution Control Association Paper No. 82-21.3. 16 pp.

Chu, L. C. and E. S. Macias. 1981. Carbonaceous Urban Aerosol—Primary or Secondary, in *Atmospheric Aerosol: Source/Air Qualtiy Relationships.* E. S. Macias and P. K. Hopke, eds. Symposium Series No. 167, American Chemical Society, Washington, D.C., pp. 251–268.

Chung, F. H. 1974a. Quantitative Interpretation of X-Ray Diffraction Patterns of Mixtures. I. Matrix-Flushing Method for Quantitative Multicomponent Analysis. *J. Appl. Cryst.* **7**:519–525.

Chung, F. H. 1974b. Quantitative Interpretation of X-Ray Diffraction Patterns of Mixtures. II. Adiabatic Principal of X-RAy Diffraction Analysis of Mixtures. *J. Appl. Cryst.* **7**:526–531.

Chung, F. H. 1975. Quantitative Interpretation of X-Ray Diffraction Patterns of Mixtures. III. Simultaneous Determination of a Set of Reference Intensities. *J. Appl. Cryst.* **8**:17–19.

Clemenson, M., T. Novakav, and S. S. Markowitz. 1980. Determination of Carbon in Atmospheric Aerosols by Deuteron Activation Analysis. *Anal. Chem.* 52:1758–1761.

Clifford, A. A. 1973. *Multivariate Error Analysis.* Applied Science Publishers, London.

Code of Federal Regulations 40, Part 50.11, Appendix B, July 1, 1975, pp. 12–16.

Conlee, C. J., P. A. Kenline, R. L. Cummins, and V. J. Konopinski. 1967. Motor Vehicle Exhaust at Three Selected Sites. *Arch. Environ. Health* **14**:429–446.

Conry, T. 1974. Airports as Trace Metal Sources: A Study of Friendship International and Washington National Airports. M.S. Thesis. University of Maryland, College Park.

Cooley, W. W. and P. R. Lohnes. 1971. *Multivariate Data Analysis.* Wiley, New York. 364 pp.

Cooper, D. W. 1983. Receptor-Oriented Source-Receptor Analysis, in *Receptor Models Applied to Contemporary Pollution Problems*, S. L. Dattner and P. K. Hopke, eds. Air Pollution Control Association, Pittsburgh, PA, pp. 296–316.

Cooper, J. A. 1979. Medford Aerosol Characterization Study: Application of Chemical Mass Balance to Identification of Major Aerosol Sources in the Medford Airshed. Interim Report to the Oregon Department of Environmental Quality, Portland.

Cooper, J. A. 1983. Receptor Model Approach to Source Apportionment of Acid Precipitation Precursors. *Proceedings of the VIth World Congress on Air Quality, Paris, France* vol. 3:223–229.

Cooper, J. A., L. A. Currie, and G. A. Klouda. 1981. Assessment of Contemporary Carbon Combustion Source Contributions to Urban Air Particulate Levels Using Carbon-14 Measurements. *Environ. Sci. Technol.* **15**:1045–1050.

Cooper, J. A. and J. G. Watson, 1980. Receptor-Oriented Methods of Air Particulate Source Apportionment. *J. Air. Pollut. Control Assoc.* **30**:1116–1125.

Cooper, J. A., J. G. Watson, and J. J. Huntzicker, 1979. Summary of the Portland Aerosol Characterization Study (PACS). Paper No. 79-29.4. Air Pollution Control Association, Pittsburgh, PA. 15 pp.

Core, J. E., J. A. Cooper, P. L. Hanrahan, and W. M. Cox. 1982. Particulate Dispersion Model Evaluation: A New Approach Using Receptor Models. *J. Air Pollut. Control Assoc.* **32**:1142–1147.

Core, J. E., P. L. Hanahan, and J. A. Cooper. 1981. Air Pollution Control Strategy Development: A New Approach Using Chemical Mass Balance Methods, in *Atmospheric Aerosol: Source/Air Quality Relationships.* E. S. Macias and P. K. Hopke, eds. Symposium Series No. 167, American Chemical Society, Washington, D.C., pp. 107–123.

Courant, R. and D. Hilbert. 1953. *Methods of Mathematical Physics*, Volume 1, 1st English Edition. Interscience, New York.

Court, J. D., R. J. Goldsack, L. M. Ferrari, and H. A. Polach. 1981. The Use of

Carbon Isotopes in Identifying Urban Air Particulate Sources. *Clean Air*, February, pp. 6–11.

Courtney, W. J., R. W. Shaw, and T. G. Dzubay. 1982. Precision and Accuracy of a β-Gauge for Aerosol Mass Determinations. *Environ. Sci. Technol.* **16**: 236–239.

Cox, W. M. and J. Clark. 1981. Ambient Ozone Concentration Patterns Among Eastern U.S. Urban Areas Using Factor Analysis. *J. Air Pollut. Control Assoc.* **31**:762–766.

Crutcher, E. R. 1983. Light Microscopy as an Analytical Approach to Receptor Modeling, in *Receptor Models Applied to Contemporary Pollution Problems*. S. L. Dattner and P. K. Hopke, eds. Air Pollution Control Association, Pittsburgh, PA, pp. 266–284.

Crutcher, E. R. and L. S. Nishimura. 1978. Estimation of Error in Quantitative Microscopic Analysis of Compounds. Proceedings of the Fourth International Symposium on Contamination Control, Washington, D.C., pp. 150–158.

Currie, L. A., S. M. Kunen, K. J. Voorhees, R. B. Murphy, and W. F. Koch. 1978. Analysis of Carbonaceous Particulates and Characterization of Their Sources of Low-Level Counting and Pyrolysis/Gas Chromatography/Mass Spectrometry. Conference on Carbonaceous Particles in the Atmosphere, CONF-7803101. U.S. Department of Energy, Lawrence Berkeley Laboratory, Berkeley, CA, pp. 36–48.

Cushing, K. M., J. D. McCain, and W. B. Smith. 1979. Experimental Determination of Sizing Parameters and Wall Losses of Five Source-Test Cascade Impactors. *Environ. Sci. Technol.* **13**:726–731.

Daisey, J. M. 1983. Receptor Source Apportionment Models for Two Polycyclic Aromatic Hydrocarbons, in *Receptor Models Applied to Contemporary Pollution Problems*. Air Pollution Control Association, Pittsburgh, PA, pp. 348–357.

Daisey, J. M., I. Hawryluk, T. J. Kneip, and F. Mukai. 1978. Mutagenic Activity in Organic Fractions of Airborne Particulate Matter. Conference on Carbonaceous Particles in the Atmosphere, CONF-7803101. U.S. Department of Energy, Lawrence Berkeley Laboratory, Berkeley, CA, pp. 187–192.

Daisey, J. M., R. J. Hershman, and T. J. Kneip. 1982. Ambient Levels of Particulate Organic Matter in New York City in Winter and Summer. *Atmospheric Environ.* **16**:2161–2168.

Daisey, J. M. and T. J. Kneip. 1981. Atmospheric Particulate Organic Matter, in *Atmospheric Aerosol: Source/Air Quality Relationships*. E. S. Macias and P. K. Hopke, eds. Symposium Series No. 167, American Chemical Society, Washington, D.C., pp. 197–221.

Daisey, J. M., M. A. Leyko, and T. J. Kneip. 1979. Source Identification and Allocation of Polynuclear Aromatic Hydrocarbon Compounds in the New York City Aerosol: Methods and Applications, in *Polynuclear Aromatic*

Hydrocarbons. P. W. Jones and P. Leber, eds. Ann Arbor Science Publishers, Ann Arbor, MI, pp. 201–215.

Dallavalle, J. M. 1948. *Micromeritics, the Technology of Fine Particles*, 2nd Edition. Pitman Publishing Corporation, New York.

Dams, R., J. A. Robbins, K. A. Rahn, and J. W. Winchester. 1970. Nondestructive Neutron Activation Analysis of Air Pollution Particulates. *Anal. Chem.* **42**: 861–867.

Dattner, S. L. 1978. Preliminary Analysis of the Use of Factor Analysis of X-Ray Fluorescence Data to Determine the Sources of Total Suspended Particulate, Draft Report. Texas Air Control Board, Austin.

Dattner, S. L. and P. K. Hopke. 1983. Receptor Models Applied to Contemporary Pollution Problems. Specialty Conference SP-48. Air Pollution Control Association, Pittsburgh, PA. 368 pp.

Dattner, S. L. and M. Jenks. 1981. Identification of Non-Emission Source Related Factors in Sets of Ambient Particulate Data. Paper No. 81-64.1. Air Pollution Control Association, Pittsburgh, PA. 13 pp.

Dattner, S. L., S. Mgebroff, G. Casuccio, and P. Janocko. 1983. Identifying the Sources of TSP and Lead in El Paso Using Microscopy and Receptor Models. Paper No. 83-49.3. Air Pollution Control Association. Pittsburgh, PA. 14 pp.

Davis, B. L. 1978. Additional Suggestions for X-Ray Quantitative Analysis of High-Volume Filter Samples. *Atmospheric Environ.* **12**:2403–2406.

Davis, B. L. 1980. "Standardless" X-Ray Diffraction Quantitative Analysis. *Atmospheric Environ.* **14**:217–220.

Davis, B. L. 1981. A Study of the Errors in X-Ray Quantitative Analysis Procedures for Aerosols Collected on Filter Media. *Atmospheric Environ.* **15**: 291–296.

Davis, B. L. and N. K. Cho. 1977. Theory and Application of X-Ray Diffraction Compound Analysis to High-Volume Filter Samples. *Atmospheric Environ.* **11**:73–85.

Davis, B. L. and L. R. Johnson. 1982. On the Use of Various Filter Substrates for Quantitative Particulate Analysis by X-Ray Diffraction. *Atmospheric Environ.* **16**:273–282.

Davis, B. L., L. R. Johnson, D. T. Griffen, W. R. Phillips, R. K. Stevens, and D. Maughan. 1981. Quantitative Analysis of Mt. St. Helens Ash by X-Ray Diffraction and X-Ray Fluorescence Spectrometry. *J. Appl. Meteorol.* **20**:922–933.

DeCesar, R. T. and J. A. Cooper. 1981. Medford Aerosol Characterization Study. Final Report to the Oregon Department of Environmental Quality, Portland.

DeCesar, R. T. and J. A. Cooper. 1983. Evaluation of Multivariate and Chemical Mass Balance Approaches to Aerosol Source Apportionment Using Synthetic Data and an Expanded PACS Data Set, in *Receptor Models Applied to*

Contemporary Pollution Problems. S. L. Dattner and P. K. Hopke, eds. Air Pollution Control Association, Pittsburgh, PA, pp. 127–140.

Draftz, R. G. 1979. Aerosol Source Characterization Study in Florida–Microscopical Analysis. Report No. EPA-600/3-79-097. U.S. Environmental Protection Agency, Research Triangle Park, NC.

Draftz, R. G. 1982. Distinguishing Carbon Aerosols by Microscopy, in *Particulate Carbon–Atmospheric Life Cycle.* G. T. Wolff and R. L. Klimish, eds. Plenum, New York, pp. 261–271.

Draftz, R. G. and K. Severin. 1980. Microscopical Analysis of Aerosols Collected in St. Louis, Missouri. Report No. EPA-600/3-80-027. U.S. Environmental Protection Agency, Research Triangle Park, NC.

Duewer, D. L., B. R. Kowalski, and J. L. Fasching. 1976. Improving the Reliability of Factor Analysis of Chemical Data by Utilizing the Measured Analytical Uncertainty. *Anal. Chem.* **48**:2002–2010.

Dunker, A. M. 1979. A Method for Analyzing Data on the Elemental Composition of Aerosols. General Motors Research Laboratories Report GMR-307 4 ENV-67, Warren, MI.

Dutot, A. L., C. Elichegaray, and R. Vie le Sage. 1983. Application de l'Analyse des Correspondances a l'Etude de la Composition Physico-Chemique de l'Aerosol Urbain. *Atmospheric Environ.* **17**:73–78.

Duval, M. M. 1980. Source Resolution Studies of Ambient Polycyclic Aromatic Hydrocarbons in the Los Angeles Atmosphere: Application of a Chemical Species Balance Method with First Order Chemical Decay. M.S. Thesis. University of Caliornia, Los Angeles.

Dzubay, T. G. 1977. *X-Ray Fluorescence Analysis of Environmental Samples.* Ann Arbor Science Publishers, Ann Arbor, MI. 310 pp.

Dzubay, T. G. 1980. Chemical Element Balance Method Applied to Dichotomous Sampler Data. *Annals N.Y. Acad. Sci.* **338**:126–144.

Dzubay, T. G. and R. K. Barbour. 1983. A Method to Improve the Adhesion of Aerosol Particles on Teflon Filters. *J. Air Pollut. Control Assoc.* **33**:692–695.

Dzubay, T. G. and R. K. Stevens. 1975. Ambient Air Analysis with Dicthotomous Sampler and X-Ray Fluorescence Spectrometer. *Environ. Sci. Technol.* **9**: 663–668.

Dzubay, T. G. and R. K. Stevens. 1983. Intercomparison of Results of Several Receptor Models for Apportioning Houston Aerosol, in *Receptor Models Applied to Contemporary Pollution Problems.* S. L. Dattner and P. K. Hopke, eds. Air Pollution Control Association, Pittsbugh, PA., pp. 60–71.

Dzubay, T. G., R. K. Stevens, and L. W. Richards. 1979. Composition of Aerosols over Los Angeles Freeways. *Atmospheric Environ.* **13**:653–659.

Dzubay, T. G., R. K. Stevens, C. W. Lewis, D. H. Hern, W. J. Courtney, J. W. Tesch, and M. A. Mason. 1982. Visibility and Aerosol Composition in Houston, Texas. *Environ. Sci. Technol.* **16**:514–525.

Eiceman, G. A., R. E. Clement, and F. W. Karasek. 1979. Analysis of Fly Ash from Municipal Incinerators for Trace Organic Compounds. *Anal. Chem.* **51**:2343–2350.

Eiceman, G. A. and V. J. Vandiver. 1983. Adsorption of Polycyclic Aromatic Hydrocarbons on Fly Ash from a Municipal Incinerator and a Coal-Fired Power Plant. *Atmospheric Environ.* **17**:461–465.

Energy Technology Consultants. 1983. Identification of the Sources of Total Suspended Particulate and Particulate Lead in the El Paso Area by Quantitative Microscopic Analysis. Report to the Texas Air Control Board, Austin, March 1983.

Environmental Protection Agency. 1979. Industrial Source Complex (ISC) Dispersion Model User's Guide. Report No. EPA-450/4-79-030. U.S. Environmental Protection Agency, Research Triangle Park, NC.

Esmen, N. A. and T. C. Lee. 1980. Distortion of Cascade Impactor Measured Size Distribution Due to Bounce and Blow-Off. *Amer. Ind. Hygiene Assoc. J.* **41**:410–419.

Exner, O. 1966. Additive Physical Properties. I. General Relationships and Problems of Statistical Nature. *Czechoslov. Chem. Commun.* **31**:3223–3251.

Fabrick, A. J. and R. C. Sklarew. 1975. Oregon/Washington Diffusion Modeling Study, Xonics, Inc.

Feely, J. A. and H. M. Liljestrand. 1983. Source Contributions to Acid Precipitation in Texas. *Atmospheric Environ.* **17**:807–814.

Feeney, P. J., T. A. Cahill, R. G. Flocchini, R. A. Eldred, D. J. Shadoan, and T. Dunn. 1975. Effects of Roadbed Configuration on Traffic Derived Aerosols. *J. Air Pollut. Control Assoc.* **25**:1145–1148.

Fisher, G. L. and D. F. S. Natusch. 1979. Size Dependence of the Physical and Chemical Properties of Fly Ash. Report No. UCD 47 2-502. U.S. Department of Energy. 47 pp.

Fordyce, J. S. 1975. Air Pollution Source Identification. Report No. NASA TMX-71704. Presented at Sources and Emissions Workshop of the Second Interagency Committee on Marine Science and Engineering Conference on the Great Lakes, Argonne, IL. 34 pp.

Foster, R. L. and P. F. Lott. 1980. X-Ray Diffractometry Examination of Air Filters for Compounds Emitted by Lead Smelting Operations. *Environ. Sci. Technol.* **14**:1240–1244.

Friedlander, S. K. 1973. Chemical Element Balances and Identification of Air Pollution Sources. *Environ. Sci. Technol.* **7**:235–240.

Friedlander, S. K. 1977. *Smoke, Dust, and Haze: Fundamentals of Aerosol Behavior.* Wiley-Interscience, New York.

Friedlander, S. K. 1981. New Developments in Receptor Modeling Theory, in *Atmospheric Aerosol: Source/Air Quality Relationships.* E. S. Macias and P. K. Hopke, eds. Symposium Series No. 167, American Chemical Society, Wahsington, D.C., pp. 1–19.

Fukasawa, T., M. Iwatsuki, S. Kawakubo, and K. Miyazaki. 1980. Heavy-Liquid

Separation and X-Ray Diffraction Analysis of Airborne Particulates. *Anal. Chem.* **52**:1784–1787.

Fukasawa, T., M. Iwatsuki, and S. P. Tillekeratne. 1983. X-Ray Diffraction Analysis of Airborne Particulates Collected by an Andersen Sampler. Compound Distribution vs. Particale Size. *Environ. Sci. Technol.* **17**: 596–602.

Full, W. E., R. Ehrlich, and J. E. Klovan. 1981. Extended *Q*-Model Objective Definition of External End Members in the Analysis of Mixtures. *Int. J. Math. Geology* **13**:331–344.

Gaarenstroom, P. D., S. P. Perone, and J. P. Moyers. 1977. Application of Pattern Recognition and Factor Analysis for Characterization of Atmospheric Particulate Composition in Southwest Desert Atmosphere. *Environ. Sci. Technol.* **11**:795–800.

Gardella, J. A. and D. M. Hercules. 1979. Surface Spectroscopic Examination of Diesel Particulates–A Preliminary Study. *Int. J. Environ. Anal. Chem.* **7**:121–136.

Gartrell, G., Jr. and S. K. Friedlander. 1975. Relating Particulate Pollution to Sources: The 1972 California Aerosol Characterization Study. *Atmospheric Environ.* **9**:279–299.

Gatz, D. F. 1975. Relative Contributions of Different Sources of Urban Aerosols: Application of a New Estimation Method to Multiple Sites in Chicago. *Atmospheric Environ.* **9**:1–18.

Gatz, D. F. 1978. Identification of Aerosol Sources in the St. Louis Area Using Factor Analysis. *J. Appl. Met.* **17**:600–608.

Gatz, D. F. 1983. Source Apportionment of Rain Water Impurities in Central Illinois. Paper No. 83-28.3. Air Pollution Control Association, Pittsburgh, PA. 16 pp.

Gerlach, R. W., L. A. Currie and C. W. Lewis. 1983. Review of the Quail Roost II Receptor Model Simulation Exercise, in *Receptor Models Applied to Contemporary Pollution Problems.* S. L. Dattner and P. K. Hopke, eds. Air Pollution Control Association, Pittsburgh, PA, pp. 96–109.

Gether, J. and H. M. Seip. 1979. Analysis of Air Pollution Data by the Combined Use of Interactive Graphic Presentation and a Clustering Technique. *Atmospheric Environ.* **13**:87–96.

Giauque, R. D. 1977. Private communication noted in Pierson and Brachaczek (1983).

Gladney, E. S. 1974, Trace Elemental Emissions from Coal-Fired Power Plants: A Study of the Chalk Point Electric Generating Station. Ph.D. Thesis. University of Maryland, College Park.

Gladney, E. S., J. A. Small, G. E. Gordon, and W. H. Zoller. 1976. Composition and Size Distribution of In-Stack Particulate Material at a Coal-Fired Power Plant. *Atmospheric Environ.* **10**:1071–1077.

Gladney, E. S., W. H. Zoller, A. G. Jones, and G. E. Gordon. 1974. Composition and Size Distribution of Atmospheric Particulate Matter in Boston Area. *Environ. Sci. Technol.* **8**:551–557.

Goldschmidt, V. M. 1958. *Geochemistry.* Academic Press, New York.

Gordon, G. E. 1980. Receptor Models. *Environ. Sci. Technol.* **14**:792–800.

Gordon, G. E., W. R. Pierson, J. M. Daisey, P. J. Lioy, J. A. Cooper, and J. G. Watson. 1984. Considerations for Design of Source Apportionment Studies. *Atmospheric Environ.* **18**:1567–1582.

Gordon, G. E., W. H. Zoller, G. S. Kowalczyk, and S. W. Rheingrover. 1981. Composition of Source Components Needed for Aerosol Receptor Models, in *Atmospheric Aerosol: Source/Air Quality Relationships.* E. S. Macias and P. K. Hopke, eds. Symposium Series No. 167, American Chemical Society, Washington, D.C., pp. 51–74.

Gordon, G. E., W. H. Zoller, E. S. Gladney, A. G. Jones, and P. K. Hopke. 1972. Neutron Activation Study of Trace Elements on Boston-Area Atmospheric Particulates. Presented to the 163rd National Meeting of the American Chemical Society.

Gordon R. J. 1976. Distribution of Airborne Polycyclic Aromatic Hydrocarbons Throughout Los Angeles. *Environ. Sci. Technol.* **10**:370–373.

Gordon, R. J. and R. J. Bryan. 1973. Patterns in Airborne Polynuclear Hydrocarbons Concentrations at Four Los Angeles Sites. *Environ. Sci. Technol.* **7**:1050–1053.

Goulding, F. S., J. M. Jaklevic, and B. W. Loo. 1981. Aerosol Analysis for the Regional Air Pollution Study—Final Report. EPA-600-/S4-8L-006. U.S. Environmental Protection Agency. Research Triangle Park, NC.

Graf, J., R. H. Snow, and R. G. Draftz. 1977. Aerosol Sampling and Analysis—Phoenix, Arizona. Report No. EPA-600/3-77-015. U.S. Environmental Protection Agency, Research Triangle Park, NC.

Greenberg, R. R. 1976. A Study of Trace Elements Emitted on Particles from Municipal Incinerators. Ph.D. Thesis. University of Maryland, College Park.

Greenberg, R. R., G. E. Gordon, W. H. Zoller, R. B. Jacko, D. W. Neuendorf, and K. J. Yost. 1978. Composition of Particles Emitted from the Nicosia Municipal Incinerator. *Environ. Sci. Technol.* **12**:1329–1332.

Greenberg, R. R., W. H. Zoller, and G. E. Gordon. 1978. Composition and Size Distribution of Particles Released in Refuse Incineration. *Environ. Sci. Technol.* **12**:566–573.

Greenberg, R. R., W. H. Zoller, and G. E. Gordon. 1981. Atmospheric Emissions from the Parkway Sewage-Sludge Incinerator. *Environ. Sci. Technol.* **15**: 64–70.

Griffin, J. J. and E. D. Goldberg. 1979. Morphologies and Origin of Elemental Carbon in the Environment. *Science* **206**:563–565.

Grimmer, G. and A. Hildebrandt. 1975. Investigations on the Carcinogenic Burden by Air Pollution in Man XII. Assessment of the Contribution of Passenger Cars to Air Pollution by Carcinogenic Polycyclic Hydrocarbons. *Zbl. Bakt. Hyg., I Abt. Orig. B* **161**:104–124.

Groblicki, P. J., G. T. Wolff, and R. J. Countess. 1981. Visibility-Reducing Species in the Denver "Brown Cloud"–I. Relationships Between Extinction and Chemical Composition. *Atmospheric Environ.* **15**:2473–2502.

Grosjean, D. 1982. Quantitative Collection of Total Inorganic Atmospheric Nitrate on Nylon Filters. *Anal. Letters* **15**(A9):785–796.

Guttman, L. 1954. Some Necessary Conditions for Common Factor Analysis. *Psychometrika* **19**:149–161.

Hadley, G. 1962. *Linear Programming.* Addison-Wesley, Reading, MA.

Hammerle, R. H. and W. R. Pierson, 1975. Sources and Elemental Composition of Aerosol in Pasadena, Calif., by Energy-Dispersive X-ray Fluorescence. *Environ. Sci. Technol.* **9**:1058–1067.

Hansen, L. D. and G. L. Fisher. 1980. Elemental Distribution in Coal Fly Ash Particles. *Environ. Sci. Technol.* **14**:1111–1117.

Hansen, L. D., L. Whiting, D. J. Eatough, T. E. Jensen, and R. M. Izatt. 1976. Determination of Sulfur (IV) and Sulfate in Aerosols by Thermonetric Methods. *Aanl. Chem.* **48**:634–638.

Harman, H. H. 1976. *Modern Factor Analysis.* 3rd ed., rev. University of Chicago Press, Chicago.

Harrison, R. M. and R. Perry. 1977. The Analysis of Tetra-Alkyl Lead Compounds and Their Significance as Urban Air Pollutants. *Atmospheric Environ.* **11**:847–852.

Hayas, A., C. F. Gonzalez, G. Pardo, and M. C. Martinez. 1982. Application of Spectral Analysis to Atmospheric Dust Pollution. *Atmospheric Environ.* **16**:1919–1922.

Heidam, N. Z. 1981. On the Origin of the Arctic Aerosol: A Statistical Approach *Atmospheric Environ.* **15**:1421–1427.

Heidam, N. Z. 1982. Atmospheric Aerosol Factor Models, Mass, and Missing Data. *Atmospheric Environ.* **16**:1923–1931.

Heindryckx, R. and R. Dams. 1974. Continental, Marine, and Anthropogenic Contributions to the Inorganic Composition of the Aerosol of an Industrial Zone. *J. Radioanal. Chem.* **19**:339–349.

Heisler, S. L., S. K. Friedlander, and R. B. Husar. 1973. The Relationship of Smog Aerosol Size and Chemical Element Distributions to Source Characteristics. *Atmospheric Environ.* **7**:633–649.

Heisler, S. L., R. C. Henry, and J. G. Watson. 1980. The Sources of the Denver Haze in November and December of 1978. Paper No. 80-58.6. Air Pollution Control Association, Pittsburgh, PA. 24 pp.

Heisler, S. L., R. C. Henry, J. G. Watson, and G. M. Hidy. 1980. The 1978 Denver Winter Haze Study. Motor Vehicles Manufacturing Association Report P-5417-1, March 1980.

Henry, R. C. 1977. The Application of Factor Analysis to Urban Aerosol Source

Identification. Proceedings of the Fifth Conference on Probability and Statistics. American Meteorological Society, Boston, pp. 134–138.

Henry, R. C. 1978. A Factor Model of Urban Aerosol Pollution: A New Method of Source Identification. Ph.D. Thesis. Oregon Graduate Center, Beaverton.

Henry, R. C. 1983. Stability Analysis of Receptor Models That Use Least-Squares Fitting, in *Receptor Modles Applied to Contemporary Pollution Problems.* S. L. Dattner and P. K. Hopke, eds. Air Pollution Control Association, Pittsburgh, PA, pp. 141–157.

Henry, R. C. and G. M. Hidy. 1979. Multivariate Analysis of Particulate Sulfate and Other Air Quality Variables by Principal Components–Part I. Annual Data from Los Angeles and New York. *Atmospheric Environ.* **13**:1581–1586.

Henry, R. C. and G. M. Hidy. 1981a. Multivariate Analysis of Particulate Sulfate and Other Air Quality Variables by Principal Components–II. Salt Lake City, Utah and St. Louis, Missouri. *Atmospheric Environ.* **16**:929–943.

Henry, R. C. and G. M. Hidy. 1981b. Authors Response to G. D. Thurston's Comments on Henry and Hidy (1979). *Atmospheric Environ.* **15**:425–426.

Hitchcock, D. R. and M. S. Black. 1978. $^{34}S/^{32}S$ Studies of Biogenic Sulfur Emissions at Wallops Island, VA, Preprints of Division of Environmental Chemistry. *American Chemical Society* **18**(2):182–185.

Hites, R. A. and J. B. Howard. 1978. Combustion Research on Characterization of Particulate Organic Matter from Flames. Report No. EPA-600-7-78-167. U.S. Environmental Protection Agency. Research Triangle Park, NC. 85 pp.

Hoffer, J. M., F. Gomez, and P. Muela. 1982. Eruption of El Chichon Volcano, Chiapas, Mexico, 28 March to 7 April 1982. *Science* **218**:1307–1308.

Hoffman, G. L. and R. A. Duce. 1971. Copper Contamination of Atmospheric Particulate Samples Collected with Gelman Hurricane Samplers. *Environ. Sci. Tech.* **5**:1134–1136.

Hollowell, C. D., P. J. Bekowics, R. D. Giauque, J. R. Wallace, and T. Novakov. 1976. Chemical and Physical Characterization of Vehicular Emissions, in *Atmospheric Aerosol Research Annual Report, 1975–76.* Lawrence Berkeley Lab. LBL-5114, Berkeley, CA, pp. 54–62.

Hollowell, C. D., R. L. Dod, R. D. Giauque, G. W. Traynor, and T. Novakov. 1977. Characterization of Winter Air Pollution Episodes, in *Atmospheric Aerosol Research Annual Report, 1976–77.* Lawrence Berkeley Lab LBL-6819, Berkeley, CA, p. 102.

Holt, B. D., P. T. Cunningham, and R. Kumar. 1981. Oxygen Isotope of Atmospheric Sulfates. *Environ. Sci. Technol.* **15**:804–808.

Holt, B. D., R. Kumar, and P. T. Cunningham. 1982. Primary Sulfates in Atmospheric Sulfates: Estimation by Oxygen Isotope Ratio Measurements. *Science* **217**:51–53.

Hopke, P. K. 1976. The Application of Multivatiate Analysis for Interpretation of the Chemical and Physical Analysis of Lake Sediments. *J. Environ. Sci. Health* **6**:367–383.

Hopke, P. K. 1978. Identification of Elements Found in Household Dusts. Presented to the 178th National Meeting of the American Chemical Society.

Hopke, P. K. 1981. The Application of Factor Analysis to Urban Aerosol Source Resolution, in *Atmospheric Aerosol: Source/Air Quality Relationships.* E. S. Macias and P. K. Hopke, eds. Symposium Series No. 167, American Chemical Society, Washington, D.C., pp. 21–49.

Hopke, P. K. 1982. Comments on Trace Element Concentrations in Summer Aerosols at Rural Sites in New York State and Their Possible Sources and Seasonal Variations in the Composition of Ambient Sulfur-Containing Aerosols in the New York Area. *Atmospheric Environ.* **16**:1279–1280.

Hopke, P. K. 1983. An Introduction to Multivariate Analysis of Environmental Data, in *Analytical Aspects of Environmental Chemistry.* D. F. S. Natusch and P. K. Hopke, eds. Wiley, New York, pp. 219–261.

Hopke, P. K., D. J. Alpert, and B. A. Roscoe. 1983. FANTASIA–A Program for Target Transformation Factor Analysis for Resolving Sources of Environmental Samples. *Computers & Chemistry* **7**:149–155.

Hopke, P. K., E. S. Gladney, G. E. Gordon, W. H. Zoller, and A. G. Jones. 1976. The Use of Multivariate Analysis to Identify Sources of Selected Elements in the Boston Urban Aerosol. *Atmospheric Environ.* **10**:1015–1025.

Hopke, P. K., R. E. Lamb, and D. F. S. Natusch. 1980. Multielemental Characterization of Urban Roadway Dust. *Environ. Sci. Technol.* **14**:164–172.

Hopke, P. K., K. G. Severin, and S. N. Chang. 1983. Application and Verification Studies of Target Transformation Factor Analysis as an Aerosol Receptor Model, in *Receptor Models Applied to Contemporary Pollution Problems.* S. L. Dattner and P. K. Hopke, eds. Air Pollution Control Association, Pittsburgh, PA, pp. 110–126.

Horst, P. 1963. *Matrix Algebra for Social Scientists.* Holt, Rinehart and Winston, New York. 517 pp.

Houck, J. E., J. E. Core, and J. A. Cooper. 1982. Source Sampling for Receptor Oriented Source Apportionment. NEA, Inc., Beaverton, OR.

Hougland, E. S. 1983. Chemical Element Balance by Linear Programming. Paper No. 83-14.7. Air Pollution Control Association, Pittsburgh, PA, 7 pp.

Huntzicker, J. J., R. S. Hoffman, and C. S. Ling. 1978. Continuous Measurement and Speciation of Sulfur-Containing Aerosols by Flame Photometry. *Atmospheric Environ.* **12**:83–88.

Husar, R. B. 1974. Atmospheric Particulate Mass Monitoring with a Radiation Detector. *Atmospheric Environ.* **8**:183–188.

Hwang, C. S. and P. K. Hopke. 1983. Time Series Analysis for Identification of Particle Emission Sources, in *Receptor Models Applied to Contemporary Pollution Problems.* S. L. Dattner and P. K. Hopke, Eds. Air Pollution Control Association, Pittsburgh, PA, pp. 317–335.

Hwang, C. S., S. G. Severin, and P. K. Hopke, Comparison of R- and Q-Mode Factor Analysis for Aerosol Mass Apportionment. *Atmospheric Environ.* **16**:345–352.

Imbrie, J. 1963. Factor and Vector Analysis Programs for Analyzing Geologic Data. Technical Report No. 6. Office of Naval Research, Evanston, IL. 83 pp.

Imbrie, J. and T. H. Van Andel. 1964. Vector Analysis of Heavy Mineral Data. *Geological Soc. of Am. Bull.* **75**:1131–1156.

Israelson, S. and C. Oluwafemi. 1975. Power and Cross-Power Spectral Studies of Electric and Meteorological Parameters Under Fair Weather Conditions in the Atmospheric Boundary Layer. *Boundary Layer Met.* **9**:461–477.

Jaklevic, J. M., R. C. Gatti, F. S. Goulding, and B. S. Loo. 1981a. A β-Gauge Method Applied to Aerosol Samples. *Environ. Sci. Technol.* **15**:680–686.

Jaklevic, J. M., R. C. Gatti, F. S. Goulding, B. W. Loo, and A. C. Thompson. 1981b. Aerosol Analysis for the Regional Air Pollution Study. Report No. EPA-600/4-81-006. U.S. Environmental Protection Agency, Research Triangle Park, NC. 45 pp.

Jaklevic, J. M., B. W. Loo, and F. S. Goulding. 1977. Photon-Induced X-Ray Fluorescence Analysis Using Energy-Dispersive Detector and Dichotomous Sampler, in *X-Ray Fluorescence Analysis of Environmental Samples*, T. G. Dzubay, ed. Ann Arbor Science Publishers, Ann Arbor, MI, pp. 3–18.

Jaklevic, J. M. and R. L. Walter. 1977. Comparison of Minimum Detectable Limits Among X-Ray Spectrometers, in *X-Ray Fluorescence Analysis of Environmental Samples*, T. G. Dzubay, ed. Ann Arbor Science Publishers, Ann Arbor, MI, pp. 63–75.

Janocko, P. B., G. S. Casuccio, S. L. Dattner, D. L. Johnson, and E. R. Crutcher. 1983. The El Paso Airshed: Source Apportionment Using Complementary Analyses and Receptor Models, in *Receptor Models Applied to Contemporary Pollution Problems*, S. L. Dattner and P. K. Hopke, eds. Air Pollution Control Association, Pittsburgh, PA, pp. 249–265.

Janossy, A. G. S., K. Kovacs, and I. Toth. 1979. Parameters for the Ratio Method by X-Ray Microanalysis. *Anal. Chem.* **51**:491–495.

Jardine, N. and R. Sibson. 1971. *Mathematical Taxonomy*. Wiley, New York.

John, W., R. Kaifer, K. Rahn, and J. J. Wesolowski. 1973. Trace Element Concentrations in Aerosols from the San Francisco Bay Area. *Atmospheric Environ.* 7:107–118.

John W. and G. Reischl. 1978. Measurement of the Filtration Efficiencies of Selected Filter Types. *Atmospheric Environ.* **12**:2015–2019.

John W., S. M. Wall, and J. J. Wesolowski. 1983. Validation of Samplers for Inhaled Particulate Matter. Report No. EPA-600/4-83-010. U.S. Environmental Protection Agency, Research Triangle Park, NC. 78 pp.

Johnson, D. L. and B. L. McIntyre. 1983. A Particle Class Balance Receptor Model for Aerosol Apportionment in Syracuse, N.Y., in *Receptor Models Applied to Contemporary Pollution Problems*, S. L. Dattner and P. K. Hopke, eds. Air Pollution Control Association, Pittsburgh, PA, pp. 238–248.

Johnson, D. L., B. McIntyre, R. Fortmann, R. K. Stevens, and R. B. Hanna. 1981. A Chemical Element Comparison of Individual Particle Analysis and Bulk Analysis Methods. *Scanning Electron Microscopy* **1**:469–476.

Johnson, D. L. and J. P. Twist. 1983. Statistical Consideration in the Employment of SAX Results for Receptor Models, in *Receptor Models Applied to Contemporary Pollution Problems*. S. L. Dattner and P. K. Hopke, eds. Air Pollution Control Association, Pittsburgh, PA, pp. 224–237.

Johnson, R. L., J. J. Shah, R. A. Cary, and J. J. Huntzicker. 1981. An Automated Thermal-Optical Method for the Analysis of Carbonaceous Aerosol, in *Atmospheric Aerosol: Source/Air Quality Relationships*, E. S. Macias and P. K. Hopke, eds. Symposium Series No. 167, American Chemical Society, Washington, D.C., pp. 223–233.

Johnson, S. C. 1967. Hierarchical Clustering Schemes. *Psychometrika* **32**: 241–254.

Joreskog, K. G., J. E. Klovan, and R. A. Reyment. 1976. *Geological Factor Analysis*. Elsevier Scientific Publishing Company, Amsterdam.

Kaakinen, J. W., R. M. Jorden, M. H. Lawasani, and R. E. West. 1975. Trace Element Behavior in Coal-Fired Power Plant. *Environ. Sci. Technol.* **9**:862–869.

Kaiser, H. F. 1959. Computer Program for Varimax Rotation in Factor Analysis. *Education and Psychological Measurement* **19**:413.

Kaiser, H. F. and S. Hunka. 1973. Some Empirical Results with Guttman's Stronger Lower Bound for the Number of Common Factors. *Education and Psych. Measurement* **33**:99–102.

Kamath, R. R. and D. N. Kelkar. 1981. Preliminary Estimates of the Primary Source Contributions to Winter Aerosols in Bombay, India. *Sci. Total Environ.* **20**:195–204.

Keith, L. H. 1984. *Identification and Analysis of Organic Pollutants in Air*. Butterworth Publishers, Woburn, MA. 486 pp.

Kelly, J. F., R. J. Lee, and S. Lentz. 1980. Automated Characteristics of Fine Particulates. *Scanning Electron Microscopy* **I**:311–322.

Kerr, P. F. 1959. *Optical Microscopy*, McGraw-Hill, New York.

Khalil, M. A. K., S. A. Edgerton, and R. A. Rasmussen. 1983. A Gaseous Tracer Model for Air Pollution from Residential Wood Burning. *Environ. Sci. Technol.* **17**:555–559.

Klein, D. H., A. W. Andren, J. A. Carter, J. F. Emery, C. Feldman, W. Fulkerson, W. S. Lyon, J. C. Ogle, Y. Talmi, R. I. Van Hook, and N. Bolton. 1975. Pathways of Thirty-Seven Trace Elements Through Coal-Fired Power Plant. *Environ. Sci. Technol.* **9**:973–979.

Kleinman, M. T. 1977. The Apportionment of Sources of Airborne Particulate Matter. Ph.D. Thesis. New York University, New York.

Kleinman, M. T., D. M. Bernstein, and T. J. Kneip. 1977. An Apparent Effect of the Oil Embargo on Total Suspended Particulate Matter and Vanadium in New York City Air. *J. Air Pollut. Control Assoc.* **27**:65–67.

Kleinman, M. T., T. J. Kneip, and M. Eisenbud. 1976. Seasonal Patterns of Airborne Particulate Concentrations in New York City. *Atmospheric Environ.* **10**:9–11.

Kleinman, M. T., M. Eisenbud, M. Lippmann, and T. J. Kneip. 1980a. The Use of

Tracers to Identify Sources of Airborne Particles. *Environment International* **4**:53–62.

Kleinman, M. T., B. S. Pasternack, M. Eisenbud, and T. J. Kneip. 1980b. Identifying and Estimating the Relative Importance of Sources of Airborne Particulate. *Environ. Sci. Technol.* **14**:62–65.

Klug, H. P. and L. E. Alexander. 1974. *X-Ray Diffraction Procedures*. Wiley, New York.

Kneip, T. J., M. Eisenbud, C. D. Strehlow, and P. C. Freudenthal. 1971. Airborne Particulates in New York City. *J. Air Pollut. Control. Assoc.* **20**:144–149.

Kneip, T. J. and P. J. Lioy. 1980. *Aerosols: Anthropogenic and Natural, Sources and Transport*. Volume 338. Annals of the New York Academy of Sciences, New York. 618 pp.

Kneip, T. J., R. P. Mallon, and M. T. Kleinman. 1983. The Impact of Changing Air Quality on Multiple Regression Models for Coarse and Fine Particle Fractions. *Atmospheric Environ.* **17**:299–304.

Knudson, E. J., D. L. Duewer, G. D. Christian, and T. V. Larson. 1977. Application of Factor Analysis to the Study of Rain Chemistry in the Puget Sound Region, in *Chemometrics: Theory and Application*. B. R. Kowalski, ed. Symposium Series No. 52, American Chemical Society, Washington, D.C. 37 pp.

Koch, R., J. Schakenback, and K. G. Severin. 1978. Identifying Sources and Quantities of Fugative Particulate Emissions in Baltimore. Third Symposium on Fugative Emissions: Measurement and Control, San Francisco, CA.

Korfmacher, W. A., E. L. Wehry, G. Mamantov, and D. F. S. Natusch. 1980. Resistance to Photochemical Decomposition of Polycyclic Aromatic Hydrocarbons Vapor-Adsorbed on Coal Fly Ash. *Environ. Sci. Technol.* **14**:1094–1099.

Kotra, J. P., D. L. Finnegan, and W. H. Zoller. 1983. El Chichon: Composition of Plume Gases and Particles. *Science* **222**:1018–1021.

Kowalczyk, G. S., C. E. Choquette, and G. E. Gordon. 1978. Chemical Element Balances and Identification of Air Pollution Sources in Washington, D.C. *Atmospheric Environ.* **12**:1143–1153.

Kowalczyk, G. S., G. E. Gordon, and S. W. Rheingrover. 1982. Identification of Atmospheric Particulate Sources in Washington, D.C. Using Chemical Element Balances. *Environ. Sci. Technol.* **16**:79–90.

Kowalski, B. R. and C. F. Bender. 1972. Pattern Recognition. A Powerful Approach to Interpreting Chemical Data. *J. Amer. Chem. Soc.* **94**:5632–5639.

Kullback, R. 1959. *Information Theory and Statistics*. Wiley-Interscience, New York.

Lance, G. N. and W. T. Williams. 1967. A General Theory of Classification Sorting Strategies. I. Hierarchical Systems. *Comput. J.* **9**:373–380.

Larsen, R. I. 1976. Air Pollution from Motor Vehicles. *Annals of the New York Academy of Sciences* **136**:275–301.

Larsen, R. I. and V. J. Konopinski. 1962. Sumner Tunnel Air Quality. *Arch. Environ. Health* **5**:597–608.

Lawson, D. R. 1980. Impaction Surface Coatings Intercomparison and Measurements with Cascade Impactors. *Atmospheric Environ.* **14**:195–199.

Lawson, C. L. and R. J. Hanson. 1974. *Solving Least Squares Problems*. Prentice-Hall, Englewood Cliffs, NJ.

Lee, R. J. and R. M. Fisher. 1980. Quantitative Characterization of Particulates by Scanning and High Voltage Electron Microscopy. National Bureau of Standards Publication 533, pp. 63–83.

Lee, R. J. and J. F. Kelly. 1980. Overview of SEM-Based Automated Image Analysis. *Scanning Electron Microscopy* **1**:303–310.

Lee, M. L., G. P. Prado, J. B. Howard, and R. A. Hites. 1977. Source Identification of Urban Airborne Polycyclic Aromatic Hydrocarbons by Gas Chromatographic Mass Spectrometry and High Resolution Mass Spectrometry. *Biomed. Mass Spectrometry* **4**:182–186.

Lee, F. S. C. and D. Schuetzle. 1983. Sampling, Extraction, and Analysis of Polycyclic Aromatic Hydrocarbons from Internal Combustion Engines, in *Polycyclic Aromatic Hydrocarbons*, A. Bjorseth, ed. Marcel Dekker, New York, pp. 27–93.

Lewis, C. W. 1981. On the Proportionality of Fine Mass Concentration and Extinction Coefficient for Bimodal Size Distributions. *Atmospheric Environ.* **15**:2639–2646.

Lewis, C. W. and E. S. Macias. 1980. Composition of Size-Fractionated Aerosol in Charleston, West Virginia. *Atmospheric Environ.* **14**:185–194.

Liljestrand, H. M. 1983. Acidic Precipitation Source Identification by Chemical Mass Balance Methods Employing Fractionation Factors, in *Receptor Models Applied to Contemporary Pollution Problems*. S. L. Dattner and P. K. Hopke, eds. Air Pollution Control Association, Pittsburgh, PA, pp. 212–223.

Liljestrand, H. M. and J. J. Morgan. 1978. Chemical Composition of Acid Precipitation in Pasadena, Calif. *Environ. Sci. Technol.* **12**:1271–1273.

Liljestrand, H. M. and J. J. Morgan. 1979. Error Analysis Applied to Indirect Methods for Precipitation Acidity. *Tellus* **31**:421–431.

Liljestrand, H. M. and J. J. Morgan, 1981. Spatial Variations of Acid Precipitation in Southern California. *Environ. Sci. Technol.* **15**:333–339.

Linton, R. W. 1979. Surface Microanalytical Techniques for the Chemical Characterization of Atmospheric Particulates, in *Monitoring Techniques for Toxic Substances from Industrial Operations*. D. Schuetzle, ed. Symposium Series No. 97, American Chemical Society, Washington, D.C., pp. 137–159.

Linton, R. W., M. E. Farmer, P. K. Hopke, and D. F. S. Natusch. 1980a. Determination of the Sources of Toxic Elements in Environmental Particles Using Microscopic and Statistical Analysis Techniques. *Environmental International* **4**:453–461.

Linton, R. W., D. F. S. Natusch, R. L. Solomon, and C. A. Evans, Jr. 1980b. Physicochemical Characterization of Lead in Urban Dusts. A Microanalytical Approach to Lead Tracing. *Environ. Sci. Technol.* **14**:159–164.

Lioy, P. J. and J. M. Daisey, 1983. The New Jersey Project on Airborne Toxic Elements and Organic Substances (ATEOS): A Summary of the 1981 Summer and 1981 Winter Studies. *J. Air Pollut. Control Assoc.* **33**:649–657.

Lioy, P. J., R. P. Mallon, M. Lippmann, T. J. Kneip, and P. J. Sampson. 1981. The Occurrence of Ozone and Sulfate in the Northeastern U.S. Under Summertime Conditions. Paper No. 81-46.3. Air Pollution Control Association, Pittsburgh, PA. 16 pp.

Lioy, P. J., P. Mallon, M. Lippmann, T. J. Kneip, and P. J. Sampson. 1982. Factors Affecting the Variability of Summertime Sulfate in a Rural Area Using Principal Components Analysis. *J. Air Pollut. Control Assoc.* **32**:1043–1047.

Liu, B. Y. H. and K. W. Lee. 1976. Efficiency of Membrane and Nucleopore Filters for Submicron Aerosols. *Environ. Sci. Technol.* **10**:345–350.

Liu, B. Y. H. and D. Y. H. Pui. 1981. Aerosol Sampling Inlets and Inhalable Particles. *Atmospheric Environ.* **15**:589–600.

Liu, C. K., B. A. Roscoe, K. G. Severin, and P. K. Hopke. 1982. The Application of Factor Analysis to Source Apportionment of Aerosol Mass. *Am. Ind. Hyg. Assoc. J.* **43**:314–318.

Lucas, J. H. 1979. A Comprehensive Study of Particulate Emissions from an Integrated Iron and Steel Plant to Aid in the Design of a Control Strategy Program, in *Air Pollution Control in the Iron and Steel Industry*. Air Pollution Control Association, Pittsburgh, PA, p. 194.

Macias, E. S. and P. K. Hopke. 1981. *Atmospheric Aerosol: Source/Air Quality Relationships*. Symposium Series No. 167, American Chemical Society, Washington, D.C. 359 pp.

Macias, E. S., C. D. Radcliffe, C. W. Lewis, and C. R. Sawicki. 1978. Proton Induced β-Ray Analysis of Atmospheric Aerosols for Carbon, Nitrogen, and Sulfur Composition. *Anal. Chem.* **50**:1130–1134.

Malinowski, E. R. 1977. Determination of the Number of Factors and the Experimental Error in a Data Matrix. *Anal. Chem.* **49**:612–617.

Malinowski, E. R. and D. G. Howery. 1980. *Factor Analysis in Chemistry*. Wiley, New York.

Malissa, H., H. Puxbaum, and B. Wopenka. 1981. Herkunftsanalyse des Atmospharischen Aerosol in Wein. Proceedings of the Second European Symposium on Physico-Chemical Behavior of Atmospheric Pollutants, Varese, Italy.

Mamuro, T. and A. Mizohata. 1978. Elemental Composition of Suspended Particles Released in Refuse Incineration. *Annual Report of the Radiation Center of Osaka Prefecture* **19**:15–19.

Mamuro, T., A. Mizohata, and T. Kubota. 1979a. Elemental Compositions of Suspended Particles Released from Iron and Steel Works. *Annual Report of the Radiation Center of Osaka Prefecture* **20**:19–28.

Mamuro, T., A. Mizohata, and T. Kubota. 1979b. Elemental Compositions of Suspended Particles Released from Various Boilers. *Annual Report of the Radiation Center of Osaka Prefecture* **20**:9–17.

Mamuro, T., A. Mizohata, and T. Kubota. 1979c. Elemental Compositions of Suspended Particles Released from Various Small Sources (I). *Annual Report of the Radiation Center of Osaka Prefecture* **20**:37–45.

Mamuro, T., A. Mizohata, and T. Kubota. 1979d. Elemental Compositions of Suspended Particles Released from Various Small Sources (II). *Annual Report of the Radiation Center of Osaka Prefecture* **20**:47–53.

Mamuro, T., A. Mizohata, and T. Kubota. 1979e. Elemental Compositions of Suspended Particles Released in Glass Manufacture. *Annual Report of the Radiation Center of Osaka Prefecture* **20**:29–35.

Mason, B. 1966. *Principles of Geochemistry*. Wiley, New York.

Massachusetts, Commonwealth of, Department of Public Health, Bureau of Environmental Sanitation. 1968. Special Report on the Investigation and Study of Air Quality in the Metropolitan (Boston) Air Pollution Control District, 1955–1965.

Massart, D. L. and L. Kaufman. 1983. *The Interpretation of Analytical Chemical Data by the Use of Cluster Analysis.* Wiley, New York.

Mayrsohn, H. and J. H. Crabtree. 1976. Source Reconciliation of Atmospheric Hydrocarbons. *Atmospheric Environ.* **10**:137–143.

Mayrsohn, H., J. H. Crabtree, M. Kuramoto, R. D. Sothern, and S. H. Mano, 1975. Source Reconciliation of Atmospheric Hydrocarbons in the South Coast Air Basin, 1974. Division of Technical Services Report No. DTS-76-2. California Air Resources Board, El Monte, July 1975.

McCrone, W. C., J. A. Brown, and I, M. Stewart. 1980. *The Particle Atlas*, Volume VI: *Electron Optical Atlas and Techniques*, 2nd Edition. Ann Arbor Science Publishers, Ann Arbor, MI. 246 pp.

McCrone, W. C. and J. G. Delly. 1973a. *The Particle Atlas*, Volume I: *Principles and Techniques*, 2nd Edition. Ann Arbor Science Publishers: Ann Arbor, MI. 296 pp.

McCrone, W. C. and J. G. Delly. 1973b. *The Particle Atlas*, Volume II: *The Light Microscopy Atlas*, 2nd Edition. Ann Arbor Science Publishers, Ann Arbor, MI. 267 pp.

McCrone, W. C. and J. G. Delly. 1973c. *The Particle Atlas*, Volume III: *The Electron Microscopy Atlas*, 2nd Edition. Ann Arbor Science Publishers, Ann Arbor, MI. 219 pp.

McCrone, W. C., J. G. Delly, and S. J. Palenik. 1979. *The Particle Atlas, Volume V: Light Microscopy Atlas and Techniques*. 2nd Edition. Ann Arbor Science Publishers, Ann Arbor, MI. 309 pp.

McFarland, A. R. and C. A. Ortiz. 1982. A 10 μm Cutpoint Ambient Aerosol Sampling Inlet. *Atmospheric Environ.* **16**:2959–2965.

Miesch, A. T. 1976. Q-Mode Factor Analysis of Geochemical and Petrologic

Data Matrices with Constant Row Sums. U.S. Geological Survey Professional Paper 574-G. U.S. Government Printing Office, Washington, D.C.

Miller, L. H. 1956. Table of Percentage Points of Kolmogorov Statistics. *J. Amer. Stat. Assoc.* **51**:113–121.

Miller, M. S., S. K. Friedlander, and G. M. Hidy. 1972. A Chemical Element Balance for the Pasadena Aerosol. *J. Colloid Interface Sci.* **39**:165–176.

Miller, Jr., R. J. 1981. *Simultaneous Statistical Inference*, 2nd Edition. Springer-Verlag, New York.

Mizohata, A. and T. Mamuro, 1979. Chemical Element Balances in Aerosol Over Sakai, Osaka. *Annual Report of the Radiation Center of Osaka Prefecture* **20**:55–69.

Morandi, M., J. M. Daisey, and P. J. Lioy. 1983. A Receptor Source Apportionment Model for Inhalable Particulate Matter in Newark, NJ. Paper No. 83-14.2. Air Pollution Control Association, Pittsburgh, PA. 16 pp.

Morgenstern, P., J. C. Goldfish, and R. L. Davis. 1970. Air Pollutant Emission Inventory for the Metropolitan (Boston) Air Pollution Control District, Walden Research Corp.

Mosteller, F. 1971. The Jackknife. *Rev. Int. Inst. Statist.* **39**:363–368.

Mosteller, F. and J. Tuckey. 1977. *Data Analysis and Regression*. Addison-Wesley, Reading, MA.

Moyers, J. L., L. E. Ranweiler, S. B. Hopf, and N. E. Korte. 1977. Evaluation of Particulate Trace Species in Southwest Desert Atmosphere. *Environ. Sci. Technol.* **11**:789–795.

Murray, L. C. and R. J. Farber. 1982. Time Series Analysis of Historical Visibility Data Base. *Atmospheric Environ.* **16**:1–10.

Mroz, E. J. 1976. The Study of the Elemental Composition of Particulate Emissions from an Oil-Fired Power Plant. Ph.D. Thesis. University of Maryland, College Park.

Nargolwalla, S. S. and E. P. Przybylowicz. 1973. *Activation Analysis with Neutron Generators*. Wiley, New York.

Nelson, E. 1979. Regional Air Pollution Study: Dichotomous Aerosol Sampling System. EPA-297 310. U.S. Environmental Protection Agency, Research Triangle Park, NC.

Nelson, J. W. 1977. Proton-Induced Aerosol Analysis: Method and Samplers, in *X-Ray Fluorescence Analysis of Environmental Samples*. T. G. Dzubay, ed. Ann Arbor Science Publishers, Ann Arbor, MI, pp. 19–34.

Nelson, P. F. and S. M. Quigley, 1982. Non-Methane Hydrocarbons in the Atmosphere of Sydney, Australia. *Environ. Sci. Technol.* **16**:650–655.

Nelson, P. F., S. M. Quigley, and M. Y. Smith. 1983. Sources of Atmospheric Hydrocarbons in Sydney: A Quantitative Determination Using a Source Reconciliation Technique. *Atmospheric Environ.* **17**:439–449.

Neustadter, H. E., J. S. Fordyce, and R. B. King. 1976. Elemental Composition of Airborne Particulates and Source Identification–Data Analysis Techniques. *J. Air Pollut. Control Assoc.* **26**:1079–1084.

Neustadter, H. E., S. M. Sidek, R. B. King, J. S. Fordyce, and J. C. Barr. 1975. The Use of Whatman-41 Filters for High Volume Air Sampling. *Atmospheric Environ.* **9**:101–109.

Noll, D. E., P. K. Mueller, and M. Imada. 1968 Visibility and Aerosol Concentration in Urban Air. *Atmospheric Environ.* 2:465–475.

O.A.Q.P.S. 1981a. Receptor Model Guideline Series, Volume I. Overview of Receptor Model Application to Particulate Source Apportionment. Report No. EPA-450/4-81-016A. U.S. Environmental Protection Agency, Research Triangle Park, NC. 56 pp.

O.A.Q.P.S. 1981b. Receptor Model Guideline Series, Volume II. Chemical Mass Balance Report No. EPA-450/4-81-016B. U.S. Environmental Protection Agency, Research Triangle Park, NC. 114 pp.

O'Connor, B. H. and J. M. Jaklevic. 1980. X-Ray Diffractometry of Airborne Particulates Deposited on Membrane Filters. *X-Ray Spectrometry* **9**: 60–65.

O'Connor, B. H. and J. M. Jaklevic. 1981. Characterization of Ambient Particulate Samples from the St. Louis Area by X-Ray Power Diffractometry. *Atmospheric Environ.* **15**:1681–1690.

Olmez, I., N. K. Aras, G. E. Gordon, and W. H. Zoller. 1974. Nondestructive Analysis for Silicon, Rubidium, and Yttrium in Atmospheric Particulate Matter. *Anal. Chem.* **46**:935–937.

Ondov, J. M. 1974. A Study of Trace Element on Particulates from Motor Vehicles. Ph.D. Thesis. University of Maryland, College Park.

Ondov, J. M. 1981. Physical and Chemical Characterization of Fine Particles in Off-Gases from Coal Combustion and Oil Shale Retorting. Report No. UCID-19158. Lawrence Livermore Laboratory, Livermore, CA. 13 pp.

Ondov, J. M., R. C. Ragaini, and A. H. Biermann. 1978. Elemental Particle-Size Emissions from Coal-Fired Power Plants: Use of an Inertial Cascade Impactor. *Atmospheric Environ.* **12**:1175–1185.

Ondov, J. M., R. C. Ragaini, and A. H. Biermann. 1979. Elemental Emissions from a Coal-Fired Power Plant. Comparison of a Venturi Wet Scrubber System with a Cold-Side Electrostatic Precipitator. *Environ. Sci. Technol.* **13**: 598–607.

Ondov, J. M., W. H. Zoller, and G. E. Gordon. 1982. Trace Element Emissions on Aerosols from Motor Vehicles. *Environ. Sci. Technol.* **16**:318–328.

Pace, T. G. 1979. An Empirical Approach for Relating Annual TSP Concentrations to Particulate Microinventory Emissions Data and Monitor Siting Characteristics. Report No. EPA-450/4-79-012. U.S. Environmental Protection Agency, Research Triangle Park, NC. 52 pp.

Paciga, J. J. and R. E. Jervis. 1976. Multielement Size Characterization of Urban Aerosols. *Environ. Sci. Technol.* **10**:1124–1128.

Parekh, P. P. and L. Husain. 1981. Trace Element Concentrations in Summer

Aerosols at Rural Sites in New York State and Their Possible Sources. *Atmospheric Environ.* **15**:1717–1725.

Parsons, M. L., S. Major, and A. R. Forster. 1983. Trace Elemental Determination by Atomic Spectroscopic Methods–State of the Art. *Applied Spectroscopy* **37**:411–418.

Pasquill, F. 1962. *Atmospheric Diffusion.* Van Nostrand, London.

Pierson, W. R. and W. W. Brachaczek. 1976. Particulate Matter Associated with Vehicles on the Road. SAE Paper 760039. *SAE Trans.* **85**:209–227.

Pierson, W. R. and W. W. Brachaczek. 1983. Particulate Matter Associated with Vehicles on the Road. II. *Aerosol Sci. Technol.* 2:1–40.

Pierson, W. R., W. W. Brachaczek, T. J. Korniski, T. J. Truex, and J. W. Butler. 1980. Artifact Formation of Sulfate, Nitrate, and Hydrogen Ion on Backup Filters: Allegheny Mountain Experiment. *J. Air Pollut. Control Assoc.* **30**: 30–34.

Pillay, K. K. S. and C. C. Thomas. 1971. Determination of the Trace Element Levels in Atmospheric Pollutants by Neutron Activation Analysis. *J. Radioanal. Chem.* 7:107–118.

Pitchford, A., M. Pitchford, W. Malm, R. Flocchini, T. A. Cahill, and E. Walther. 1981. Regional Analysis of Factors Affecting Visual Air Quality. *Atmospheric Environ.* **15**:2043–2054.

Price, J. H., V. C. Anselmo, S. L. Dattner, J. M. Jenks, and S. Mgebroff. 1982. Cost Effective Measurement of Elemental Concentrations in Aerosol in Texas by X-Ray Fluorescence Analysis. Paper No. 82-39.5. Air Pollution Control Association, Pittsburgh, PA, 16 pp.

Prinz, B. and H. Stratmann. 1968. The Possible Use of Factor Analysis in Investigating Air Quality. *Staub-Reinhalt Luft* **28**:33–39.

Que Hee, S. S., V. N. Finelli, F. L. Fricke, and K. A. Wolnik. 1982. Metal Content of Stack Emissions, Coal and Fly Ash from Some Eastern and Western Power Plants in the U.S.A. As Obtained by ICP-AES. *Intern. J. Environ. Anal. Chem.* **13**:1–18.

Rabinowitz, M. B. and G. W. Wetherill. 1972. Identifying Sources of Lead Contamination by Stable Isotope Techniques. *Environ. Sci. Technol.* **6**:705–709.

Radian Corporation. 1983. A Study to Characterize Ambient Air Quality and Assess Emission Source Contributions to Ambient Air Pollution Concentrations for El Paso County, Austin, Texas. Draft Final Report (DNC 82-144-771-04), March 1983.

Raybold, R. L. and R. Byerly, Jr. 1972. Investigation of Products of Tire Wear. National Bureau of Standards Report 10834. 39 pp.

Redman, T. C. and A. R. Zinmeister. 1982. Comments on a Quantitative Determination of Sources in the Boston Urban Aerosol. *Atmospheric Environ.* **16**:1567.

Rheingrover, S. G. and G. E. Gordon. 1984. Wind-Trajectory Methods for

Determining Compositions of Particles from Major Air Pollution Sources, submitted to *Environ. Sci. Technol.* November 1984.

Roberts, J., H. Watters, F. Austin, and M. Crooks. 1979. Particulate Emissions for Paved Roads in Seattle and Tacoma Non-Attainment Areas. Puget Sound Air Pollution Control Agency, Seattle, WA.

Robinson, J. W., L. Rhodes, and D. K. Wolcott. 1975. The Determination and Identification of Molecular Lead Pollutants in the Atmosphere. *Anal. Chem. Acta* **78**:474–478.

Robson, C. D. and K. E. Foster. 1962. Evaluation of Air Particulate Sampling Equipment. *Amer. Ind. Hygiene Assoc. J.* **24**:404.

Rohbock, E., H.-W. Georgh, and J. Muller. 1980. Measurements of Gaseous Lead Alkyls in Polluted Atmospheres. *Atmospheric Environ.* **14**:89–98.

Rom, J. J. 1972. Maintenance, Calibration, and Operation of Isokinetic Source Sampling Equipment. Report No. APTD-0576, U.S. Environmental Protection Agency, Research Triangle Park, NC.

Roscoe, B. A. and P. K. Hopke, 1981a. Comparison of Weighted and Unweighted Target Transformation Rotations in Factor Analysis. *Computers & Chem.* **5**:1–7.

Roscoe, B. A. and P. K. Hopke, 1981b. Error Estimation of Factor Loadings and Scores Obtained with Target Transformation Factor Analysis. *Anal. Chim. Acta* **132**:89–97.

Roscoe, B. A., P. K. Hopke, S. L. Dattner, and J. M. Jenks. 1982. The Use of Principal Components Factor Analysis to Interpret Particulate Compositional Data Sets. *J. Air Pollut. Control Assoc.* **32**:637–642.

Roscoe, B. A., C. Y. Chen, and P. K. Hopke. 1984. Comparison of the Target Transformation Factor Analysis of Coal Composition Data with X-Ray Diffraction Analysis. *Anal. Chim. Acta* **160**:121–134.

Rosen, H., A. D. A. Hansen, R. L. Dod, and T. Novakov. 1980. Soot in Urban Atmospheres: Determination by an Optical Absorption Technique. *Science* **208**:741–744.

Rozett, R. W. and E. M. Petersen. 1975. Methods of Factor Analysis of Mass Spectra. *Anal. Chem.* **47**:1301–1308.

Schwartz, G. P., J. M. Daisey, and P. J. Lioy. 1981. Effect of Sampling Duration on the Concentration of Particulate Organics Collected on Glass Fiber Filters. *Am. Ind. Hygiene Assoc. J.* **42**:258–263.

Sedlacek, W. A., G. H. Heiken, E. J. Mroz, E. S. Gladney, D. R. Perrin, R. Leifer, I. Fisenne, L. Hinchliffe, and R. L. Chuan. 1981. Physical and Chemical Characteristics of Mt. St. Helens Airborne Debris. Los Alamos Scientific Laboratory Report LA-UR-81-110, Los Alamos, NM. 34 pp.

Sedlacek, W. A., G. Heiken, W. H. Zoller, and M. S. Germani. 1982. Aerosols from the Soufriere Eruption Plume of 17 April 1979. *Science* **216**: 1119–1121.

Serth, R. W. and T. W. Hughes. 1980. Polycyclic Organic Matter (POM) and Trace

Element Contents of Carbon Black Vent Gas. *Environ. Sci. Technol.* **14**: 298–301.

Severin, K. G., B. A. Roscoe, and P. K. Hopke. 1984. The Use of Factor Analysis in Source Determination of Particulate Emissions. *Particulate Science and Technology* **1**:183–192.

Schmitt, B. F., C. Segebade, and H. U. Fusban. 1980. Waste Incinerator Ash–A Versatile Environmental Reference Material. *J. Radioanal. Chem.* **60**:99–109.

Shah, J. J., J. J. Huntzicker, J. A. Cooper, and J. G. Watson. 1981a. Source of Visibility Degradation in Portland, Oregon. Paper No. 81-54.4. Air Pollution Control Association, Pittsburgh, PA. 15 pp.

Shah, J. J., J. J. Huntzicker, T. J. Kneip, and J. M. Daisey. 1981b. Investigation of the Sources of Carbonaceous Aerosol in New York City by Multiple Linear Regression. Paper No. 81-64.4. Air Pollution Control Association, Pittsburgh, PA. 14 pp.

Shah, J. J., R. L. Johnson, E. K. Heyerdahl, and J. J. Huntzicker. 1982. Carbonaceous Aerosol at Urban and Rural Sites in the United States. Presented at the 75th Annual Meeting of the Air Pollution Control Association, New Orleans, LA. 8 pp.

Shirahata, H., R. W. Elias, C. C. Patterson, and M. Koide. 1980. Chronological Variations in Concentrations and Isotopic Compositions of Anthropogenic Atmospheric Lead in Sediments of a Remote Subalpine Pond. *Geochim. Cosmoschim. Acta* **44**:149–162.

Shum, Y. S. and W. D. Loveland. 1974. Atmospheric Trace Element Concentrations Associated with Agricultural Field Burning in the Williamette Valley of Oregon. *Atmospheric Environ.* **8**:645–655.

Sievering, H., M. Dave, D. Dolske, and P. McCoy. 1980. Trace Element Concentrations over Midlake Michigan as a Function of Meteorology and Source Region. *Atmospheric Environ.* **14**:39–53.

Silvey, S. D. 1969. Multicollinearity and Imprecise Estimation. *J. Royal Statistical Society, Series B* **31**:539–552.

Slanina, J., J. H. Baard, W. L. Zijp, and W. A. H. Asman. 1983. Tracing the Sources of the Chemical Composition of Precipitation by Cluster Analysis. *Water, Air, Soil Pollut.* **20**:41–45.

Small, J. A. 1976. An Elemental and Morphological Characterization of the Emissions from the Dickerson and Chalk-Point Coal-Fired Power Plants. Ph.D. Thesis. University of Maryland, College Park.

Small, M. 1979. Composition of Particulate Trace Elements in Plumes from Industrial Sources. Ph.D. Thesis. University of Maryland, College Park.

Small, M., M. S. Germani, A. M. Small, W. H. Zoller, and J. L. Moyers. 1981a. Airborne Plume Study of Emissions from the Processing of Copper Ores in Southeastern Arizona. *Environ. Sci. Technol.* **15**:293–298.

Small, M., M. S. Germani, W. H. Zoller, and J. L. Moyers. 1981b. Fractionation of Elements During Copper Smelting. *Environ. Sci. Technol.* **15**:299–304.

Smith, R. D., J. A. Campbell, and K. K. Nielson. 1979. Characterization and Formation of Submicron Particles in Coal-Fired Plants. *Atmospheric Environ.* **13**:607–617.

Sneath, P. H. A. and R. R. Sokal. 1973. *Numerical Taxonomy.* Freeman, San Francisco.

Spicer, C. W., J. E. Howes, Jr., T. A. Bishop, L. H. Arnold, and R. K. Stevens. 1982. Nitric Acid Measurement Methods: An Intercomparison. *Atmospheric Environ.* **16**:1487–1500.

Spurny, K. R., J. P. Lodge, Jr., E. R. Frank, and D. C. Sheesley. 1969. Aerosol Filtration by Means of Nucleopore Filters: Structural and Filtration Properties. *Environ. Sci. Technol.* **3**:453–468.

Stafford, M. A. and H. M. Liljestrand. 1983. Ambient Particulate Matter Contributions to Acidic Precipitation, in *Receptor Models Applied to Contemporary Pollution Problems*, S. L. Dattner and P. K. Hopke, eds., Air Pollution Control Association, Pittsburgh, PA, pp. 200–211.

Stevens, R. K. and T. G. Pace. 1983. Status of Source Apportionment Methods: Quail Roost II, in *Receptor Models Applied to Contemporary Pollution Problems*, S. L. Dattner and P. K. Hopke, eds. Air Pollution Control Association, Pittsburgh, PA, pp. 46–59.

Stiles, D. C. 1983. Evaluation of an S^2 Sampler for Receptor Modeling of Woodsmoke Emissions. Paper No. 83-54.6. Air Pollution Control Association, Pittsburgh, PA, 16 pp.

Stolzenburg, T. R., A. W. Andren, and M. R. Stolzenburg. 1982. Source Reconciliation of Atmospheric Aerosols. *Water, Air Soil Pollution* **17**:75–85.

Taback, H. J., A. R. Brienza, J. Macko, and N. Brunetz. 1979. Fine Particle Emissions from Stationary and Miscellaneous Sources in the South Coast Air Basin. Report No. KVB5806-7 83. KVB Inc., Tustin, CA.

Tanner, R. L., J. S. Gaffney, and M. F. Phillips. 1982. Determination of Organic and Elemental Carbon in Atmospheric Aerosol Samples by Thermal Evolution. *Anal. Chem.* **54**:1627–1630.

Tanner, R. L. and B. P. Leaderer, 1981. Seasonal Variations in the Composition of Ambient Sulfur-Containing Aerosols in the New York Area. *Atmospheric Environ.* **15**:569–580.

Taylor, S. R. 1964. Abundance of Chemical Elements in the Continental Crust: A New Table. *Geochim. Cosmoschim Acta* **28**:1273–1285.

Ter Haar, G. L., D. L. Lenane, J. N. Hu, and M. Brandt. 1972. Composition, Size and Control of Automotive Exhaust Particulates. *J. Air Pollut. Control Assoc.* **22**:39–46.

Thomae, S. C. 1977. Size and Composition of Atmospheric Particles in Rural Areas Near Washington, D.C. Ph.D. Thesis. University of Maryland, College Park.

Thompson, A. C., J. M. Jaklevic, B. H. O'Connor, and C. M. Morris. 1982. X-Ray Power Diffraction System for Chemical Speciation of Particulate Aerosol Samples. *Nucl. Intr. Methods* **198**:539–546.

Thompson, A. C., L. R. Johnson, and J. M. Jaklevic. 1983. Quantitative X-Ray Power Diffraction Analysis of Air Particulate Samples, in *Receptor Models Applied to Contemporary Pollution Problems.* S. L. Dattner and P. K. Hopke, eds. Air Pollution Control Association, Pittsburgh, PA, pp. 72–83.

Thurston, G. D. 1981. Discussion of "Multivariate Analysis of Particulate Sulfate and Other Air Quality Variables by Principal Components—Part I. Annual Data from Los Angeles and New York" by R. C. Henry and G. M. Hidy. *Atmospheric Environ.* **15**:424–425.

Thurston, G. D. and J. D. Spengler. 1981. An Assessment of Fine Particulate Sources and Their Interaction with Meteorological Influences via Factor Analysis. Paper No. 81-64.4. Air Pollution Control Association, Pittsburgh, PA. 16 pp.

Thurston, G. D. and J. D. Spengler. 1982. Source Contributions to Inhalable Particulate Matter in Metropolitan Boston, MA. Paper No. 82-21.5. Air Pollution Control Association, Pittsburgh, PA. 16 pp.

Torres, L., M. Frikha, J. Mathieu, and M. L. Riba. 1983. Preconcentration of Volatile Sulphur Compounds on Solid Sorbents. *Intern. J. Environ. Anal. Chem.* **13**:155–164.

Trijonis, J. 1982. Analysis of Ambient Lead Data Near the ASARCO Smelter in El Paso. Sante Fe Research Corporation, Sante Fe, NM.

Turekian, K. K. 1971. Elements, Geochemical Distribution of, in *McGraw-Hill Encyclopedia of Science and Technology*, 3rd Edition, Volume 4. McGraw-Hill, New York.

Van Craen, M. J., E. A. Denoyer, D. F. S. Natusch, and F. Adams. 1983. Surface Enrichment of Trace Elements in Electric Steel Furnace Dust. *Environ. Sci. Technol.* **17**:435–439.

Verner, S. S., ed. 1980. *Sampling and Analysis of Toxic Organics in the Atmosphere.* ASTM Special Technical Publication 721. American Society for Testing and Materials, Philadelphia PA. 192 pp.

Vinogradov, A. P. 1959. *The Geochemistry of Rare and Dispersed Chemical Elements in Soils*, 2nd Edition. Consultants Bureau, Inc., New York.

Voorhees, K. J., S. M. Kunen, S. L. Durfee, L. A. Currie, and G. A. Klouda. 1981. Characterization of Airborne Particulates by Pyrolysis/Mass Spectrometry and Carbon-14 Analysis. *Anal. Chem.* **53**:1463–1465.

Vossler, T., D. L. Anderson, N. K. Aras, J. M. Phelan, and W. H. Zoller. 1981. Trace Element Composition of the Mount St. Helens Plume: Stratospheric Samples from the 18 May Eruption. *Science* **211**:827–830.

Waggoner, A. P. and R. E. Weiss. 1980. Scattering Extinction in Ambient Aerosol. *Atmospheric Environ.* **14**:623–626.

Watson, J. G. 1979. Chemical Element Balance Receptor Model Methodology for Assessing the Source of Fine and Total Suspended Particulate Matter in Portland, Oregon. Ph.D. Thesis. Oregon Graduate Center, Beaverton.

Watson, J. G. 1981. Receptor Models Relating Ambient Suspended Particulate Matter to Sources. Report No. EPA-600/2-81-039. U.S. Environmental Protection Agency, Research Triangle Park, NC. 88 pp.

Watson, J. G. 1983. Overview of Receptor Model Principles, in *Receptor Model Applied to Contemporary Pollution Problems.* S. L. Dattner and P. K. Hopke, eds. Air Pollution Control Association, Pittsburgh, PA, pp. 6–17.

Wedding, J. B., A. R. McFarland, and J. E. Cermak. 1977. Large Particle Collection Characteristics of Ambient Aerosol Samplers. *Environ. Sci. Technol.* **11**:387–390.

Wedding, J. B., and M. A. Weigand. 1982. Design, Frabrication, and Testing of Ambient Aerosol Sampler Inlets. Report No. EPA-600/3-82-039. U.S. Environmental Protection Agency, Research Triangle Park, NC. 24 pp.

Wedding, J. B., M. Weigand, W. John, and S. Wall. 1980. Sampling Effectiveness of the Inlet to the Dichotomous Sampler. *Environ. Sci. Technol.* **14**: 1367–1370.

Wedepohl, K. H. 1971. *Geochemistry.* Holt, Rinehart and Winston, New York.

Weiss, R. E., T. V. Larson, and A. P. Waggoner. 1982. *In Situ* Rapid-Response Measurement of $H_2SO_4/(NH_4)_2SO_4$ Aerosols in Rural Virginia. *Environ. Sci. Technol.* **16**:525–532.

Whitby, K. T. and B. Cantrell. 1976. Atmospheric Aerosols–Characteristics and Measurement. Paper No. 29.1. International Conference on Environmental Sensing and Assessment, Volume 2. 6 pp.

Whitby, K. T., R. B. Husar, and B. Y. U. Liu. 1972. The Aerosol Size Distribution of Los Angeles Smog. *J. Colloid Interface Sci.* **39**:177–204.

White, W. H. and P. T. Roberts. 1977. On the Nature and Origins of Visibility Reducing Aerosols in the Los Angeles Air Basin. *Atmospheric Environ.* **11**:803–812.

Williamson, H. J., W. D. Balfour, and C. E. Schmidt. 1983. Weighted Ridge Regression, A Statistical Technique for Increased Source Resolution in Chemical Element Balance Analysis. Submitted to *Atmospheric Environ.*

Williamson, H. J. and D. A. DuBose. 1983. Receptor Model Technical Series, Volume III. User's Manual for Chemical Mass Balance Model. Report No. EPA-450/4-83-0L4. U.S. Environmental Protection Agency, Research Triangle Park, NC.

Winchester, J. W. and G. D. Nifong. 1971. Water Pollution in Lake Michigan by Trace Elements from Pollution Aerosol Fallout. *Water, Air Soil Pollut.* **1**:50–64.

Wine, R. L. 1976. *Beginning Statistics.* Winthrop Publishers, Cambridge, MA.

Yarmartino, R. J. 1983. Formulation and Application of a Hybrid Source Receptor Model, in *Receptor Models Applied to Contemporary Pollution Problems.* S. L. Dattner and P. K. Hopke, eds. Air Pollution Control Association, Pittsburgh, PA, pp. 285–295.

Yarmartino, R. J. and D. J. Lamich. 1979. The Formulation and Application of a Source Finding Algorithm. Fourth Symposium of Turbulence, Diffusion and Air Pollution, Jan. 15–18, Reno, NV. Published by the American Meteorological Society, Boston, MA, pp. 84–88.

Zinmeister, A. R. and T. C. Redman. 1978. A Time Series Analysis of Aerosol Composition Measurements. Department of Statistics Technical Report M477, Florida State University, Tallahassee. 48 pp.

Zinsmeister, A. R. and T. C. Redman. 1980. A Time Series Analysis of Aerosol Composition Measurements. *Atmospheric Environ.* **14**:201–245.

Zoller, W. H. and G. E. Gordon. 1970. Instrumental Activation Analysis of Atmospheric Pollutants Utilizing Ge(Li) γ-Ray Detectors. *Anal. Chem.* **42**:257–265.

APPENDIX

SELECTED SOURCE PROFILES

In this appendix there is a tabulation of composition profiles for a number of sources. The major compilations from which these tables are derived are the report by Taback et al. (1979) as listed by Cass and McRae (1981) and the thesis of Watson (1979). A number of other sources are also included. In the Taback listings in Cass and McRae, there is generally a size fractionation of the profiles. In general, the 0–10 μm size range profile has been quoted here since it appeared that there were more values given. In the Taback et al. profiles a value of <5500 μg/g indicates that there was an indication of an unquantifiable value between 1000 and 100,000 μg/g while <500 μg/g values indicate a value between 100 and 1000 μg/g. The presence of a blank value indicates that no information was provided on that element. A dash indicates that the element was sought but no detectable concentration was found, although a detection limit was not provided. For details of the sampling and analysis methods, the original reference should be consulted.

This listing is not intended to be all inclusive but to provide a reasonably large fraction of the available data. Other compilations are being developed and more source data will become available in the near future.

Soil (Rural) ($\mu g/g$)

	Vinogradov (1959)	Bowen (1966)	Watson Fine (1979)	Watson Coarse (1979)	Friedlander (1973)	Taback et al. (1979)	Thomae (1977)
Carbon (total)	20,000	20,000					
Carbon (organic)			55,000	14,100			
Carbon (elemental)			2,600	11,000			
N	1,000	1,000					
F	200	200		20			
Na	6,300	6,300	5,300	12,500	25,000		3,200
Mg	6,300	6,300	13,900	11,700	14,000		5,600
Al	71,300	71,000	114,000	73,000	82,000	80,000	59,900
Si	330,000	330,000	259,000	321,000	200,000	200,000	350,000
P	800	650					
S	850	700	560				
Cl	100	100	100	100			
K	13,600	14,000	10,400	22,000	15,000	20,000	
Ca	13,700	13,700	7,300	14,000	15,000	20,000	4,900
Ti	4,600	5,000	8,000	7,300	4,000		
V	100	100	290	140	60	500	140
Cr	200	100	310	60			59
Mn	850	850	185	81	1,100	5,500	990
Fe	38,000	38,000	69,100	41,600	32,000	30,000	37,800

Ni	40	40	80		40	500	
Cu	20	20	180		80	500	
Zn	50	50	300		50	500	
As	5	6					82
Se	0.01	0.2					4.6
Br	5	5	100				0.6
Rb	100	100				7	
Sr	300	300					
Zr	300	300					
Cd	(0.5)	0.06					
Sn	(10)	10					
Sb	–	(2–10?)					
I	5	5					0.60
Cs	(5)	6					
Ba	500	500					3.0
La	(40)	30			600	500	520
Ce	(50)	50					56
Fu	–	–					71
Hf	(6)	6					2.2
Ta	–	–					13.5
Hg	0.01	0.33					1.3
Pb	10	10	210	250	200	500	
Th	6	5					
U	1	1					8.2

Crustal Rock ($\mu g/g$)

	Goldschmidt (1958)	Taylor (1964)	Mason (1966)	Wedepohl (1971)	Turekian Model A (1971)	Turekian Model B (1971)
Carbon (total)	320	200	200	320	–	–
Carbon (organic)						
Carbon (elemental)						
N	–	20	20	20	20	20
F	800	625	625	720	650	460
Na	28,300	23,600	28,300	24,500	31,900	23,200
Mg	20,900	23,300	20,900	13,900	33,000	27,700
Al	81,300	82,300	81,300	78,300	77,400	80,000
Si	277,200	281,500	277,200	305,400	311,400	272,000
P	1,200	1,050	1,050	810	820	1,010
S	520	260	260	310	300	300
Cl	480	130	130	320	210	190
K	25,900	20,900	25,900	28,200	29,500	16,800
Ca	36,300	41,500	36,300	28,700	25,700	50,600
Ti	4,400	5,700	4,400	4,700	4,400	8,600
V	150	135	135	95	98	170
Cr	200	100	100	70	48	96
Mn	1,000	950	950	690	670	1,000
Fe	50,000	56,300	50,000	35,000	34,300	58,000
Ni	100	75	75	44	37	72
Cu	70	55	55	30	32	58
Zn	80	70	70	60	63	82
As	5	1.8	1.8	1.7	1.7	2.0
Se	0.09	0.05	0.05	0.09	0.05	0.05
Br	2.5	2.5	2.5	2.9	3.0	4.0
Rb	280	90	90	120	120	70
Sr	150	375	375	290	300	450
Zr	220	165	165	160	170	140
Cd	0.18	0.2	0.1	0.15	0.18	
Sn	40	2	2	3	2.1	1.5
Sb	(1)	0.2	0.2	0.2	0.3	0.2
I	0.3	0.5	0.5	0.5	0.5	0.5
Ca	3.2	3	3	2.7	3.1	1.6
Ba	430	425	425	590	610	380
La	18.3	30	30	44	58	50
Ce	41.6	60	60	75	74	82
Eu	1.06	1.2	1.2	1.4	1.6	2.2
Hf	4.5	3	3	3	4	4
Ta	2.1	2	2	3.4	3.3	2.4
Hg	0.5	0.08	0.08	0.03	0.03	0.02
Pb	16	12.5	13	15	15	10
Th	11.5	9.6	7.2	11	12	5.8
U	4	2.7	1.8	3.5	2.9	1.6

Volcanic Emissions ($\mu g/g$ Except as Noted)

	El Chichon[a]	El Chichon[b] 3/28–4/2/82	El Chichon[b] 4/3–4/7/82	Mt. St. Helens[c, d] 17.7 km	Mt. St. Helens[e] May 20, 1980	Mt. St. Helens[e] May 21, 1981	Mt. St. Helens[f] Average Value	Soufriere[g, h] 1.8 km
Carbon (total)								
Carbon (organic)								
Carbon (elemental)								
N								
F								
Na	68,000 ± 3,000	22,000	20,500	96	30,800	36,100	21,900	1,520
Mg	<20,000	10,900	13,000				6,200	2,540
Al	146,000 ± 2,000	91,500	91,500	260	85,200	90,500	95,300	15,400
Si		277,000	263,000	1100			310,000	
P		440	440				2,000	
S				5.8				
Cl	6,300 ± 900			2.2	1,700	9,000		960
K	79,000 ± 3,000	26,000	24,000	46	39,800	–	14,600	
Ca	54,000 ± 7,000	44,300	49,000	130	21,900	45,300	36,400	
Ti	10,000 ± 3,000	4,200	4,800	16			2,300	880
V	110 ± 20			0.18	32	84		18
Cr	<10					230		150
Mn	1,800 ± 500	770	770	1.3	530	660	600	6,340
Fe	31,000 ± 2,000	42,000	49,000	110	35,400	82,500	35,700	
Ni								
Cu					150	290		
Zn	80 ± 10			0.39				16

Volcanic Emissions ($\mu g/g$ Except as Noted) (*Continued*)

	El Chichon[a]	El Chichon[b] 3/28–4/2/82	El Chichon[b] 4/3–4/7/82	Mt. St. Helens[c, d] 17.7 km	Mt. St. Helens[e] May 20, 1980	Mt. St. Helens[e] May 21, 1981	Mt. St. Helens[f] Average Value	Soufriere[g, h] 1.8 km
As	19 ± 1			0.0043	4.9			<60
Se	7 ± 2			0.00077	2.4	8.7		<0.15
Br	100 ± 50							~60
Rb	290 ± 10			0.14				
Sr	1,300 ± 100							
Zr	260 ± 50							
Cd				0.099				
Sn								
Sb	3.4 ± 4.0			0.00092				0.61
I	<20							4.1
Cs	13 ± 1							0.38
Ba	1,900 ± 100			1.3	530			130
La	85 ± 2			0.046	26			
Ce	78 ± 4			0.11	56			12
Eu	1.5 ± 0.1			0.0039	1.3			0.152
Hf	7.3 ± 0.3			0.014				0.41
Ta	1.5 ± 0.2			0.0014				0.051
Hg	<5							
Pb								
Th	26.1 ± 1			0.0031	6.0			0.54
U								0.58

[a]Kotra, Finnegan and Zoller (1983).
[b]Hoffer, Gomez and Muela (1982).
[c]Vossler et al. (1981).
[d]$\mu g/m^3$.
[e]Sedlacek et al. (1981).
[f]Davis et al. (1981).
[g]Sedlacek et al. (1982).
[h]ng/m^3.

Vegetative Sources ($\mu g/g$)

	Flour Processing[a]		Grain Elevator[a]		Rice[b] Dryer	Coffee[b] Carob Dryer	Wood[b] Sawing	Wood[b] Sanding	Feed and[b] Grain Operations	Livestock[b] Dust
	Fine	Coarse	Fine	Coarse						
Carbon (total)					–	240,000	420,000	410,000	300,000	20,000
Carbon (organic)	253,000	118,000	278,000	257,000	–	230,000	390,000	350,000		
Carbon (elemental)	2,800	2,300	8,000	9,500						
NO_3	<6,000		<3,000	<1,000	–	<5,500	<5,500			
F	<2,000	2,000	<1,000	210						
Na	1,000	600	1,000	1,150						
Mg	<10,000	<20,000	3,000	2,900						
Al	2,000	500	12,000	11,800						80,000
Si	<1,000	<1,000	43,500	65,500	58,890				150,000	200,000
P										
S	31,500	4,100	1,450	2,150	–	1,350	<1,250		<1,250	
Cl	3,400	6,100	3,300	1,700	3,060					
K	<300	<1,000	16,000	19,500	10,940	–			<5,500	20,000
Ca	2,400	1,800	7,600	6,750	10,000	<500		<500	<5,500	20,000
Ti	<500	500	490	550	–	480	<500			<5,500
V	<100	<100	7.5	7.5	1,220					<500
Cr	<100	<100	<8	<10	<5,500	–	<500			
Mn	<100	100	320	280	<5,500	–	<500			<5,500
Fe	300	180	7,800	8,000	45,560	<500	<5,500	<500	<500	30,000

Vegetative Sources (μg/g) (*Continued*)

	Flour Processing[a]		Grain Elevator[a]		Rice[b] Dryer	Coffee[b] Carob Dryer	Wood[b] Sawing	Wood[b] Sanding	Feed and[b] Grain Operations	Livestock[b] Dust
	Fine	Coarse	Fine	Coarse						
Ni	<100	<100	<10	<10	<5,500	<500	<500			<500
Cu	500	270	230	<100	<500	<500	<500		<500	<500
Zn	<100	<100	130	150	–	<500	<500			
As					–					
Se						480				
Br	<100	70	70	20	<500	<500	<5,500			
Rb					–					
Sr					–	<500		<500		
Zr										
Cd						480				
Sn							<500			
Sb										
I										
Cs										
Ba					–		<500			
La										
Ce										
Eu										
Hf										
Ta										
Hg										
Pb	<500	<100	<40	<50			<500			<500
Th										
U										

[a] Watson (1979).
[b] Taback et al. (1979).

Agricultural Burning ($\mu g/g$)

	Slash Burn[a]		Simulated Field[a] Burning		Agricultural[b] Burning	Forest[b] Fire	Field[a] Burning	Structural[b] Fire	Field[c,d] Burning
	Fine	Coarse	Fine	Coarse					
Carbon (total)					300,000	300,000		300,000	
Carbon (organic)	594,000	474,000	47,200	401,000	100,000	100,000		10,000	
Carbon (elemental)	35,000	49,000	44,000	80,000					
NO_3^-	51,000	10,000	20,000	35,000					
F		1,900	3,200	8,000					
Na	6,500	1,600	3,300	5,300			250,000		14,400
Mg									
Al	14,000	2,300	4,500	9,600			290,000		16,700
Si	8,900	54,000	4,900	18,000	100,000	100,000		100,000	
P									
S	4,300	1,900	16,000	28,000				75	
Cl	55,500	9,300	99,000	93,000	20,000	20,000	170,000	16,300	9,840
K	6,000	14,800	65,000	90,000	50,000	50,000	90,000		5,200
Ca	10,700	1,300	9,200	19,000	10,000	10,000		16,000	
Ti			700				41,000		2,330
V							1,100		70
Cr			120				1,100		62
Mn	1,200	90	470	710			11,000		650
Fe	1,900	5,400	540	1,300	20,000	20,000	16,000	25,500	9,100
Ni				170					

Agricultural Burning ($\mu g/g$) *(Continued)*

	Slash Burn[a]		Simulated Field[a] Burning		Agricultural[b] Burning	Forest[b] Fire	Field[a] Burning	Structural[b] Fire	Field[c,d] Burning
	Fine	Coarse	Fine	Coarse					
Cu	900	140	540	160			<900		<50
Zn		390							
As									28
Se									
Br	5,300	230	450	790			2,300		130
Rb									
Sr									
Zr									
Cd									
Sn									
Sb									50
I									
Cs									
Ba									<30
La									16
Ce									19
Eu									1.8
Hf									2.2
Ta									27
Hg			350						
Pb									
Th									
U									

[a] Watson (1979).
[b] Taback et al. (1979).
[c] Shum and Loveland (1974).
[d] ng/m^3.

Incinerators—Waste (μg/g Except as Noted)

	Nicosia[a]	Alexandria[b]	SWRC #1[b]	Berlin[c]	Osaka[d]
Carbon (total) (%)				2.5 ± 0.1	
Carbon (organic)					
Carbon (elemental)					
N					
F (%)				0.15 ± 0.03	
Na (%)	8.2 ± 1.3	9.8 ± 2.8	6.5 ± 1.3	1.65 ± 0.04	12
Mg (%)	2.8 ± 0.8	0.6 ± 0.25	0.36 ± 0.07	1.54 ± 0.04	
Al (%)	0.5	1.6 ± 0.8	2.1 ± 1.0	8.28 ± 0.07	0.42
Si (%)				14.0 ± 0.1	
P (%)				0.3 ± 0.1	
S (%)					13
Cl (%)	27 ± 3	20 ± 5	14 ± 3	1.28 ± 0.02	27
K (%)				3.0 ± 0.6	20
Ca (%)				10.2 ± 0.2	1.1
Ti	500	2900 ± 1400	3600 ± 2000	7100 ± 10	900
V					27
Cr	105 ± 17	490 ± 350	870 ± 370	420 ± 50	850
Mn	270 ± 80	1500 ± 1400	410 ± 110	1440 ± 50	330
Fe	3300 ± 1500	9000 ± 3300	7100 ± 1400	78,900 ± 1600	6100
Ni	79 ± 29	200 ± 80	170 ± 70	126 ± 4	
Cu	1700 ± 300	2000 ± 1200	1500 ± 500	750 ± 30	3600
Zn (%)	11.4 ± 2.8	12 ± 6	13 ± 5	1.2 ± 0.1	2.6
As	200 ± 90	210 ± 100	310 ± 160	88 ± 8	150
Se	49 ± 37	23 ± 21	39 ± 29	23 ± 3	
Br	880 ± 390	2600 ± 1400	920 ± 520	141 ± 6	820
Rb				140 ± 2	260
Sr				778 ± 38	
Zr				154 ± 3	
Cd	1500 ± 400	1100 ± 400	1900 ± 700	250 ± 30	500
Sn (%)	1.29 ± 0.16	1.07 ± 0.15	1.08 ± 0.11	3240 ± 150	0.30
Sb	1600 ± 800	2400 ± 2400	2400 ± 1100	220 ± 12	610
I				24 ± 1	
Cs	8.5 ± 3.3	3.1 ± 1.7	5.5 ± 2.2	16 ± 3	12
Ba	220 ± 130			4200 ± 300	
La	2.9 ± 0.6	3.8 ± 2.6	4.6 ± 1.1	34 ± 4	
Ce				57 ± 4	
Eu				0.23 ± 0.03	
Hf				3.6 ± 0.1	
Ta				1.85 ± 0.14	
Hg				0.3 ± 0.1	
Pb (%)	6.9 ± 1.0	9.7 ± 2.6	7.7 ± 1.1	0.64 ± 0.02	1.7
Th	0.37 ± 0.28	1.8 ± 1.4	3	11.3 ± 0.3	
U				4.2 ± 0.2	

[a]Greenberg et al. (1978).
[b]Greenberg, Zoller, and Gordon (1978).
[c]Schmitt, Segebade, and Fusbaum (1980).
[d]Mamuro and Mizohata (1978).

Sewage Sludge Incinerators

	Laurel, MD[a, b] ($\mu g/m^3$)	4 U.S. Cities[c] (% Total Mass)				Osaka[e] ($\mu g/g$)
		Site O	Site P	Site Q[d]	Site R	
Carbon (total)						
Carbon (organic)						
Carbon (elemental)						
N						
F						
Na	38 ± 18	2.76	0.57	1.96	0.34	6,800
Mg	22 ± 16	0.72	0.86	1.01	0.03	
Al	72 ± 7	1.02	3.05	3.07	0.05	4,900
Si		4.69	8.13	5.33	0.91	
P		4.69	3.36	4.14	0.84	
S		5.25	1.80	14.42	3.83	4,700
Cl	10	1.80	0.26	0.93	2.91	68,000
K		1.71	0.89	1.07	0.18	8,700
Ca	300 ± 220	6.20	5.09	8.76	0.23	17,000
Ti	7.7 ± 14.7	0.22	0.88	0.65	0.02	910
V	0.41 ± 0.18	0.28	0.07	0.18	—	96
Cr	0.65 ± 0.54	0.77	0.94	0.28	0.33	2,100
Mn	1.7 ± 0.5	0.21	0.16	0.46	0.01	840
Fe	170 ± 250	2.91	13.75	4.52	0.34	53,000
Ni		—	0.24	—	—	

Cu	3.9 ± 4.4	—	1.58	0.28	0.77	4,400
Zn	7.1 ± 10.7	3.58	1.72	2.69	9,700	
As						78
Se	0.67 ± 0.67		0.04	0.36	0.04	
Br	3.2 ± 4.4		0.11	0.11	0.26	460
Rb						180
Sr						
Zr						
Cd	0.81 ± 0.48	0.33	0.07	0.13	2.77	800
Sn		1.42	2.39	0.59	1.16	
Sb	0.13 ± 0.08	0.05	0.05	—	0.06	36
I						
Cs	5.5 ± 4.5					8.4
Ba	34 ± 27	0.08	0.56	0.40	0.02	
La						
Ce	0.39 ± 0.18					
Eu						
Hf						
Ta	0.0003					
Hg						
Pb	9.1 ± 3.8	4.07	2.27	2.26	3.14	7,900
Th						
U						

[a]Greenberg, Zoller, and Gordon (1981).
[b]Mean and standard deviations of 5 samples.
[c]Bennet and Knapp (1982).
[d]Fluidized bed combustor.
[e]Mamuro and Mizohata (1978).

Wood or Residential Fuel Combustion ($\mu g/g$)

	Wood[a]	Hog Fuel Boiler[b]		Fireplace Burn[b]		Choked Wood Stove[b]	
		Fine	Coarse	Fine	Coarse	Fine	Coarse
Carbon (total)							
Carbon (organic)			111,000	459,000	420,000	491,000	
Carbon (elemental)			43,000	129,000	97,000	83,000	
NO_3		5,100	2,000	1,300	<10,000	330	
F		1,500	1,100	3,800	<4,000	4,300	16,000
Na	150,000	24,000	6,900	370	3,200	600	3,700
Mg			4,600	<100	<10,000	<7,500	<20,000
Al	1,000	2,400	2,800	2,400	1,300	180	1,800
Si		7,600	9,600	240	2,200	<50	980
P							
S	–	88,000	5,900	370	<1,300	138	<100
Cl	230,000	95,000	8,300	6,100	20,000	940	11,000
K	52,000	224,000	13,000	5,300	450	530	<100
Ca	40,000	56,000	52,000	550	3,600	370	5,600
Ti	100	<600	970	<40	<200	<30	<200
V	–	<20	55	<0.6	<0.6	13	90
Cr	39	150	73	<8	2,800	<10	<200
Mn	1,000	5,100	2,900	15	230	120	180
Fe	2,300	12,600	9,200	23	130	35	<400
Ni	270	67	<50	4.8	<10	<40	<200
Cu	240	1,200	230	190	480	25	900
Zn	670	7,300	390	1,200	130	<100	<100
As	84						
Se							
Br	1,800	550	13	70	150	23	<100
Rb	91						
Sr							
Zr							
Cd							
Sn							
Sb	3.2						
I							
Cs	1.8						
Ba	660						
La							
Ce							
Eu							
Hf							
Ta							
Hg							
Pb	350	4,200	200	110	180	<50	<100
Th							
U							

[a]Mamuro et al. (1979b).
[b]Watson (1979).
[c]Stiles (1983).
[d]10% moisture.

Domestic Burning[b]								
Fine	Coarse	Pine[c,d]	Polar[c,d]	Sweet[c] Gum	Oak[c,d]	Oak[c,e]	Oak[c,f]	Fireplace[g]
298,000	389,000	387,000	332,000	330,000	292,000	445,000	430,000	300,000
37,000	8,400	45,000	70,000	86,000	72,000	16,000	51,000	100,000
8,800								
2,800	21,000							
960	7,900							
<90								
1,500	4,400							
5,000	2,800							100,000
2,800	7,700	1,000	36,000	39,000	16,000	2,000	15,000	<1,350
13,500	75,000	1,000	5,000	4,000	18,000	2,000	7,000	<5,500
16,600	20,000	10,000	169,000	135,000	77,000	15,000	68,000	40,000
2,100	24,000							50,000
<10	<400							
<10	<500							
18	160							
30	860							
710	2,100							40,000
<500	<2,000							
140	1,500							
500	<1,000							
80	160							
18,000	<1,000							

[e]30% moisture.
[f]50% moisture.
[g]Taback et al. (1979).

Construction Materials Sources ($\mu g/g$)

	Rock Crusher[a]		Rock[b] Crusher	Asphalt Production[a]		Asphalt[b] Roofing Manufacture	Asphalt[b] Concrete Batch Plant	Calcination[b] of Gypsum	Cement[b] Production	Limestone[b] Kiln	Brick[b] Grinding
	Fine	Coarse		Fine	Coarse						
Carbon (total)						240,000	106,370	23,820	149,350		
Carbon (organic)				19,900	50,000	230,000		15,450	42,610		
Carbon (elemental)				8,400	10,400			4,130[c]	61,520[c]	400,000[c]	
NO_3	4,100	<1,000		<1,000	<1,000		–	<500	<5,500		<500
F				<100	<100						
Na	19,500	11,200		14,900	18,000						
Mg	14,300	8,000		16,800	13,300						
Al	80,700	79,900		86,600	63,900						
Si	299,400	327,100	400,000	271,300	246,500				10,000		270,000
P									59,000		
S	660	610	<1,250	2,200	900	56,600	<1,250	136,000			<1,250
Cl	<100	<100		<100	<100	<500		1,380			
K	14,600	9,400		11,400	5,700	<5,500	–	3,680	20,000		10,000
Ca	32,700	27,200		28,600	23,200	30,000	87,000	123,180	207,390	300,000	<5,500
Ti	5,600	4,900		8,600	8,000		<5,500		<500		<5,500
V	146	120		210	350						
Cr	164	110		320	1,150		<5,500	1,750	<500		
Mn	810	2,900		2,000	1,800	<5,500		<500	<500		<500
Fe	45,900	37,900	<5,600	70,400	39,500	20,000	<5,500	<5,500	<5,500	20,000	20,000

Ni	39	37	<500	69	110	<5,500		1,750	<500		<500
Cu	530	65	<500	148	1,700	<500		<500	<500		<500
Zn	102	61		152	4,200	<5,500		<500	<500		<500
As							—	180			<500
Se						<5,500		310	<500	100,000	
Br	38	<100		80	200	<500		130			
Rb									320		<500
Sr								<500			
Zr						—					<500
Cd						<500	—	320	320		
Sn											
Sb											
I											
Cs						120,000					
Ba						<500	—	180	320		<500
La											
Ce											
Eu											
Hf											
Ta											
Hg											
Pb	61	32		<100	<200	<500		320	<500		
Th											
U											

[a]Watson (1979).
[b]Taback et al. (1979).
[c]Carbonate carbon.

Coal-Fired Boilers and Power Plants (μg/g Except Where Noted)

	Stack[a] Sample	Stack[b] After ESP	Stack[c] (No Controls)	House[c] (Coal Boiler)	Stack After[d]		Stack Samples After ESP[e,f]	
					Scrubber	ESP	Western 1a	Western 1b
Carbon (total)								
Carbon (organic)								
Carbon (elemental)								
N								
F								
Na	9,700	11,300	7,500	2,500			8.3	37
Mg			7,400	5,400			45	95
Al	26,000	76,000	33,000	11,500	74,000	92,000	344	733
Si							—	15
P							22	15
S	15,000							
Cl	11,000		3,500	25,500				
K	10,000	24,000	14,800	3,500			110	130
Ca	23,000	32,000	8,500	5,400			230	490
Ti	2,600	10,000	2,300	400			23	46
V	130	1,180	380	87				
Cr	400	900	250	53			2.9	0.69
Mn	310	430	330	125			2.8	48
Fe	18,000	150,000	53,000	7,800	49,000	74,000	72	150
Ni	180	650	1,000	220			0.32	0.54
Cu	1,300		2,800	560	290	320	0.30	2.7
Zn	3,000	5,900	2,700	1,450	600	370	1.1	1.6
As	260	440	160	160	280	150		
Se			48	52	440	62		
Br	160		120	670				
Rb	51	55	70	25	28	56		
Sr					2,500	2,500	2.6	5.3
Zr					80	160		
Cd		51	120	250				
Sn								
Sb	81	36	45	29	22	18		
I			28	200				
Cs	8.1	27	4.1	1.4				
Ba	1,100	750	1,300	630			4.1	8.1
La	12	42	90	17				
Ce		120	195	28				
Eu		1.3	9.3	3.7				
Hf		5.0	5.2	1.4				
Ta		1.8						
Hg			22	14.4				
Pb	2,000	190			340	130	0.62	1.1
Th		26	8.0	1.8				
U								

[a]Mamuro et al. (1979b).
[b]Klein et al. (1975).
[c]Block and Dams (1976).
[d]Kaakinen et al. (1975).
[e]Que Hee et al. (1982).
[f]μg/m³.

Stack Samples After ESP[e,f]							
Western 2	Eastern	Stack[g] After ESP	Stack[h] After ESP	Stack[h] After Scrubber	Stack A^i (<1 μm)	Stack B^i (<1 μm)	Stack[j,f] After ESP
					3,000	4,000	40,400
							36,200
							404,000
280	490	2,080	18,000	15,600	11,400	20,000	40,400
320	1000	6,800	11,100	10,300	18,000	23,000	36,200
1,130	23,300	130,000	130,000	62,400	130,000	110,000	404,000
15	4,200				230,000	220,000	
14	180				13,000	<3,000	
					5,000	8,000	
		260	–	3,400	<530	<480	
170	2,200	16,900	7,400	2,500	4,000	14,000	22,800
710	1,700	11,700	27,600	71,700	47,000	60,000	70,000
72	1,000	7,800	7,250	3,950	26,000	8,000	17,500
		390	323	860	900	340	497
3.3	23	210	78	1,230	265	230	126
11	26	390	335	2,110	1,110	950	959
350	12,300	91,000	30,000	21,900	56,000	35,000	87,800
0.78	17	260	–	1,220	90	190	
2.2	16	520	–	1,060	650	500	
4.5	65	650	364	5,100	680	210	506
		780	124	480	74	79	118
		195	47	12,000	15	68	
		520	54	90	<1	6.2	
			45	–	40	62	163
27	98		694	870	2,300	1,600	1,320
			281	225	1,400	280	788
		34	2.1	–	<8	<10	
					60	51	
		33	17.5	127	9	23	19.5
		520			<10	<17	
			3.3	1.3	4.0	5.0	11.1
200	140	1,200	6,560	9,000	2,700	12,000	5,870
		78	73	25	105	71	209
		156	130	44	170	180	380
					5.7	2.0	4.12
					27		34
			2.6	1.6	9.5	3.2	7.7
					<5	<5	
1.8	1.8	520			80	250	
		26	30	9.4		33	86.5
			27	48			41.9

[g]Kowalczyk, Choquette, and Gordon (1978).
[h]Ondov, Ragaini, and Biermann (1979).
[i]Smith, Campbell, and Nielson (1979).
[j]Ondov, Ragaini, and Biermann (1978).

Glass Manufacturing Emissions ($\mu g/g$)

	Melting[a] Furnace	Group I[b,c] Boro-silicate	Lamp Glass	Opal Glass	Glass Pots
Carbon (total)					
Carbon (organic)					
Carbon (elemental)					
NO_3	<5,500				
F					
Na		71,000	42,000	56,000	31,000
Mg					
Al		4,000	2,600	51,000	79,000
Si		–	46,000	–	–
P					
S	580,000	84,000	84,000	1,600	57,000
Cl		5,600	6,300	2,100	13,000
K	30,000	38,000	27,000	23,000	–
Ca		3,500	2,100	100,000	36,000
Ti					
V		74	190	–	180
Cr	<5,500	120	150	–	490
Mn		210	510	90	670
Fe	<5,500	3,600	3,700	–	43,000
Ni	<500	1,600	–	–	610
Cu					
Zn		770	1,600	50,000	3,900
As	20,000	5,800	3,800	–	420
Se	40,000	80	30,000	15	–
Br		–	6.2	–	120
Rb		180	97	–	–
Sr					
Zr					
Cd		–	48,000	–	87
Sn					
Sb		690	2,400	5,300	40
I					
Cs		6.3	10	–	–
Ba					
La					
Ce					
Eu					
Hf					
Ta					
Hg					
Pb	<5,500	7,300	46,000	400	5,400
Th					
U					

[a]Taback et al. (1979).

[b]Mamuro et al. (1979e).

[c]Emissions were not mixed with fuel combustion emissions.

Group II[b,d]						Group III[b,e]
Opal Glass	Opal Glass	Glass Vessels	Sauce Bottles	Optical[b] Glass	Crucible[b]	Mixed (Mean Value)
140,000	150,000	150,000	44,000	5,100	540	2,200
7,900	11,000	3,600	670	760	9,000	1,600
150,000	–	–	–	–	30,000	–
52,000	130,000	170,000	140,000	–	51,000	92,000
3,200	–	–	–	3,900	3,400	–
14,000	36,000	27,000	49,000	16,000	–	770
670	5,600	4,600	2,800	1,800	–	1,100
1,200	2,400	1,500	1,500	2.2	280	4,900
140	57	3,400	380	–	250	940
100	130	260	26	22	82	96
1,200	650	10,000	2,100	–	3,100	5,200
440	1,600	1,300	710	–	730	1,800
3,800	15,000	660	230	510	780	210
17,000	51,000	580	150	340	3,100	84
13	110	1,300	48	–	9.0	100
930	–	17	6.7	–	–	3.3
140	300	130	570	–	34	–
–	–	–	99	–	–	210
77	110	17	42	9,500	9.0	20
9.5	12	4.7	7.3	–	–	1.3
–	–	–	3,700	900,000	740	740

[d] Emissions were mixed with heavy oil combustion.

[e] Emissions were primarily heavy oil combustion.

Oil-Fired Boilers and Power Plants (μg/g)

	Heating[a] Furnace	Oil[a,b] Boiler	Indonesian[a] Oil	Waste[b] Oil	Pitch[b]	Miscellaneous[b]
Carbon (total)						
Carbon (organic)						
Carbon (elemental)						
NO_3						
F						
Na	2,900	10,000	22,000	15,000	150,000	8,400
Mg				7,100	–	21,000
Al	2,700	2,100	5,400	3,300	4,500	2,200
Si						
P						
S	130,000	96,000	110,000	150,000	250,000	91,000
Cl	570	920	12,000	36,000	–	68,000
K	630	850	1,400	42,000	–	–
Ca	1,300	850	8,600	40,000	13,000	180,000
Ti	270	740	1,100	130	–	–
V	1,400	9,200	670	19	12,000	9.1
Cr	560	210	420	210	150	630
Mn	280	120	140	740	300	160
Fe	11,000	4,600	13,000	10,000	17,000	9,800
Ni	2,400	4,900	24,000	66	15,000	140
Cu	2,800	–	–	1,500	–	1,900
Zn	800	400	1,300	8,500	660	84,000
As	120	23	250	–	34	–
Se	39	48				
Br	22	8.5	390	2,200	–	3,100
Rb	–	–	–	170	–	–
Sr						
Zr						
Cd	–	240	–	120	–	28
Sn						
Sb	58	6.9	29	32	20	66
I						
Cs	11	–	–	1.6	0.53	–
Ba		920	–	7,100	–	20,000
La						
Ce						
Eu						
Hf						
Ta						
Hg						
Pb	2,100	330	3,800	19,000	1,000	20,000
Th						
U						

[a]Mamuro et al. (1979a).

[b]Mamuro et al. (1979b).

[c]Watson (1979).

Kerosene[b]	Residual Oil[c] Fine	Residual Oil[c] Coarse	Distillate Oil[c] Fine	Distillate Oil[c] Coarse	Power[d] Plant	Industrial[e] Boiler	Residual[e] Fuel Boiler
						204,140	203,090
	70,000	176,000	180,000	96,000			141,860
	31,000	139,000	178,000	76,000			
	6,500	<150,000	10,000	<20,000		<500	<500
3,800	35,000	9,000	3,200	6,900	37,100		
					12,000		
6,000	5,300	14,700	3,100	5,900	1,300		
	9,600	37,000	2,700	2,900			
14,000	133,000	21,000	69,300	<100		126,000	74,200
2,600	<3,300	7,400	12,000	16,000	37,100		
–	2,800	<1,000	180	<120	1,300	4,170	<5,500
3,600	15,800	35,600	5,000	5,000	250,000	<5,500	97,080
450	1,100	1,400	<3,000	<2,000	77		<500
150	34,400	17,900	50	<100	70,000		<5,500
240	470	540	<500	<1,000	180	<5,500	<5,500
260	460	270	140	130	300	<500	<500
15,000	29,700	14,700	1,200	2,300	8,400	31,670	20,620
1,200	53,600	13,900	90	<200	12,000	<5,500	49,070
–	750	860	1,700	1,100	2,500	<500	<500
1,400	4,000	500	290	<50	4,900		<5,500
31					84	380	<500
					100	<500	<500
45	130	<100	260	<100	160		480
–							
						<500	<500
					8.4		<500
24					20		
					–		
–							
47,000					5,800	<500	5,330
1.3					50		
					40		
–	1,100	<500	5,400	<1,000	1,200		<500
					5		

[d]Kowalczyk, Choquette, and Gordon (1978).

[e]Taback et al. (1979).

— Ferrous Metal-Related Sources ($\mu g/g$)

	Electric[a] Furnace Dust	Medium[b] Steel Furnace	Induction[b] Furnace	Special[b] Steel Furnace	Stainless[b] Steel Furnace	Nickel[b] Sulfide Furnace	Electric[b] Steel Furnace
Carbon (total)							
Carbon (organic)							
Carbon (elemental)							
NO_3							
F							
Na	1,000 ± 500	20,000	20,000	11,000	–	26,000	14,000
Mg	20,000 ± 5,000						
Al	7,000 ± 1,500	6,200	13,000	13,000	20,000	890	10,000
Si	30,000 ± 4,000	–	21,000	–	–	17,000	–
P	1,400 ± 300						
S	2,000 ± 300	84,000	–	27,000	–	120,000	48,000
Cl	14,500 ± 2,500	42,000	25,000	30,000	14,000	830	34,000
K	10,000 ± 2,000	15,000	32,000	9,100	–	20,000	13,000
Ca	47,000 ± 4,500	22,000	15,000	70,000	12,000	82	45,000
Ti	300 ± 60	920	4,800	1,300	–	–	1,000
V	410 ± 150	58	59	180	350	5.9	130
Cr	1,500 ± 300	3,000	240	3,300	4,600	1,200	3,200
Mn	49,000 ± 8,000	40,000	17,000	16,000	4,100	150	22,000
Fe		270,000	57,000	120,000	44,000	3,700	160,000
Ni		3,700	980	2,800	10,000	2,700	2,900
Cu	2,400 ± 300	5,300	1,200	2,600	–	13,000	3,700
Zn	64,000 ± 9,000	210,000	29,000	25,000	15,000	1,400	52,000
As		180	120	77	–	150	100
Se		51	7.7	–	260	720	51
Br	340 ± 70	280	210	100	180	–	140
Rb			110	77	–	320	77
Sr	30 ± 10						
Zr							
Cd		500	120	160	4,300	–	250
Sn							
Sb		170	31	66	230	13	90
I							
Cs			12	–	–	26	–
Ba	95 ± 12						
La							
Ce							
Eu							
Hf							
Ta							
Hg							
Pb	16,000 ± 3,000	52,000	5,400	7,600	11,000	6,400	14,000
Th							
U							

[a]Van Craen et al. (1983). [b]Mamuro et al. (1979a). [c]Watson (1979).

Cupola[b]	Electric[a] Arc Furnace Fine	Electric[a] Arc Furnace Coarse	Ferromanganese[c] Furnace Fine	Ferromanganese[c] Furnace Coarse	Steel[d] Sinter	Open[d] Hearth Furnace	Basic[d] Oxygen Furnace	Electric[e] Steelworks
					100,000	200,000	200,000	
			90,000	160,000		200,000	200,000	
			15,000	30,000				
			57,000	12,000		<5,500	<5,500	
								24,000
13,000	12,600	14,000	31,000	3,400				
	65,000							
								10,000
11,000	6,500		6,400	1,800				3,700
240,000	50,000	66,000	9,900	2,900				14,000
								2,600
23,000	6,250	3,250	10,500	1,750	48,000	100,000	100,000	
8,900	18,500	38,000	4,200	8,500	280,000			
30,000	9,200	5,600	105,000	12,000	190,000	50,000	50,000	
10,000	62,000	69,000	13,000	5,700	<5,500	<5,500	<5,500	115,000
600	2,000	1,100	460	<88				2,000
89	630	580	240	90		<5,500	<5,500	
520	21,000	21,000	420	230	30,000	20,000	20,000	2,100
45,000	87,000	95,000	173,000	27,500	<5,500	<5,500	<5,500	35,400
150,000	319,500	308,000	21,000	3,800	140,000	110,000	110,000	224,000
350	7,100	6,200	<50	<200		<5,500	<5,500	
2,600	2,800	4,300	360	740	20,000	<5,500	<5,500	
8,300	12,000	12,000	5,800	350				241,000
130					<5,500	<500	<500	
18					<500			
99	<1,000	<1,000	<1,600	<7,500	<5,500	<500	<500	
220					<5,500	<500	<500	
					<500			
					<500			
—					<5,500	<500	<500	300
						<500	<500	
370						<500	<500	
					<500			
20					<5,500			
2,300	7,600	6,400	450	<500	100,000	<5,500	<5,500	270,000

[d] Cass and McRae (1981).
[e] Alary et al. (1983).

Aluminum Production Sources ($\mu g/g$)

	Al Melting[a] Furnace	Al Processing[b] Stack		Al Processing[b] Roof Vent		Alumina Handling[b]		Al[c] Foundry
		Fine	Coarse	Fine	Coarse	Fine	Coarse	
Carbon (total)								130,000
Carbon (organic)		39,000	<10,000	50,000	120,000			130,000
Carbon (elemental)		23,000	16,000	<10,000	58,000			
NO_3		4,100	<20,000	<20,000	<9,000	5,900	<20,000	<5,500
F		60,000	42,000	<6,000	5,300	1,200	<5,000	
Na	30,000	41,000	24,000	55,000	6,600	2,400	15,000	
Mg		28,000	27,000	<20,000	12,000	1,350	<10,000	
Al	100,000	267,000	313,000	258,000	141,000	361,000	22,900	
Si		3,400	970	7,600	1,300	2,300	2,100	
P								
S	180,000	14,000	4,300	15,000	1,300	610	4,000	40,000
Cl	30,000	13,300	12,000	52,000	4,200	3,600	8,500	110,000
K	42,000	2,200	<1,000	26,000	<500	680	<120	<5,500
Ca	3,900	3,300	8,100	12,000	2,900	4,800	2,800	30,000
Ti	–	400	760	470	780	150	130	<500
V	320	640	400	780	180	25	15	
Cr	510	<100	160	200	<10	165	–	<500
Mn	150	110	<200	<200	<100	490	540	<5,500
Fe	15,000	4,500	3,800	7,100	1,700	3,100	360	60,000
Ni	4,600	1,900	2,100	1,500	330	180	–	<5,500

Cu	3,600	440	1,400	1,900	400	200	760	<500
Zn	7,500	150	90	<10	<100	120	–	<500
As	90							
Se	–							
Br	200	370	1,200	<50	<100	<10	200	
Rb	–							
Sr								
Zr								
Cd	2,300							
Sn	–							<500
Sb	65							
I								
Cs	22							
Ba	–							
La	–							
Ce	–							
Eu								
Hf	3.8							
Ta								
Hg	33							
Pb	6,000	120	<1,000	<1,000	300	<500	–	<500
Th								
U								

[a] Mamuro et al. (1979c).
[b] Watson (1979).
[c] Taback et al. (1979).

	Plume Samples[a] Smelter Number					Ore[b]
	1 (ng/m^3)	2 (ng/m^3)	3 (ng/m^3)	4 (ng/m^3)	5 (ng/m^3)	Smelter 2 ($\mu g/g$)
Carbon (total)						
Carbon (organic)						
Carbon (elemental)						
N						
F						
Na	550 ± 80	700 ± 300	230 ± 70	210 ± 50	240 ± 40	17,000 ± 4,000
Mg	370 ± 260	<800	<90	210 ± 80	<200	7,600 ± 700
Al	220 ± 290	1,000 ± 800	480 ± 120	350 ± 180	170 ± 100	62,000 ± 2,000
Si						
P						
S	280,000 ± 32,000	120,000 ± 24,000	54,000 ± 12,000	45,000 ± 7,000	<6,000,000	<50,000
Cl						
K	1,300 ± 240	900 ± 450	720 ± 110	290 ± 90	510 ± 190	47,000 ± 14,000
Ca	1,600 ± 300	780 ± 500	1,700 ± 110	1,200 ± 210	100 ± 800	11,700 ± 500
Ti	180 ± 50	700 ± 100	<80	<20	<10	3,600 ± 800
V	59 ± 8	4.8 ± 1.2	2.4 ± 1.8	3.2 ± 0.6	0.60 ± 0.31	69 ± 2
Cr	6.5 ± 2.6	35 ± 15	5.9 ± 2.3	77 ± 70	14 ± 5	67 ± 2
Mn	39 ± 3	28 ± 7	11 ± 6.5	6.8 ± 1.5	3.2 ± 0.9	190 ± 10
Fe	1,800 ± 100	2,500 ± 500	2,700 ± 100	1,100 ± 100	1,100 ± 100	53,000 ± 1,000
Ni						
Cu	3,200 ± 300	6,800 ± 1,300	9,500 ± 900	4,200 ± 420	2,000 ± 200	5,500 ± 500

Zn	340 ± 110	4,500 ± 400	22,000 ± 2,000	7,200 ± 5,100	7,600 ± 600	540 ± 20
As	2,000 ± 100	2,000 ± 100	1,900 ± 100	1,400 ± 100	4,300 ± 100	5.6 ± 0.4
Se	920 ± 10	880 ± 20	49 ± 2	170 ± 10	42 ± 10	11 ± 1
Br	17 ± 4	33 ± 10	10 ± 6	<50	<10	
Rb						
Sr						
Zr						
Cd						6 ± 2
Sn						
Sb	80 ± 5	140 ± 10	210 ± 10	58 ± 5	370 ± 20	5.9 ± 0.2
I	6 ± 6	6.8 ± 3.5	47 ± 5	53 ± 6	2 ± 2	8.4 ± 0.2
Cs	0.45 ± 0.20	1.5 ± 0.2	0.41 ± 0.21	0.28 ± 0.15	0.34 ± 0.20	
Ba	14 ± 8	<10	50 ± 50	<3	<7	
La	1.9 ± 0.3	1.8 ± 0.4	0.76 ± 0.07	0.27 ± 0.07	0.50 ± 0.30	45 ± 1
Ce	1 ± 1	0.12 ± 0.02	2 ± 1	0.6 ± 0.5	<0.3	78 ± 5
Eu						1.49 ± 0.04
Hf						
Ta						
Hg						
Pb	5,100 ± 1,000	3,000 ± 600		2,600 ± 500	1,900 ± 400	50 ± 5
Th	0.34 ± 0.02	0.31 ± 0.20	<0.2	<0.1	<0.1	
U						

[a] Small et al. (1981a).

[b] Small et al. (1981b).

Nonferrous Metal-Related Sources ($\mu g/g$)

	Cd Red[a] and Cd 0 Furnace	Drying[a] Furnace for Cd Red	Mixing[a] and Powdering Machine for Cd Yellow	Ni–Cd[a] Electrode Production	Brass[a] Induction Furnace	Brass[a] Reverberatory Furnace	Induction[a] Furnace for Be–Cu	Induction[a] Furnace for Zn–Cu
Carbon (total)								
Carbon (organic)								
Carbon (elemental)								
N								
F								
Na	16,000	24,000	5,200	1,400	6,100	5,200	5,800	15,000
Mg								
Al	14,000	10,000	6,300	2,500	7,100	500	6,900	4,600
Si								
P								
S	–	–	–	–	26,000	–	–	–
Cl	100,000	63,000	–	11,000	39,000	37,000	110,000	40,000
K	–	–	–	–	3,300	6,100	–	–
Ca	–	–	–	–	2,800	3,200	4,500	6,900
Ti	–	–	–	2,800	340	–	–	–
V	170	55	–	14	25	55	–	–
Cr	1,700	1,200	300	–	500	–	330	410
Mn	720	550	440	220	820	12	2,800	140
Fe	29,000	11,000	9,000	44,000	5,100	1,200	–	–
Ni	1,700	–	–	250,000	–	–	–	–
Cu	1,700	1,300	–	–	29,000	24,000	6,000	790
Zn	5,600	2,000	1,200	1,900	500,000	230,000	41,000	19,000
As	180	–	–	–	25	93	–	–
Se	400	220	–	–	–	–	–	–
Br	110	100	–	840	92	240	6,800	60
Rb	–	–	–	–	–	–	–	–
Sr								
Zr								
Cd	32,000	2,900	14,000	170,000	1,500	1,800	–	–
Sn	–	–	–	–	–	–	–	–
Sb	110	35	–	–	32	310	17	14
I								
Cs	–	1.0	–	–	–	–	–	–
Ba	–	–	–	–	–	–	–	–
La	–	–	30	–	–	–	–	–
Ce	–	–	–	520	–	–	–	–
Eu								
Hf	120	–	–	–	–	–	–	–
Ta								
Hg	–	–	–	–	110	–	23	44
Pb	10,000	–	–	–	20,000	44,000	6,200	4,900
Th								
U								

[a]Mamuro et al. (1979c).

Induction[a] Furnace for Cr–Cu	Zn[a] Melting Furnace	Zn[a] Chromate Pulverizer	Drying[a] Furnace for Cr Yellow	Cr[a] Yellow Pulverizer	Reverberatory[a] Furnace for Pb	Blast[a] Furnace for Pb	Minimum[a] Furnace
24,000	2,200	1,700	34,000	1,400	28,000	10,000	2,400
24,000	3,600	470	29,000	2,600	2,200	3,000	1,100
45,000	–	–	430,000	–	83,000	97,000	26,000
12,000	59,000	4,900	38,000	–	26,000	15,000	4,900
–	–	–	–	–	–	–	570
30,000	690	1,100	40,000	–	–	4,700	6,500
–	–	310	–	–	–	–	–
100	27	9.5	200	36	25	27	11
3,600	160	95,000	98,000	1,800	550	920	230
410	300	–	810	–	98	540	42
13,000	3,900	–	33,000	3,900	12,000	11,000	4,900
170	–	–	5,100	–	2,000	4,000	24
86,000	1,500	180	1,000	–	–	380	–
8,200	150,000	530,000	5,200	660	1,800	2,400	190
260	120	–	300	250	1,500	2,600	52
19	–	–	–	–	–	5,900	–
73	44	10	180	39	54	72	64
29	–	–	–	–	–	–	580
–	420	240	–	–	470	950	–
–	–	–	–	–	2,900	–	2,600
68	15	2.9	730	86	5,200	7,600	230
1.8	–	–	–	–	–	–	1.1
–	–	–	47,000	1,300	–	–	450
3.9	–	–	–	14	–	–	0.25
–	260	130	–	26	–	–	–
–	19	–	–	–	–	–	–
–	400	120	99	–	22	–	3.8
2,600	3,400	2,500	420,000	8,000	160,000	110,000	16,000

Carbonaceous Industrial Sources (μg/g)

	Carbonaceous[a] Manufacture		Car Shredder[a]		Carbide Furnace[a]		Paint Spray[b] Booth (Water)
	Fine	Coarse	Fine	Coarse	Fine	Coarse	
Carbon (total)							500,000
Carbon (organic)	136,000	46,000	77,000	310,000	73,000	45,000	400,000
Carbon (elemental)	550,000	240,000	6,000	40,000	12,000	36,000	
NO_3	16,000				5,700	<5,000	<500
F	500	<3,000	<1,000	–	<1,000	<2,000	
Na	1,200	2,600	1,800	–	9,200	4,300	
Mg	7,800	<4,000	<2,000	–	24,000	12,500	
Al	3,500	28,000	4,700	19,000	5,800	8,600	
Si	35,000	120,000	8,700	–	25,000	29,000	
P							
S	48,000	4,100	4,700	–	16,000	5,900	4,000
Cl	17,000	9,900	6,800	87,000	10,500	9,600	
K	7,900	7,200	2,700	–	12,500	4,800	
Ca	1,400	5,700	4,800	55,000	297,000	258,000	<5,500
Ti	140	330	<500	–	<200	<200	30,000
V	70	120	<10	–	60	87	
Cr	90	40	40	2,000	<100	<100	<500
Mn	350	290	88	780	420	360	
Fe	330	2,700	57,600	37,000	5,400	3,300	<5,500
Ni	40	170	310	<10	220	210	<500
Cu	190	550	1,000	9,900	200	390	<500
Zn	40	<100	21,000	8,700	150	<100	<500
As							
Se							
Br	170	110	4,900	84,000	<100	<100	<500
Rb							
Sr							
Zr							
Cd							<500
Sn							
Sb							
I							
Cs							
Ba							
La							
Ce							
Eu							
Hf							
Ta							
Hg							
Pb	<100	<100	4,900	<50	80	<200	
Th							
U							

[a]Watson (1979).

[b]Taback et al. (1979).

Paint Spray[b] Booth (Oil)	Boric Acid[b] Manufacture	Chemical[b] Fertilizer (Urea)	Kraft Recovering Boiler[a]		Sulfite Recovery Boiler[a]	
			Fine	Coarse	Fine	Coarse
550,000		330,000				
550,000		310,000	158,000	17,000		
			18,000	2,200		
	<5,500	<5,500			<500	<1,000
			<5,000	<1,000	<100	<400
			53,000	127,000	18,000	1,100
				6,300	3,600	900
			2,800	2,500		400
			<1,300	1,500	2,900	250
4,000	4,000	10,100	33,000	117,000	108,000	5,300
	5,440	109,380	29,000	18,000	5,600	<1,200
		<5,500	4,000	15,000	230,000	9,200
<5,500	<500	<5,500	3,600	<5,000	<20,000	<500
30,000			<2,000	60	<100	<800
			<10	10	<10	<10
	10		4,800	2,800	<10	<100
	10	<500	520	300	540	52
<5,500	<5,500	600	18,400	12,000	660	64
	<500	<500	2,200	1,400	<10	
	<500	<500	600	210	160	110
	<500	<500	<100	690	170	<100
		490				
		<500	560	1,300	<10	<100
	<500	<500				
	10		<200	130	<100	<100

Miscellaneous Industrial Sources (μg/g)

	Drying[a] Furnace for Aggregate	Drying[a] Furnace for L-ABS	Calcining[a] Furnace for Barite	Coking[a] Still	Powder[a] Metallurgy of Ultrahard Metal	Drying[a] Furnace for Sand Mold	Rotary[a] Drying Oven for Thenardite	Reaction[a] Pot for Rock Phosphate and H_2SO_4
Carbon (total)								
Carbon (organic)								
Carbon (elemental)								
N								
F								
Na	6,000	22,000	6,400	4,300	13,000	13,000	400,000	15,000
Mg								
Al	70,000	7,300	2,800	3,400	5,600	7,100	190	8,100
Si								
P								
S	7,500	44,000	83,000	–	240,000	–	360,000	–
Cl	810	4,000	5,500	3,800	32,000	46,000	–	13,000
K	14,000	–	49,000	–	–	12,000	2,000	–
Ca	81,000	2,200	–	4,000	6,300	13,000	5,000	–
Ti	2,800	30,000	–	–	530	–	83	–
V	72	54	320	75	24	26	–	63
Cr	55	530	210	74	1,200	850	140	210
Mn	830	1,100	63	80	98	1,200	24	–
Fe	53,000	28,000	6,200	6,400	8,900	17,000	430	4,800
Ni	–	1,200	330	–	–	–	–	–
Cu	–	2,700	440	–	–	1,000	–	–
Zn	430	2,600	21,000	320	430	15,000	150	290
As	26	89	1,700	60	–	49	–	38,000
Se	6.4	62	–	19	39	–	–	–
Br	9.2	110	100	76	–	170	–	23,000
Rb	55	–	780	–	–	–	–	–
Sr								
Zr								
Cd	–	–	510	50	–	–	–	–
Sn	–	–	–	–	–	–	–	–
Sb	3.6	97	1,200	5.3	30	130	7.3	27
I								
Cs	4.6	–	42	0.80	–	11	–	–
Ba	340	1,800	120,000	850	–	–	1,200	–
La	19	9.3	–	4.8	–	–	–	–
Ce	31	10	0.78	4.0	–	–	–	–
Eu								
Hf	3.4	–	–	–	–	–	–	–
Ta	0.49	60	–	–	–	–	–	–
Hg	0.62	–	360	11	5.6	76	21	270
Pb	120	2,400	56,000	950	3,600	5,800	300	150,000
Th	13	3.8	–	3.6	–	–	–	–
U								

[a]Mamuro et al. (1979d).

Pulverizer[a] for Rock Phosphate	Drying[a] Furnace for Mercury Chloride	Drying[a] Furnace for Mercury Amide Chloride	Reaction[a] Pot for Metallic Soap	Dissolution[a] Vessel for Ferric Chloride	Adsorption[a] Facility for Cl_2 and HCl	Sandblast[a]	Polishing[a] Machine for Asbestos
7,100	6,700	11,000	2,800	89,000	3,700	8,000	5,900
12,000	4,800	4,900	1,100	1,900	23,000	7,500	27,000
30,000	–	–	–	–	–	8,800	–
–	96,000	580,000	7,100	220,000	110,000	28,000	16,000
23,000	–	–	–	–	–	–	–
230,000	6,200	–	–	–	–	8,200	36,000
460	–	–	–	–	–	4,000	4,500
120	28	–	3.8	–	–	31	74
–	720	420	210	170	9,100	5,800	620
260	300	230	23	–	–	1,400	580
12,000	6,200	3,600	990	–	2,200	63,000	21,000
–	–	–	66	–	–	3,200	–
–	–	–	–	–	–	6,900	–
500	720	440	1,100	–	970	4,700	4,900
330	–	4.1	–	–	–	99	58
–	–	–	–	–	–	440	–
100	–	29	13	5,800	4,900	47	110
–	–	–	–	–	–	–	–
–		–	–	–	–	–	–
–	–	–	–	–	–	1,300	–
6.0	–	5.8	8.4	12	4.0	110	42
–	–	–	–	–	–	–	–
–	–	–	–	–	–	–	–
39	–	–	–	–	–	11	–
100	–	460	–	–	–	110	–
–	–	–	–	–	–	41	–
–	–	–	–	–	–	–	–
–	86,000	4,200	1.3	28	20	3,700	–
–	4,300	820	35,000	–	–	5,800	–
–	–	–	–	4.4	–	20	–

Automobile-Related Sources: Direct Emission ($\mu g/g$)

	Catalyst[a] Equipped	Noncatalyst[a] Equipped	Unleaded Fuel[b]		Leaded Fuel[b]		Leaded[c] Fueled
			Fine	Coarse	Fine	Coarse	
Carbon (total)							467,000
Carbon (organic)	211,000	501,000	500,000 ± 5,000	215,000 ± 30,000	280,000 ± 240,000	63,000	
Carbon (elemental)	179,000	44,000	18,000 ± 2,000	58,000 ± 6,000	45,000 ± 21,000	5,000	
NO_3			6,300 ± 4,500	17,000 ± 6,400	830 ± 600	12,000	
F							
Na							
Mg							
Al	1,200	430	1,200 ± 600	710 ± 560	430 ± 120	2,500	
Si	5,100	750	5,100 ± 1,300	6,400 ± 700	750 ± 350	1,600	
P							
S	125,000	425	23,000 ± 2,800	8,400 ± 900	5,200 ± 1,400	7,900	
Cl		54,000	13,200 ± 6,900	8,900 ± 3,000	4,900 ± 2,000	8,000	
K	400		400 ± 230	1,300 ± 1,500	<120	500	
Ca	1,700		1,700 ± 400	4,800 ± 4,200	250 ± 28	1,500	
Ti			–	800 ± 100	15 ± 14	<1,000	
V			<6	–	<5	<5	
Cr			<8	590 ± 450	<10	<10	
Mn	150		150 ± 70	–	6.0 ± 1.4	60	
Fe	1,100	2,500	1,100 ± 400	6,300 ± 8,000	2,500 ± 3,100	3,200	
Ni	150		150 ± 50	<100	<50	<100	
Cu	240	40	240 ± 160	440 ± 50	38 ± 4	440	
Zn	800	210	800 ± 100	<50	210 ± 100		

As						
Se						
Br	82,000	44,000 ± 16,000	3,000 ± 1,400	6,500 ± 700	4,200	
Rb						
Sr						
Zr						
Cd						
Sn						
Sb						
I						
Cs						
Ba						
La						
Ce						
Eu						
Hf						
Ta						
Hg						
Pb	211,000	116,000 ± 40,000	7,000 ± 1,300	20,000 ± 2,000	148,000	132,000
Th						
U						

[a]Cass and McRae (1981).
[b]Watson (1979).
[c]Ter Haar et al. (1972).

Automobile-Related Sources: Tire Wear Particles ($\mu g/g$)

	Raybold and Byerly (1972)	Ondov (1974)				
		Sample 1	Sample 2	Sample 3	Sample 4	Sample 5
Carbon (total)						
Carbon (organic)						
Carbon (elemental)						
N						
F						
Na		550 ± 50	440 ± 40	100 ± 10	170 ± 20	140 ± 10
Mg		110 ± 15	–	430 ± 60	600 ± 90	–
Al		160 ± 20	–	200 ± 20	440 ± 40	–
Si						
P						
S		9,000 ± 1,000	–	14,000 ± 1,000	11,000 ± 1,000	–
Cl		200 ± 20	850 ± 90	140 ± 10	130 ± 10	95 ± 10
K						
Ca		860 ± 90	–	170 ± 20	67 ± 7	–
Ti		23 ± 2	69 ± 7	210 ± 20	140 ± 10	92 ± 9
V		0.66 ± 0.07	–	4.7 ± 0.5	0.72 ± 0.07	–
Cr						
Mn		1.3 ± 0.1	9.0 ± 0.9	1.3 ± 0.1	3.1 ± 0.1	1.1 ± 0.1
Fe						
Ni						
Cu		3.8 ± 0.8	–	4.5 ± 0.8	2.6 ± 0.5	–
Zn	7,600 ± 700	11,000 ± 1,000	8,600 ± 900	11,000 ± 1,000	11,000 ± 1,000	11,000 ± 1,000
As		–	0.013 ± 0.001	0.053 ± 0.005	0.26 ± 0.03	0.11 ± 0.01
Se						
Br		6	–	5.1 ± 1.0	2.3 ± 0.4	1.9 ± 0.4
Rb						
Sr						
Zr						
Cd						
Sn						
Sb		–	0.39 ± 0.04	0.12 ± 0.01	0.69 ± 0.07	0.52 ± 0.05
I						
Cs						
Ba		8.1 ± 1.6	170 ± 30	1.5 ± 0.3	4.4 ± 0.8	6.7 ± 1.3
La		0.93 ± 0.09	9.2 ± 0.9	3.0 ± 0.3	8.0 ± 0.8	1.6 ± 0.2
Ce						
Eu						
Hf						
Ta						
Hg						
Pb						
Th						
U						

Ondov (1974)						Taback et al. (1979)
Sample 6	Sample 7	Sample 8	Sample 9	Sample 10	Sample 11	
						870,000
500 ± 50	1,100 ± 100	88 ± 9	660 ± 70	1,300 ± 100	860 ± 90	
170 ± 30	540 ± 70	370 ± 60	810 ± 120	160 ± 20	360 ± 60	
240 ± 20	1,900 ± 200	190 ± 20	860 ± 90	1,600 ± 200	1,100 ± 100	
17,000 ± 2,000	13,000 ± 1,000	14,000 ± 1,000	75,000 ± 1,000	15,000 ± 1,000	11,000 ± 1,000	
1,100 ± 100	1,500 ± 200	77 ± 8	890 ± 90	1,700 ± 200	1,300 ± 100	
330 ± 30	150 ± 15	71 ± 7	220 ± 20	200 ± 20	81 ± 8	
210 ± 20	14 ± 2	30 ± 3	300 ± 30	160 ± 20	560 ± 60	
0.43 ± 0.04	1.0 ± 0.1	5.1 ± 0.5	18 ± 2	1.2 ± 0.1	0.79 ± 0.08	
1.2 ± 0.1	3.4 ± 0.3	1.3 ± 0.1	10 ± 1	1.7 ± 0.2	1.8 ± 0.2	
3.0 ± 0.6	7.9 ± 1.6	8.9 ± 1.9	3.4 ± 0.5	<0.72	2	
12,000 ± 1,000	11,000 ± 1,000	8,700 ± 900	11,000 ± 1,000	6,700 ± 700	10,000 ± 1,000	10,000
0.056 ± 0.006	0.039 ± 0.004	0.011 ± 0.001	0.017 ± 0.002	0.073 ± 0.007	0.024 ± 0.002	
5.4 ± 1.0	–	5.5 ± 1.0	–	–	6.2 ± 1.2	
0.24 ± 0.02	0.065 ± 0.007	0.12 ± 0.01	0.93 ± 0.09	0.33 ± 0.03	0.27 ± 0.03	
1.6 ± 0.3	4.3 ± 0.8	0.96 ± 0.20	3.0 ± 0.6	3.6 ± 0.8	–	
3.6 ± 0.4	0.82 ± 0.08	1.0 ± 0.1	11 ± 1	2.0 ± 0.2	3.0 ± 0.3	

Automobile-Related Sources: Brake Lining Material (μg/g)

	Lynch (1968)	Ondov (1974) Sample 1	Ondov (1974) Sample 2	Ondov (1974) Sample 3
Carbon (total)	283,000			
Carbon (organic)	233,000			
Carbon (elemental)				
N				
F				
Na		240 ± 20	400 ± 20	
Mg	82,500	136,000 ± 14,000	140,000 ± 10,000	
Al		1,600 ± 200	2,700 ± 200	
Si	154,000			
P				
S				
Cl		1,600 ± 200	1,300 ± 100	
K				
Ca	55,000	36,000 ± 4000	1,900 ± 200	
Sc		4.2 ± 0.3	2.6 ± 0.2	4.2 ± 0.3
V		8.5 ± 0.8	7.4 ± 0.6	
Cr		750 ± 30	660 ± 50	840 ± 60
Mn		400 ± 40	450 ± 20	
Fe		28,000 ± 2,000	23,000 ± 200	35,000 ± 3,000
Co		61 ± 3	56 ± 4	53 ± 4
Cu		<2	–	
Zn		2,800 ± 300	22 ± 2	2,100 ± 200
As		13 ± 1	–	–
Se		0.18 ± 0.06	0.41 ± 0.06	–
Br		–	<2	
Rb				
Sr		410 ± 40	–	
Zr				
Cd				
Sn				
Sb		5.1 ± 0.5	<0.31	–
I				
Cs		–	–	0.28 ± 0.03
Ba		31 ± 3	20 ± 2	22,000 ± 1,000
La		1.4 ± 0.1	0.19 ± 0.02	2.2 ± 0.2
Ce		7.8 ± 0.3	0.46 ± 0.04	10 ± 1.5
Eu				
Hf		0.35 ± 0.03	–	0.28 ± 0.02
Ta		0.12 ± 0.03	–	–
Hg				
Pb				
Th				0.37 ± 0.03
U				

Ondov (1974)					Taback et al. (1979)
Sample 4	Sample 5	Sample 6	Sample 7	Sample 8	
					160,000
					140,000
					215,000
					80,000
3.3 ± 0.3	3.0 ± 0.03	17 ± 2	7.1 ± 0.7	6.2 ± 0.15	
640 ± 40	600 ± 50	2,500 ± 200	920 ± 60	910 ± 80	
30,000 ± 3,000	25,000 ± 3,000	116,000 ± 10,000	29,000 ± 1,000	34,000 ± 3,000	
44 ± 3	35 ± 3	180 ± 10	57 ± 4	61 ± 3	
6,200 ± 600	3,500 ± 300	6,000 ± 400	1,800 ± 200	2,400 ± 200	
–	–	–	–	–	
–	–	–	–	–	
0.16 ± 0.04	<0.07	3.2 ± 0.3	2.3 ± 0.2	2.1 ± 0.1	
–	0.12 ± 0.01	–	0.26 ± 0.03	–	
36,000 ± 2,000	12,000 ± 1,000	150 ± 60	3,500 ± 200	28 ± 5	
1.4 ± 0.2	0.6 ± 0.05	3.9 ± 0.6	1.0 ± 0.1	0.41 ± 0.04	
8.3 ± 1.0	20 ± 0.2	11 ± 2	1.9 ± 0.2	0.77 ± 0.07	
0.65 ± 0.05	1.8 ± 0.1	0.74 ± 0.10	0.13 ± 0.01	0.11 ± 0.01	
–	–	0.4 ± 0.2	–	–	
0.17 ± 0.02	2.7 ± 0.2	0.93 ± 0.20	1.2 ± 0.1	0.62 ± 0.06	

Automobile-Related Sources: New Lubricating Oils[a] (μg/g Except as Noted)

	Multiweight HD[b]	SAE 30	SAE 30 (Premium)	Multiweight HD	Multiweight HD
Carbon (total)					
Carbon (organic)					
Carbon (elemental)					
N					
F					
Na	28 ± 3	63 ± 6	6.0 ± 0.6	5.3 ± 0.5	7.4 ± 0.7
Mg	5.8 ± 1.2	7.5 ± 1.6	3.3 ± 0.3	9.7 ± 1.0	940 ± 90
Al	0.21 ± 0.02	0.23 ± 0.02	0.028 ± 0.003	0.47 ± 0.05	0.42 ± 0.04
Si					
P					
S	5,000 ± 500	5,800 ± 600	1,400 ± 100	4,800 ± 500	3,900 ± 400
Cl	410 ± 40	23 ± 2	250 ± 30	130 ± 10	240 ± 20
K					
Ca	3,100 ± 300	310 ± 30	2,200 ± 200	1,800 ± 200	140 ± 20
Sc (ng/g)	0.15 ± 0.03	0.27 ± 0.06	<0.37	1.6 ± 0.2	<0.30
V	0.027 ± 0.003	0.017 ± 0.002	0.0023 ± 0.0002	0.045 ± 0.005	0.0089 ± 0.0009
Cr					
Mn	0.33 ± 0.03	0.035 ± 0.004	0.12 ± 0.01	0.12 ± 0.01	0.24 ± 0.02
Fe	2.6 ± 0.5	0.60 ± 0.30	2.9 ± 0.5	3.4 ± 0.6	1.8 ± 0.9
Co (ng/g)	0.70 ± 0.30	1.2 ± 0.6	1.1 ± 0.2	2.1 ± 0.4	1.7 ± 0.8
Cu (ng/g)	200 ± 40	0.40 ± 0.08	4.8 ± 0.8	110 ± 20	150 ± 20
Zn	1,500 ± 200	62 ± 6	880 ± 90	850 ± 90	1,100 ± 100
As					
Se (ng/g)	<50	32 ± 6	<44	<76	<80
Br	0.79 ± 0.08	0.093 ± 0.010	0.24 ± 0.02	0.91 ± 0.09	0.65 ± 0.07
Rb					
Sr					
Zr					
Cd					
Sn					
Sb (ng/g)	30 ± 3	<2.1	12 ± 2	6.0 ± 3.0	10 ± 5
I					
Cs					
Ba	<2.0	85 ± 9	2.9 ± 0.6	4,400 ± 400	10 ± 2
La (ng/g)	41 ± 4	<3.0	6.9 ± 0.7	12 ± 1	–
Ce					
Eu					
Hf					
Ta					
Hg					
Pb					
Th					
U					

[a]From Ondov (1974).

[b]HD signifies high detergent oil.

SAE 30 HD	SAE 30	Multiweight HD	Multiweight HD	Multiweight HD	SAE 30
2.0 ± 0.2	7.5 ± 0.8	12 ± 1	6.6 ± 0.7	4.7 ± 0.5	8.4 ± 0.8
440 ± 100	28 ± 6	5.5 ± 1.0	400 ± 80	1,300 ± 200	12 ± 4
0.039 ± 0.004	0.50 ± 0.05	0.14 ± 0.01	0.17 ± 0.02	0.16 ± 0.02	0.29 ± 0.03
1,800 ± 200	11,000 ± 1,000	4,900 ± 500	3,300 ± 300	5,800 ± 600	4,500 ± 500
2.0 ± 0.2	134 ± 10	160 ± 20	10 ± 1	27 ± 3	150 ± 20
78 ± 8	1,400 ± 100	1,300 ± 100	1,100 ± 100	31 ± 6	1,400 ± 100
<0.25	0.75 ± 0.15	1.5 ± 0.3	0.40 ± 0.20		
0.010 ± 0.001	0.12 ± 0.01	0.029 ± 0.003	0.023 ± 0.002	16 ± 2	59 ± 6
0.013 ± 0.001	0.072 ± 0.007	0.075 ± 0.008	0.30 ± 0.03	0.017 ± 0.002	0.037 ± 0.004
<0.4	2.9 ± 0.6	1.6 ± 0.8	1.9 ± 0.8		
3.0 ± 0.6	<3.1	2.0 ± 0.4	1.8 ± 0.9		
<2.6	120 ± 20	190 ± 40	190 ± 40	90 ± 20	75 ± 16
33 ± 3	99 ± 10	830 ± 80	1,300 ± 100	2,100 ± 200	1,100 ± 100
18 ± 2	<330	<200	<50		
3.2 ± 0.3	1.1 ± 0.1	1.6 ± 0.2	0.12 ± 0.01	0.088 ± 0.009	1.1 ± 0.1
1.0 ± 1.0	1,800 ± 200	2.1 ± 0.5	17 ± 3		
64 ± 6	2,300 ± 200	3,000 ± 300	10 ± 1	13 ± 2	4,200 ± 400
–	–	–	–		

(Continued on page 310)

Automobile-Related Sources: New Lubricating Oils[a] (μg/g Except as Noted) *(Continued)*

	Multiweight	SAE 10	Multiweight	SAE 30	SAE 30	SAE 20
Carbon (total)						
Carbon (organic)						
Carbon (elemental)						
N						
F						
Na	1.1 ± 0.1	3.5 ± 0.4	170 ± 20	3.5 ± 0.4	32 ± 3	10 ± 1
Mg	2.3 ± 0.4	790 ± 80	14 ± 3	1.1 ± 0.3	10 ± 2	3.1 ± 0.3
Al	0.060 ± 0.006	0.16 ± 0.02	0.21 ± 0.02	0.079 ± 0.008	0.070 ± 0.007	0.14 ± 0.01
Si						
P						
S	1,600 ± 200	4,200 ± 400	4,700 ± 500	2,000 ± 200	6,000 ± 600	2,200 ± 200
Cl	48 ± 5	24 ± 2	520 ± 50	41 ± 4	500 ± 50	270 ± 30
K						
Ca	0.77 ± 0.15	150 ± 20	760 ± 80	31 ± 3	3,400 ± 300	230 ± 50
Sc (ng/g)						
V	6.9 ± 0.7	12 ± 1	22 ± 2	1.9 ± 0.2	30 ± 3	24 ± 2
Cr						
Mn	<0.008	0.13 ± 0.01	0.096 ± 0.010	0.012 ± 0.001	0.34 ± 0.03	0.024 ± 0.002
Fe						
Co (ng/g)						
Cu (ng/g)	47 ± 8	79 ± 16	250 ± 50	51 ± 10	59 ± 12	170 ± 30
Zn	<28	1.9 ± 0.2	1,200 ± 100	280 ± 30	2,300 ± 200	270 ± 30
As						
Se (ng/g)						
Br	2.6 ± 0.3	0.19 ± 0.02	0.87 ± 0.09	0.19 ± 0.02	0.50 ± 0.05	0.63 ± 0.06
Rb						
Sr						
Zr						
Cd						
Sn						
Sb (ng/g)						
I						
Cs						
Ba	0.38 ± 0.12	350 ± 40	2,100 ± 200	20 ± 2	0.92 ± 0.18	340 ± 60
La (ng/g)						
Ce						
Eu						
Hf						
Ta						
Hg						
Pb						
Th						
U						

Diesel Powered Vehicles (μg/g)

	Truck[a]		Train[a]		Jet[b] Aircraft	Diesel[c] Particles
	Fine	Coarse	Fine	Coarse		
Carbon (total)					960,000	679,000
Carbon (organic)	325,000	410,000	653,000	230,000	259,000	
Carbon (elemental)	680,000	235,000	214,000	140,000		128,000
NO_3	7,200	7,200	25,000	33,000		
F	<5,000					
Na	3,700	7,200	4,700	2,100		
Mg			2,250			
Al	3,400	15,000	2,200	129,000		
Si	1,700	4,300	9,800	102,000		
P						
S	11,200	4,000	5,000	8,000		20,000
Cl	16,900	17,000	7,400	20,000		
K	<100	<120	370			
Ca	8,400	17,000	6,300	12,700		
Ti	<2,000	<2,000	98	<2,000		
V	100	<100	6.3	<10		
Cr	<300	<100	50			
Mn	270	90	58	170		
Fe	13,200	17,400	1,800	<4,000		
Ni	<200	<200	160	<500		
Cu	7,300	3,000	67	300		
Zn	2,300	740	<100	<1,000		
As						
Se						
Br	310	300	<50	230		
Rb						
Sr						
Zr						
Cd						
Sn						
Sb						
I						
Cs						
Ba						
La						
Ce						
Eu						
Hf						
Ta						
Hg						
Pb	950	<1000	510			
Th						
U						

[a]Watson (1979).

[b]Cass and McRae (1981).

[c]Gardella and Hercules (1979).

Mixed Motor Vehicle Sources: Gasoline and Diesel Powered (Normalized to Pb = 1.0)*

	Summer Tunnel[a]		Los Angeles Freeway[b]		Baltimore Harbor[c] Tunnel 1973	Caldecott Tunnel[d] 1975–1977	Det. & Canada Tunnel[e] Allegheny Tunnel 1971–1974	Allegheny Tunnel Tuscarara Tunnel 1975–1979
	1961	1963	1972	1976				
Carbon (total)						1.0 (0.3–2.3)	9	14
Carbon (organic)								
Carbon (elemental)								
N								0.3
F								0.1
Na								
Mg					0.075 ± 0.03			0.2
Al			0.12	~0[g]	0.07 ± 0.02			0.2
Si				~0.15[g]				0.2
P								0.02
SO_4^{2-}	0.16	0.9		<0		0.059 ± 0.025	0.43	0.7
Cl				0.24 ± 0.04	0.22 ± 0.04			0.06
K			0.007					0.02
Ca			0.12	~0.08[g]	0.11 ± 0.045			0.2
Ti	0.0093	0.02			<0.04			0.002
V					<0.015			0.00005
Cr	0.00094	~0.0003			<0.015			0.0004
Mn	0.0019				0.003 ± 0.001			0.012[h]
Fe	0.487	~0.1	0.19	~0.1[g]	0.08 ± 0.04	~0.01		0.09
Ni		~0.0006			0.008 ± 0.0006			~0.0005
Cu	0.0049	~0.004	0.005		0.007 ± 0.003	~0.0005		0.01
Zn	0.017		0.024		0.017 ± 0.004	~0.005, 0.01	0.009	0.02
As								
Se					0.000081 ± 0.000034			

Br			0.30	0.47 ± 0.06	0.39 ± 0.05	0.55	0.36	0.3
Rb								
Sr								0.002
Zr								0.0006
Cd	0.00023				0.0024 ± 0.0006			<0.0003
Sn	0.00048							<0.004
Sb	0.00039				0.0014 ± 0.0003			0.00005
I								0.0004
Cs								0.000015
Ba					0.015 ± 0.004		0.016	0.018
La					<0.00006			<0.00006
Ce					<0.00011			0.00009
Eu								
Hf					<0.00009			<0.00002
Ta					<0.000014			<0.00001
Hg								0.00001
Pb	1.0	1.0	1.0	1.0	1.0	1.0	1.0	1.0[i]
Th					<0.000055			<0.00005
U								

*This table reproduced from Pierson and Brachaczek (1983). Used with permission.

[a]1961: From Larsen and Konopinski (1962) and Larsen (1966).
1963: From Conlee et al. (1967).

[b]1972: From Cahill and Feeney (1973) and Feeney et al. (1975).
1976: From Dzubay et al. (1979).

[c]From Ondov et al. (1982) and Ondov (1974).

[d]From Hollowell et al. (1976, 1977) and Giauque (1977).

[e]From Pierson and Brachaczek (1976).

[f]Traffic approximately 85–90% gasoline-powered vehicles, the rest diesel trucks. Calculated as long-term average percent divided by long-term average percent Pb (rather than average of the experiment-by-experiment ratios to Pb).

[g]From only one set of samples, data of Figure 2 of Dzubay et al. (1979); assumed to be representative.

[h]Changing year to year, owing to changes in MMT usage.

[i]Declining year to year in mg/km.

Marine ($\mu g/g$)

	Kowalczyk et al. (1978)	Watson (1979)	Taback et al. (1979)
Carbon (total)			
Carbon (organic)			
Carbon (elemental)			
NO_3		37,500	
F			
Na	310,000	109,000	310,000
Mg	40,300		40,000
Al	0.29	19,700	
Si		49,000	
P			
S		12,700	22,500
Cl	561,000	150,000	550,000
K	11,200	13,000	10,000
Ca	11,800	10,900	10,000
Ti	0.03	1,480	
V	0.06	<2	
Cr	0.0015	550	
Mn	0.06	330	
Fe	0.30	7,600	
Ni	0.06	<4	
Cu	0.09	<5	
Zn	0.03	680	
As	0.09		
Se	0.12		
Br	1,900	750	
Rb			
Sr			
Zr			
Cd	0.003		
Sn			
Sb	0.015		
I	1,800		
Cs			
Ba	0.87		
La	0.009		
Ce	0.012		
Eu			
Hf			
Ta			
Hg			
Pb	0.001	<20	
Th	0.001		
U			

INDEX